SFL
QC985.5.N7 C65 2008
v.2
Cold region atmospheric and
hydrologic studies : the
Mackenzie GEWEX experience

Ming-ko Woo
Editor

Cold Region Atmospheric and Hydrologic Studies
The Mackenzie GEWEX Experience

Volume 2: Hydrologic Processes

Ming-ko Woo
Editor

Cold Region Atmospheric and Hydrologic Studies

The Mackenzie GEWEX Experience

Volume 2: Hydrologic Processes

with 162 Figures

Mr. Ming-ko Woo
McMaster University
School of Geography
and Earth Sciences
1280 Main Street West
Hamilton ON L8S 4K1
Canada

Email: woo@mcmaster.ca

Cover photograph: Ice jam formed during spring breakup at the mouth of Liard River near Fort Simpson

Library of Congress Control Number: 2007931626

ISBN   978-3-540-74927-1 Springer Berlin Heidelberg New York

This work is subject to copyright. All rights are reserved, whether the whole or part of the material is concerned, specifically the rights of translation, reprinting, reuse of illustrations, recitation, broadcasting, reproduction on microfilm or in any other way, and storage in data banks. Duplication of this publication or parts thereof is permitted only under the provisions of the German Copyright Law of September 9, 1965, in its current version, and permission for use must always be obtained from Springer-Verlag. Violations are liable to prosecution under the German Copyright Law.

Springer is a part of Springer Science+Business Media
springer.com
© Springer-Verlag Berlin Heidelberg 2008

The use of general descriptive names, registered names, trademarks, etc. in this publication does not imply, even in the absence of a specific statement, that such names are exempt from the relevant protective laws and regulations and therefore free for general use.

Cover design: deblik, Berlin
Production: Almas Schimmel
Typesetting: Camera-ready by the editor

Printed on acid-free paper   32/3180/as 5 4 3 2 1

Editor's Note

The circumpolar areas are crucially important to the global environment as they play the dual roles of being the harbinger of and a contributor to climate change. A rising clamor for environmental information on such areas is evident in scientific and media publications. Yet, relative to the temperate zones, the cold regions remain under-studied. There exists a need to improve knowledge of the cold region sciences, and atmospheric and hydrologic sciences are of no exception. This book was prepared with the purpose of sharing the experience of a team of Canadian researchers with the practitioners and stakeholders, residents and environmental managers, educators and students of the cold regions.

A good understanding of the physical processes and enhanced capability to model the regional atmospheric and the hydrologic conditions are the keys to environmental preservation and resource sustainability in the circumpolar areas. However, the cold environment presents a challenge to atmospheric and hydrologic research. The low temperatures test the endurance of field workers and their instruments; the distance of these areas from large urban centers raises the cost of logistics and reduces the availability of measured data. In 1994, a group of researchers from academic and government institutions embarked upon a major collaborative program, using the large and complex Mackenzie River Basin in northern Canada as the focal point to study the atmospheric and hydrologic phenomena of the cold regions. The goals of this project, known as the Mackenzie GEWEX Study or MAGS, were: (1) to understand and model the high-latitude energy and water cycles that play roles in the climate system, and (2) to improve our ability to assess the changes to Canada's water resources that arise from climate variability and anthropogenic climate change.

The Mackenzie Basin has many of the physical attributes common to cold regions worldwide. Results from MAGS investigation including the enhanced understanding of the atmospheric and hydrologic processes, the improved and developed models, and the data collected using novel and conventional methods, are largely applicable to other circumpolar areas. This publication is a documentation of the research outcomes of MAGS. For a study that spanned over a decade (1994–2005), MAGS yielded an immense amount of information, too extensive to be accommodated within one publication. Thus, this book is divided into two volumes. Volume I of this book, entitled "Atmospheric Processes of a Cold Region: the Mackenzie GEWEX Study Experience", concentrates on the atmospheric investigations. Volume II is a complementary report on the hydrologic aspect. To provide continuity with Volume I, a synopsis of its chapters is presented. This information is provided in the spirit of the Mackenzie GEWEX Study which emphasizes the integration of atmospheric and hydrologic research.

Acknowledgements are in order for the funding and in-kind contributions of various institutions. The Natural Sciences and Engineering Research Council of Canada (NSERC) has given continued support to the university investigators and their students to carry out multiple years of research. Environment Canada offered unwavering support through the former Atmospheric Environment Service, the National Water Research Institute and the Prairie and Northern Region. The Departments of Indian and Northern Affairs, and Natural Resources of Canada, and several Canadian universities (Alberta, McGill, McMaster, Quebec, Saskatchewan, Toronto, Waterloo and York) also participated in the program. Other sources of support are acknowledged in individual chapters.

MAGS investigators have been most enthusiastic in sharing their knowledge by authoring chapters in this publication. This collective work benefits further from the reviewers of various book chapters and from members of the advisory group for ensuring its high standard. I am particularly thankful to members of our production team: Michael and Laurine Mollinga, Robin Thorne and Laura Brown. Their dedication and efficiency contributed tremendously to shaping the materials into the final book format.

Ming-ko Woo

May 2007

Table of Contents

Editor's note .... *v*
Table of Contents .... vi*i*
List of Contributors .... *xi*
List of Acronyms .... *xix*

1. Synopsis of Atmospheric Research under MAGS
   Ming-ko Woo 1

2. MAGS Contribution to Hydrologic and Surface Process Research
   Ming-ko Woo and Wayne R. Rouse 9

3. Analysis and Application of 1-km Resolution Visible and Infrared
   Satellite Data over the Mackenzie River Basin
   Normand Bussières 39

4. On the Use of Satellite Passive Microwave Data
   for Estimating Surface Soil Wetness in the Mackenzie River Basin
   Robert Leconte, Marouane Temimi, Naira Chaouch,
   François Brissette and Thibault Toussaint 59

5. Studies on Snow Redistribution by Wind and Forest,
   Snow-covered Area Depletion and Frozen Soil Infiltration
   in Northern and Western Canada
   John W. Pomeroy, Donald M. Gray and Phil Marsh 81

6. Snowmelt Processes and Runoff at the Arctic Treeline:
   Ten Years of MAGS Research
   Philip Marsh, John Pomeroy, Stefan Pohl, William Quinton,
   Cuyler Onclin, Mark Russell, Natasha Neumann, Alain Pietroniro,
   Bruce Davison and S. McCartney 97

7. Modeling Maximum Active Layer Thaw
   in Boreal and Tundra Environments using Limited Data
   Ming-ko Woo, Michael Mollinga and Sharon L. Smith 125

8. Climate-lake Interactions
   Wayne R. Rouse, Peter D. Blanken, Claude R. Duguay,
   Claire J. Oswald and William M. Schertzer 139

9. Modeling Lake Energy Fluxes in the Mackenzie River Basin using Bulk Aerodynamic Mass Transfer Theory
   Claire J. Oswald, Wayne R. Rouse and Jacqueline Binyamin 161

10. The Time Scales of Evaporation from Great Slave Lake
    Peter Blanken, Wayne Rouse and William Schertzer 181

11. Interannual Variability of the Thermal Components and Bulk Heat Exchange of Great Slave Lake
    William M. Schertzer, Wayne R. Rouse, Peter D. Blanken, Anne E. Walker, David C.L. Lam and Luis León 197

12. Flow Connectivity of a Lake–stream System in a Semi-arid Precambrian Shield Environment
    Ming-ko Woo and Corrinne Mielko 221

13. Hydrology of the Northwestern Subarctic Canadian Shield
    Christopher Spence and Ming-ko Woo 235

14. Recent Advances toward Physically-based Runoff Modeling of the Wetland-dominated Central Mackenzie River Basin
    William L. Quinton and Masaki Hayashi 257

15. River Ice
    Faye Hicks and Spyros Beltaos 281

16. Regression and Fuzzy Logic Based Ice Jam Flood Forecasting
    Chandra Mahabir, Claudine Robichaud, Faye Hicks and Aminah Robinson Fayek 307

17. Impact of Climate Change on the Peace River Thermal Ice Regime
    Robyn Andrishak and Faye Hicks 327

18. Climate Impacts on Ice-jam Floods in a Regulated Northern River
    Spyros Beltaos, Terry Prowse, Barrie Bonsal, Tom Carter, Ross MacKay, Luigi Romolo, Alain Pietroniro and Brenda Toth 345

19. Trends in Mackenzie River Basin Streamflows
    Donald H. Burn and Nicole Hesch 363

20. Re-scaling River Flow Direction Data
    from Local to Continental Scales
    *Lawrence W. Martz, Alain Pietroniro, Dean A. Shaw,
    Robert N. Armstrong, Boyd Laing and Martin Lacroix* 371

21. Lessons from Macroscale Hydrologic Modeling: Experience
    with the Hydrologic Model SLURP in the Mackenzie Basin
    *Robin Thorne, Robert N. Armstrong, Ming-ko Woo
    and Lawrence W. Martz* 397

22. Development of a Hydrologic Scheme for Use in Land Surface
    Models and its Application to Climate Change
    in the Athabasca River Basin
    Ernst Kerkhoven and Thian Yew Gan 411

23. Validating Surface Heat Fluxes and Soil Moisture Simulated by the
    Land Surface Scheme CLASS under Subarctic Tundra Conditions
    *Lei Wen, David Rodgers, Charles A. Lin, Nigel Roulet,
    and Linying Tong* 435

24. The MAGS Integrated Modeling System
    E.D. Soulis and Frank R. Seglenieks 445

25. Synthesis of Mackenzie GEWEX Studies
    on the Atmospheric–Hydrologic System of a Cold Region
    Ming-ko Woo 475

Index 497

List of Contributors to Volume II

Robyn Andrishak
Department of Civil and Environmental Engineering
University of Alberta, Edmonton, AB
Canada T6G 2W2

Robert N. Armstrong
Department of Geography, University of Saskatchewan
9 Campus Drive, Saskatoon, SK
Canada S7N 5A5

Spyros Beltaos
National Water Research Institute
867 Lakeshore Road, Burlington, ON
Canada L7R 4A6

Jacqueline Binyamin
School of Geography and Earth Sciences
McMaster University, Hamilton, ON
Canada L8S 4K1

Peter Blanken
Department of Geography, 260 UCB
University of Colorado
Boulder, Colorado
USA 80309-0260

Barrie Bonsal
National Water Research Institute at NHRC
11 Innovation Blvd., Saskatoon, SK
Canada S7N 3H5

François Brissette
Department of Construction Engineering, École de technologie supérieure
1100 Notre Dame Street West, Montreal, QC
Canada H3C 1K3

Donald H. Burn
Department of Civil and Environmental Engineering
University of Waterloo, 200 University Avenue West, Waterloo, ON
Canada N2L 3G1

Tom Carter
National Water Research Institute at NHRC
11 Innovation Blvd., Saskatoon, SK
Canada S7N 3H5

Naira Chaouch
Department of Construction Engineering, École de technologie supérieure
1100 Notre Dame Street West, Montreal, QC
Canada H3C 1K3

Bruce Davison
National Water Research Institute at NHRC
11 Innovation Blvd., Saskatoon, SK
Canada S7N 3H5

Claude R. Duguay
Department of Geography
University of Waterloo, Waterloo, ON
Canada N2L 3G1

Thian Yew Gan
Department of Civil and Environmental Engineering
University of Alberta, Edmonton, AB
Canada T6G 2W2

Donald M Gray
Division of Hydrology, University of Saskatchewan
Saskatoon, SK, Canada S7N 0W0
[Deceased]

Masaki Hayashi
Department of Geology and Geophysics
University of Calgary, Calgary, AB
Canada T2N 1N4

Nicole Hesch
Department of Civil and Environmental Engineering
University of Waterloo, 200 University Avenue West, Waterloo, ON
Canada N2L 3G1

Faye Hicks
Department of Civil and Environmental Engineering
University of Alberta, Edmonton, AB
Canada T6G 2W2

Ernst Kerkhoven
Department of Civil and Environmental Engineering
University of Alberta, Edmonton, AB
Canada T6G 2W2

Martin Lacroix
Water & Climate Impacts Research Centre, NWRI
Department of Geography, University of Victoria
PO Box 3050 STN CSC, Victoria, BC
Canada V8W 3P5

Boyd Laing
Department of Geography, University of Saskatchewan
9 Campus Drive, Saskatoon, SK
Canada S7N 5A5

David C.L. Lam
National Water Research Institute
867 Lakeshore Road, Burlington, ON
Canada L7R 4A6

Robert Leconte
Department of Construction Engineering, École de technologie supérieure
1100 Notre Dame Street West, Montreal, QC
Canada H3C 1K3

Luis León
Computing and Information Science
University of Guelph, Guelph, ON
Canada, N1G 2W1

Charles A. Lin
Department of Atmospheric and Oceanic Sciences
McGill University, Montreal, QC
Canada H3A 2K6

Ross MacKay
National Water Research Institute at NHRC
11 Innovation Blvd., Saskatoon, SK
Canada S7N 3H5

Chandra Mahabir
River Engineering, Alberta Environment, Edmonton, AB
Canada T6B 2X3
[Formerly at University of Alberta]

Philip Marsh
National Water Research Institute at NHRC
11 Innovation Blvd., Saskatoon, SK
Canada S7N 3H5

Lawrence W. Martz
Department of Geography, University of Saskatchewan
9 Campus Drive, Saskatoon, SK
Canada S7N 5A5

S. McCartney
[Formerly at National Water Research Institute]

Corrinne Mielko
School of Geography and Earth Sciences
McMaster University, Hamilton, ON
Canada L8S 4K1

Michael Mollinga
School of Geography and Earth Sciences
McMaster University, Hamilton, ON
Canada L8S 4K1

Natasha Neumann
University of British Columbia Okanagan, Kelowna, BC
Canada V1V 1V7

Cuyler Onclin
National Water Research Institute at NHRC
11 Innovation Blvd., Saskatoon, SK
Canada S7N 3H5

Claire J. Oswald
Department of Geography
University of Toronto, Toronto, ON
Canada M5S 3G3

Alain Pietroniro
National Water Research Institute at NHRC
11 Innovation Blvd., Saskatoon, SK
Canada S7N 3H5

Stefan Pohl
National Water Research Institute at NHRC
11 Innovation Blvd., Saskatoon, SK
Canada S7N 3H5

John W Pomeroy
Centre for Hydrology, University of Saskatchewan
117 Science Place, Saskatoon, SK
Canada S7N 5C8

Terry Prowse
Water & Climate Impacts Research Centre, NWRI
Department of Geography, University of Victoria
PO Box 3050 STN CSC, Victoria, BC
Canada V8W 3P5

William L. Quinton
Cold Regions Research Centre
Wilfrid Laurier University, Waterloo, ON
Canada N2L 3C5

Claudine Robichaud
River Engineering, Alberta Environment, Edmonton, AB
Canada T6B 2X3
[Formerly at University of Alberta]

Aminah Robinson Fayek
Department of Civil and Environmental Engineering
University of Alberta, Edmonton, AB
Canada T6G 2W2

David Rodgers
Ontario Storm Prediction Centre
4905 Dufferin St., Downsview, ON
Canada M3H 5T4

Luigi Romolo
Southern Regional Climate Center, Louisiana State University
E328 Howe-Russell Geoscience Complex
Baton Rouge, LA 70803, USA

Nigel Roulet
Department of Geography
McGill University, Montreal, QC
Canada H3A 2K6

Wayne Rouse
School of Geography and Earth Sciences
McMaster University, Hamilton, ON
Canada L8S 4K1

Mark Russell
National Water Research Institute at NHRC
11 Innovation Blvd., Saskatoon, SK
Canada S7N 3H5

William M. Schertzer
National Water Research Institute
867 Lakeshore Road, Burlington, ON
Canada L7R 4A6

Frank R. Seglenieks
Department of Civil and Environmental Engineering
University of Waterloo, 200 University Avenue West, Waterloo, ON
Canada N2L 3G1

Dean A. Shaw
Department of Geography, University of Saskatchewan
9 Campus Drive, Saskatoon, SK
Canada S7N 5A5

Sharon L. Smith
Geological Survey of Canada, Natural Resources Canada
601 Booth Street, Ottawa, ON
Canada K1A 0E8

E.D. (Ric) Soulis
Department of Civil and Environmental Engineering
University of Waterloo, 200 University Avenue West, Waterloo, ON
Canada N2L 3G1

Christopher Spence
National Water Research Institute at NHRC
11 Innovation Blvd., Saskatoon, SK
Canada S7N 3H5

Marouane Temimi
Department of Construction Engineering, École de technologie supérieure
1100 Notre Dame Street West, Montreal, QC
Canada H3C 1K3

Robin Thorne
School of Geography and Earth Sciences
McMaster University, Hamilton, ON
Canada L8S 4K1

Linying Tong
Meteorological Service of Canada, CMC, Environment Canada
2121 Rte TransCanadienne, Doval, QC
Canada H9P 1J3

Brenda Toth
National Water Research Institute at NHRC
11 Innovation Blvd., Saskatoon, SK
Canada S7N 3H5

Thibault Toussaint
Department of Construction Engineering, École de technologie supérieure
1100 Notre Dame Street West, Montreal, QC
Canada H3C 1K3

Anne Walker
Science and Technology Branch, Environment Canada
4905 Dufferin Street, Downsview, ON
Canada M3H 5T4

Lei Wen
Department of Atmospheric and Oceanic Sciences
McGill University, Montreal, QC
Canada H3A 2K6

Ming-ko Woo
School of Geography and Earth Sciences
McMaster University, Hamilton, ON
Canada L8S 4K1

List of Acronyms

AL	Aleutian Low
AMSR	Advanced Microwave Scanning Radiometer
AO	Arctic Oscillation
AVHRR	Advanced Very High Resolution Radiometer
BALTEX	Baltic Sea (GEWEX) Experiment
BERMS	Boreal Ecosystem Research and Monitoring Sites
BOREAS	Boreal Ecosystem-Atmosphere Study
CAGES	Canadian GEWEX Enhanced Study (Sept. 1998 to July 1999)
CANGRID	Environment Canada's gridded monthly surface climate dataset
CCCma	Canadian Centre for Climate modelling and analysis
CCRS	Canadian Centre for Remote Sensing
CERES	Clouds and Earth's Radiant Energy System
CLASS	Canadian Land Surface Scheme
CMC	Canadian Meteorological Centre (of Environment Canada)
CRCM	Canadian Regional Climate Model
CRYSYS	Cryosphere System in Canada
CSE	Continental Scale Experiment (of GEWEX)
DEM	Digital Elevation Model
DIAND	Department of Indian Affairs and Northern Development
DYRESM	A 1-D Dynamic Reservoir Model
EC	Environment Canada
ECMWF	European Centre for Medium-range Weather Forecasts
ELCOM	A 3-D hydrodynamic model
ENSO	El Niño and Southern Oscillation
ERA	European Reanalysis of Global Atmospheric data (from ECMWF)
ERA-40	40-year Global Reanalysis data from ECMWF
GAME	GEWEX Asian Monsoon Experiment
GAPP	GEWEX Americas Prediction Project (formerly GCIP)
GCIP	GEWEX Continental-scale International Project
GCM	Global Climate Model; or General Circulation Model
GEM	Global Environmental Multi-scale Model
GEWEX	Global Energy and Water Cycle Experiment
GFDL	Geophysical Fluid Dynamics Laboratory
GHP	GEWEX Hydrometeorology Panel

GRACE	Gravity Recovery and Climate Experiment
GPCP	Global Precipitation Climatology Project
GRDC	Global Runoff Data Center
GSC	Geological Survey of Canada
GTOPO-30	Global 30 Arc-Second Elevation Data Set
HYDAT	Hydrometric Data from Environment Canada
ISBA	Interactions Soil-Biosphere-Atmosphere (land surface scheme)
ISCCP	International Satellite Cloud Climatology Project
LiDAR	Light Detection And Ranging
MAGS	Mackenzie GEWEX Study
MBIS	Mackenzie Basin Impact Study
MC2	Mesoscale Compressible Community Model
MEC	Modèle Environnemental Communautaire (of CMC)
MODIS	Moderate-Resolution Imaging Spectroradiometer
MRB	Mackenzie River Basin
MSC	Meteorological Service of Canada
NARR	North American Regional Reanalysis (from NCEP)
NAO	North Atlantic Oscillation
NASA	National Aeronautics and Space Administration
NCAR	National Center for Atmospheric Research
NCEP	National Center for Environmental Prediction
NESDIS	National Environmental Satellite, Data and Information Service (NOAA)
NOAA	National Oceanographic and Atmospheric Administration
NRC	National Research Council of Canada
NRCan	Natural Resources Canada
NSIDC	National Snow and Ice Data Center
NSERC	Natural Sciences and Engineering Research Council
NWP	Numerical Weather Prediction
NWRI	National Water Research Institute
NWT	Northwest Territory, Canada
PBSM	Prairie Blowing Snow Model
PDO	Pacific Decadal Oscillation
PIEKTUK	York University blowing snow model
PNA	Pacific North American Oscillation
RadarSat	Canadian Space Agency satellite designed to study polar regions
RCM	Regional Climate Model
RFE	Regional Finite Element model

RIVJAM	River ice jam model
SAR	Synthetic Aperture Radar
ScaRaB	Scanner for Earth Radiation Budget
SPOT	Satellite Probatoire d'Observation de la Terre
SEF	Canadian Global Spectral Forecast model
SLURP	Semi-distributed Land Use-Based Runoff Processes (hydrological model)
SMMR	Scanning Multichannel Microwave Radiometer
SRES	Special Report on Emissions Scenarios
SSM/I	Special Sensor Microwave Imager
TOPAZ	TOPgraphic PArameterZation
WATCLASS	A coupled model of WATFLOOD and CLASS
WATFLOOD	University of Waterloo river basin model
WCRP	World Climate Research Program
WEBS	Water and Energy Budget Study
WSC	Water Survey of Canada

Chapter 1

Synopsis of Atmospheric Research under MAGS

Ming-ko Woo

Cold regions present a challenge to atmospheric and hydrologic research. Their low temperatures test the endurance of field workers and their instruments; their distance from large urban centers raises the cost of logistics and reduces the availability of measured data. Yet, these regions are sensitive to climate variability and change and are prone to accelerated human activities. Improved understanding of the physical processes and enhanced capability to model the regional atmospheric and hydrologic conditions are the keys to their environmental preservation and resource sustainability.

Volume I of this book documents the atmospheric research on the cold environment, principally of the Mackenzie River Basin (MRB) in northern Canada. Volume II is a complementary report on the hydrologic aspect of this cold region. To provide continuity with Volume I, a synopsis of its chapters is presented. This information is provided in the spirit of the Mackenzie GEWEX Study (MAGS) that emphasizes the integration of atmospheric and hydrologic research.

The first volume is entitled "Atmospheric Processes of a Cold Region: the Mackenzie GEWEX Study Experience". All its chapters are listed in the Appendix. The Introduction chapter (Woo et al., Chap. 1) presents the physical setting of the Mackenzie Basin and places its cold environment in the context of the continental water and energy balances. In 1994, an interdisciplinary study began in Canada to participate in the international Global Energy and Water Cycle Experiment (GEWEX) under the auspices of the World Climate Research Program. The goals of the MAGS were (1) to understand and model the high-latitude energy and water cycles that play roles in the climate system, and (2) to improve our ability to assess the changes to Canada's water resources that arise from climate variability and anthropogenic climate change.

Szeto et al. (Chap. 2) provide a synthesis of atmospheric research contributions of MAGS. Major scientific accomplishments include improved understanding of the large-scale atmospheric processes that control the transport of water and energy into and out of the MRB; enhanced understanding of the interactions of large-scale atmospheric flows with physical features of the Basin to affect its weather and climate; and applications of research results to address climate issues in the Basin and other cold regions. This overview is a companion paper to Woo and Rouse (Chap. 2 in this Volume) which surveys the hydrologic process research undertaken under MAGS.

Most parts of the MRB are located in a center of high pressure in the winter. Ioannidou and Yau (Chap. 3) showed that MRB anticyclones are deep, warm core structures developed through an amplification of the climatological semi-permanent ridge over western North America. They are also sensitive to variation of the Pacific–North American and the Arctic Oscillations, manifested in a weaker or stronger meridional/zonal orientation of the anticyclonic activities. These anticyclones are transformed into cold core structures as they move to eastern Canada. The weather of the MRB is also strongly affected by the pressure pattern and storm systems over the North Pacific. As the Pacific airflow crosses the lofty western Cordillera and descends the mountains, adiabatic warming associated with subsidence governs the heat budget (Szeto, Chap. 4). Anomalously warm or cold winters are related to the strength of this incursion of the North Pacific air. A large winter temperature variability is linked to the interaction between this airflow and the regional environment, notably the Cordillera. On top of this variability is a significant winter warming trend in the last several decades (Cao et al., Chap. 5). Warming events are largely due to the advection of warm air from west and south of the Basin and through adiabatic descent induced by topography and by the pressure system, especially when low-level temperature inversion occurs.

Vapor flux into the MRB comes from the Pacific and Arctic Oceans. These sources also bring moisture to its southern neighbor, the South Saskatchewan Basin in the prairies, though the prairies also receive moisture flux from the Gulfs of Mexico and California, and from the Hudson Bay (Liu et al., Chap. 6). Although there are moisture exchanges between the two basins, topography and surface properties give rise to differences in their vertical profiles of moisture transport. In the MRB, there are occasions when extreme summer rainfall (>100 mm) is fed by moisture that can be traced back to the Gulf of Mexico (Brimelow and Reuter, Chap. 7). In these cases, rapid lee cyclogenesis over Alberta (associated with a 500-hPa cutoff low) and the Great Plains Low Level Jet act in unison to transport moisture to southern MRB. The transport time is 6–10 days.

Within the MRB, local evaporation is also a moisture source for precipitation (Szeto, Chap. 8). Compared with several other basins, the MRB has a precipitation recycling ratio that is below that of the Amazon (30%) but similar to the Mississippi and the Lena basins (23–25%). The ratio is expectedly higher in the summer than in the cold season, and about half of the summer precipitation in the downstream regions of these basins is derived from local evaporation.

Closely associated with precipitation is the cloud system, both features being linked to synoptic forcing conditions in the MRB (Hudak et al., Chap. 9). Intense observation at Fort Simpson (1998–99) near the center of the Basin, including cloud radar sampling, showed the common occurrence of multi-layered clouds and a reduction in precipitation due to sublimation beneath or between cloud layers. Both observation and modeling of clouds may be improved. Ground measurement of precipitation has allowed a set of precipitation–elevation relationships to be developed on the lee slopes of the Alberta foothills (Smith, Chap. 10). The monthly accumulated precipitation is linearly correlated with elevation when precipitation exceeds 70% of the long term average and this altitudinal increase depends on the total precipitation observed. Dupilka and Reuter (Chap. 11) noted the linear relationship between maximum snowfall and cloud base temperature. For Alberta exclusive of the mountains, it was found that the snow amount depends roughly linearly on the 850 hPa temperature.

Passive microwave data are well suited to snow cover application and Derksen et al. (Chap. 12) reviewed the performance of snow water equivalent retrieval algorithm at high latitudes. There is recent progress in understanding the impact of subgrid land cover variability, validating algorithm performance in northern boreal forest, and the development of algorithm for the boreal forest and tundra. Snow cover extent in the MRB was analyzed using satellite data, ground observation, and Canadian Regional Climate Model (CRCM) simulations (Derksen et al., Chap. 13). Consistent with air temperature trends, there are significant decreases in spring snow cover duration and earlier snow disappearance, but no significant trends for snow cover onset in the fall. An east–west gradient in the MRB shows earlier melt in the mountains, little change or slight increase in the spring snow cover in the northeast, and a cyclical behavior in the south that is linked to the Pacific–North America and Pacific Decadal Oscillation. A zone with high winter snow water equivalent (>100 mm) occurs across the northern fringe of the boreal forest. CRCM simulations indicate its correspondence with the mean monthly patterns of the 850 hPa frontogenesis forcing.

The winter snow cover is subject to blowing snow transport and sublimation. Blowing snow events are rare in the boreal forest but common in the tundra (Déry and Yau, Chap. 14). A parameterization based on the PIEKTUK-D model and the ERA-15 data provides a first-order estimate that indicates surface and blowing snow sublimation would deplete 29 mm yr^{-1} or about 7% of the annual precipitation of the MRB. Gordon and Taylor (Chap. 15) also parameterized the sublimation of blowing snow at six locations in the Basin. They predicted that up to 12% of the total snow precipitation in the Basin may be removed by this process, though these amounts could be an overestimate.

Assessment of the water and energy budget of the MRB is a major goal of MAGS. Szeto et al. (Chap. 16) provided a comprehensive climatology using remotely sensed and ground observations and several sets of reanalyzed and modeled data. The magnitude of the residuals in balancing the budgets is often comparable to the budget terms themselves and the spread of the budget estimates from different datasets is also large. However, the water budget closure for the Basin was within 10% of the measured runoff. Discussion of the sources of error and level of uncertainty suggests areas of improvement for the observation and the models.

Moisture flux convergence for the MRB showed significant yearly variations (Schuster, Chap. 17). Within the decade of 1990–2000, the 1994–95 water year had the lowest flux convergence. For some years the maximum flux convergence was at 850 hPa, implying that there was a large overall moisture flux into the MRB. However, usually the largest moisture flux convergence was at 700 hPa and this was particularly prominent in 1994–95, the year with record low discharge for the Mackenzie River.

The abundance of lakes of various sizes plays an important role in the energy and water cycling of the MRB. Rouse et al. (Chap. 18) used results of measured data to study the sensitivity of energy and moisture exchanges between lakes and the atmosphere. They found that were the Basin devoid of lakes, there would be a positive water balance for wet and average years and only a small negative balance for the driest years. If all the lakes in the region were large or medium-sized, there would be a regional evaporation increase by 8–10%.

Satellite observation and modeling are a major source of information for the study of energy budget of a vast domain. Trishchenko et al. (Chap. 19) used satellite information to examine the seasonal and interannual variability of albedo of the MRB for 1985–2004. They found a seasonal change in the broadband albedo from 0.11±0.03 in summer to 0.4–0.55 in winter, but no systematic trends because of large interannual variability, uncertainties in sensor properties, atmospheric correction, and retrieval procedures. Guo et al. (Chap. 20) compared the net surface radiation fluxes from satellite observation with those from the CRCM. They found that the latter overestimates short-wave fluxes at the top of atmosphere and underestimates the surface absorbed solar fluxes. CRCM also underestimates the outgoing long-wave fluxes at the top of atmosphere in winter. Differences were noted between the cloud fields simulated by the model and deduced from satellite measurements. Clouds and wildfire aerosols have strong influences on the radiation budgets of the MRB. Guo et al. (Chap. 21) investigated their radiative forcing at the top of atmosphere and at the surface. Overall, the cloud forcing could impact the short-wave and long-wave radiation budgets by 30–50%, both at the top of atmosphere and at the surface of the MRB region.

Most wildfires in northern Canada are started by lightning. Thunderstorms and associated lightning also play an important role in the cycling of water and energy during the warm season over the boreal and subarctic ecosystem. Kochtubajda et al. (Chap. 22) used observational datasets and model-derived products to characterize these storms and to examine their impacts on the forests and polar bear habitats. Fires peak in July but much of the burning occurs in June. The region of maximum lightning activity varies in space and time, and there is evidence that smoke from fires enhances the probability of positive cloud-to-ground lightning flashes. A tree-structured regression method successfully predicts the probability of lightning. Under a future warmer climate, more severe fire weather, more area burned, more ignition, and longer fire season are expected (Flannigan et al., Chap. 23). The burned area may increase by 25–300% and the fire season may lengthen by 30–50 days over a large of the Northwest Territories in Canada.

A final section presents contributions to atmospheric modeling. Ritchie and Delage (Chap. 24) analyzed the predictions of water and energy fluxes over the MRB by the Canadian Global Spectral Forecast model (SEF) which was used by the Canadian Meteorological Centre for weather forecasting. They found that the accumulation of precipitation-minus-evaporation is highly sensitive to initial conditions. Within SEF, the replacement of the force-restore scheme by the Canadian Land Surface Scheme (CLASS) improves the predicted energy and water budgets for the MRB, and better initialization of the CLASS variables can greatly affect its performance. CLASS was also used in the MAGS version of the CRCM, tailored for use over North America (MacKay et al., Chap. 25). Its simulation of the MRB climate was evaluated against surface observations. The model yielded an annual precipitation bias of 13% and a surface temperature cold bias of <1°C, suggesting that the physics package developed by a coarse-resolution GCM can be used in high resolution regional climate model with minimum modification.

Volume I presents the primary atmospheric processes that occur in northwestern North America, particularly in the domain of the MRB. Many of the research methods employed, including ground based observations, remote sensing techniques and modeling approaches, are shared by hydrologic studies in MAGS. Most of the atmospheric processes summarized in this Chapter are related to or even responsible for the hydrologic activities discussed in Volume II.

Appendix

List of chapters in Volume I

Woo MK, Rouse WR, Stewart RE, Stone JMR (Chapter 1) The Mackenzie GEWEX Study: a contribution to cold region atmospheric and hydrologic sciences

Szeto KK, Stewart RE, Yau MK, Gyakum J (Chapter 2) The Mackenzie Climate System: a synthesis of MAGS atmospheric research

Ioannidou L, Yau MK (Chapter 3) Climatological analysis of the Mackenzie River Basin anticyclones: structure, evolution and interannual variability

Szeto KK (Chapter 4) Variability of cold-season temperatures in the Mackenzie Basin

Cao Z, Stewart RE, Hogg WD (Chapter 5) Extreme winter warming over the Mackenzie Basin: observations and causes

Liu J, Stewart RE, Szeto KK (Chapter 6) Water vapor fluxes over the Canadian Prairies and the Mackenzie River Basin

Brimelow JC, Reuter GW (Chapter 7) Moisture sources for extreme rainfall events over the Mackenzie River Basin

Szeto KK, Liu J, Wong A (Chapter 8) Precipitation recycling in the Mackenzie and three other major river basins

Hudak D, Stewart RE, Rodriguez P, Kochtubajda B (Chapter 9) On the cloud and precipitating systems over the Mackenzie Basin

Smith CD (Chapter 10) The relationship between monthly precipitation and elevation in the Alberta foothills during the Foothills Orographic Precipitation Experiment

Dupilka ML, Reuter GW (Chapter 11) On predicting maximum snowfall

Derksen C, Walker A, Toose P (Chapter 12) Estimating snow water equivalent in northern regions from satellite passive microwave data

Derksen C, Brown R, MacKay M (Chapter 13) Mackenzie Basin snow cover: variability and trends from conventional data, satellite remote sensing, and Canadian Regional Climate Model simulations

Déry SJ, Yau MK (Chapter 14) Recent studies on the climatology and modeling of blowing snow in the Mackenzie River Basin

Gordon M, Taylor PA (Chapter 15) On blowing snow and sublimation in the Mackenzie River Basin

Szeto KK, Tran H, MacKay M, Crawford R, Stewart RE (Chapter 16) Assessing water and energy budgets for the Mackenzie River Basin

Schuster M (Chapter 17) Characteristics of the moisture flux convergence over the Mackenzie River Basin for the 1990–2000 water-years

Rouse WR, Binyamin J, Blanken PD, Bussières N, Duguay CR, Oswald CJ, Schertzer WM, Spence C (Chapter 18) The influence of lakes on the regional energy and water balance of the central Mackenzie River Basin

Trischenko AP, Khlopenkov KV, Ungureanu C, Latifovic R, Luo Y, Park WB (Chapter 19) Mapping of surface albedo over Mackenzie River Basin from satellite observations

Guo S, Leighton HG, Feng J, MacKay M (Chapter 20) Comparison of solar radiation budgets in the Mackenzie River Basin from satellite observations and a regional climate model

Guo S, Leighton HG, Feng J, Trischenko A (Chapter 21) Wildfire aerosol and cloud radiative forcing in the Mackenzie River Basin from satellite observations

Kochtubajda B, Flannigan MD, Gyakum JR, Stewart RE, Burrows WR, Way A, Richardson E, Stirling I (Chapter 22) The nature and impacts of thunderstorms in a northern climate

Flannigan MD, Kochtubajda B, Logan KA (Chapter 23) Forest fires and climate change in the Northwest Territories

Harold Ritchie and Yves Delage (Chapter 24) The Impact of CLASS in MAGS Monthly Ensemble Predictions

MacKay M, Bartlett P, Chan E, Verseghy D, Soulis ED, Seglenieks FR (Chapter 25) The MAGS Regional Climate Modeling System: CRCM-MAGS

Woo MK (Chapter 26) Synopsis of hydrologic research under MAGS

Chapter 2
MAGS Contribution to Hydrologic and Surface Process Research

Ming-ko Woo and Wayne R. Rouse

Abstract The Mackenzie GEWEX Study (MAGS) research contributed to advancement in our knowledge on hydrologic and surface processes common to all cold regions. These include the accumulation, sublimation and ablation aspects of the snow in boreal forest and tundra areas; infiltration into and thawing of frozen soil; breakup of river ice and the associated floods. Additionally, there are several land surface features distinctive to the Mackenzie River Basin, including lakes and wetlands, mountainous topography, Precambrian Shield and organic terrain. Hydrologic knowledge on these landscapes was gained through field research, conceptualization and modeling effort. Most of these studies were carried out at a local scale that allows understanding of the physical processes through intense field and modeling investigations.

1 Introduction

As for all cold regions, snow, frost and ice exert considerable influence on the hydrology of the Mackenzie River Basin (MRB). Additionally, there are several land surface features distinctive to the Basin. They include lakes and wetlands, mountains, Precambrian Shield and organic terrain. The surface and hydrologic processes are important in the water balance through evaporation, storage and runoff. Latent heat exchanges are particularly relevant because of their roles in ground freeze-thaw, ice formation and breakup, snow sublimation and melt, and evaporation. The significant research results arising from the Mackenzie GEWEX Study (MAGS) that deal with these processes and features are addressed in this synthesis.

2 Snow Processes

The MRB is snow-covered for five to eight months each year (Hydrological Atlas of Canada 1978), with the duration increasing northward towards the Arctic coast and westward at high elevations (Fig. 1). Winter precipitation comes mainly as snowfall but rainfall can occur occasionally, particularly in the south. The snow cover is subject to sublimation loss and redistribution by drifting. Snowmelt in the spring is a major hydrologic event that triggers extensive runoff, infilling of lakes and ponds, inundation of wetlands together with river ice breakup and the flooding of riparian zones. Problems associated with the measurement of snowfall have been discussed extensively (Goodison 1978; Yang et al. 1999). The distribution of snowfall (Fig. 1) has to be interpreted with caution because of possible inaccuracies, especially due to gauge undercatch. Research carried out under MAGS has emphasized snow ablation rather than accumulation. Snow sublimation and melt were the main processes considered.

2.1 Snow Sublimation and Redistribution

Sublimation of snow intercepted on the tree canopy reduces snow accumulation on the forest floor. Tree-weighing experiments conducted in northern Saskatchewan (Fig. 2) and southern Yukon clearly demonstrated the loss of intercepted snow to sublimation, melt and shedding from the tree canopy (Pomeroy et al. 1998). Leaf area, canopy closure, vegetation type, time since snowfall, snowfall amount and the existing snow load determine the efficiency with which snow is intercepted. Cold coniferous forest canopies can store more than half of the cumulative snowfall and much of this can be lost by sublimation. Parviainen and Pomeroy (2000) developed a boundary layer model that provides a reasonable approximation of sublimation losses and the within-canopy energetics, enabling an evaluation of the role of sublimation from a boreal forest.

On the scale of a forest, Pomeroy et al. (2002) provided simple yet physically appropriate equations for estimating snow accumulation beneath forest canopies, relating canopy properties (leaf area index or canopy density) to either snowfall or snow accumulation in the clearings. For sites where mid-winter melt, snow redistribution by wind, and surface sublimation are infrequent or limited, this approach can be used to estimate the seasonal snow accumulation. Comparisons with data and results from eastern Europe and Siberia suggest that these findings are transferable.

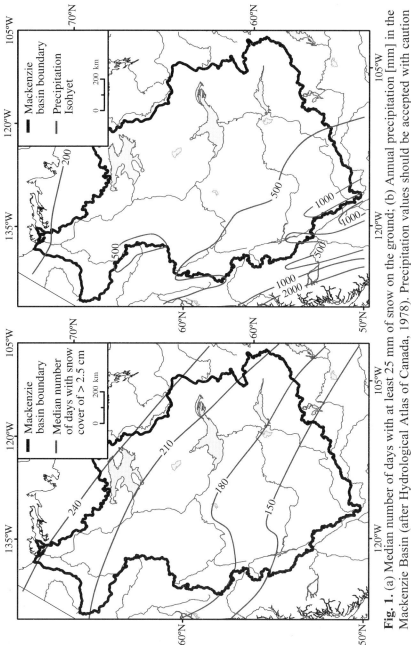

Fig. 1. (a) Median number of days with at least 25 mm of snow on the ground; (b) Annual precipitation [mm] in the Mackenzie Basin (after Hydrological Atlas of Canada, 1978). Precipitation values should be accepted with caution as snowfall is usually underestimated

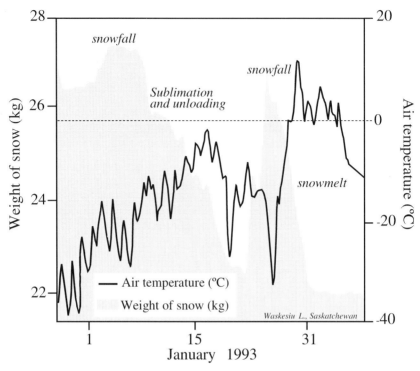

Fig. 2. Weight of snow on a 9-m tall black spruce tree as snow accumulated, sublimated, melted and slid off its canopy. Experiment was carried out in a forest in northern Saskatchewan. (Modified after Pomeroy and Gray 1995)

Snow redistribution by wind is an important process in the open terrain. Accompanying blowing snow events is the loss to sublimation by the transported snow. However, there are large discrepancies among various blowing snow models concerning the rates of sublimation loss. Essery et al. (1999), using their Prairie Blowing Snow Model (PBSM), suggested that 47% of the snow in a small tundra basin (Trail Valley Creek; see Marsh et al. 2007 for description of the basin) sublimated over a winter period. For the same study site and year, Déry and Yau (2001) utilized a version of the PIEKTUK model and indicated that the near-surface relative humidity quickly approached saturation with respect to ice; consequently sublimation becomes limited. On a regional scale, Déry and Yau (2002, 2007) used reanalysis data at a resolution of 2.5° to study the effects of surface sublimation and blowing snow on the surface mass balance. They found that surface sublimation removes 29 mm yr^{-1} snow water equivalent (SWE), or about 7% of the annual precipitation of the MRB. Taylor and

Wilson (2003) presented an example (Fig. 3) that showed sublimation as estimated by the PIEKTUK model could be an order of magnitude lower than by the PBSM.

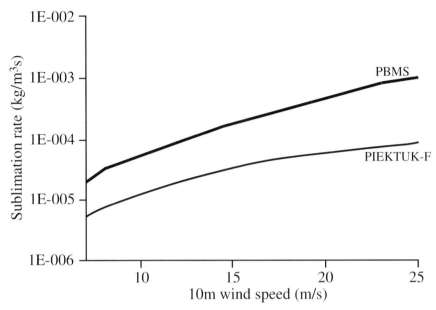

Fig. 3. Comparison of snow sublimation calculated using the Prairie Blowing Snow Model and the PIEKTUK model, showing discrepancies mainly due to different assumptions of air saturation during blowing snow events. (After Taylor and Wilson 2003)

The lack of agreement between two modeling approaches hinges upon the role of advective process that hinders the rise of relative humidity during a blowing event, thus preventing the attainment of the self-limiting condition for sublimation. Perhaps such differences in the evaluation of sublimation may be reconciled by considering that the air quickly becomes saturated under a wet regime, such as over sea ice, but under a dry regime there is always an overlying pool of dry air that can be entrained into the boundary layer to prevent saturation (Liston and Sturm 2004).

In the open tundra environment, blowing snow is strongly affected by the terrain which therefore plays a controlling role in the relocation of snow. Studies in Trail Valley Creek basin indicate that roughly two-thirds of the SWE accumulated as snowdrifts in shrub tundra that covers only one-third of the landscape while one-third of the SWE remained on the upland tundra that occupies two-thirds of the basin (Pomeroy et al. 2007).

Such drifts result in delayed runoff due to percolation through the deep snow, which needs recognition in runoff model development (Marsh et al. 2002).

2.2 Snowmelt

Snowmelt studies were conducted in the forest and tundra environments in MRB and in Wolf Creek basin in Yukon Territory. In terms of snowmelt around a single tree in subarctic woodland, Woo and Giesbrecht (2000) observed and modeled the decrease in SWE close to the tree trunk. They also measured reduced short-wave radiation caused by shadows, but enhanced long-wave radiation to the snow from the tree. Giesbrecht and Woo (2000) extended the results to an open woodland where the spruce trees occupy only 2% of a slope. They found that due to low sun angles, elongated tree shadows of the low-density canopy affect a disproportionately large area (Fig. 4). As the shadow zone receives only diffuse but not direct beam radiation, the contrast between the shadow and no-shadow areas produces spatial melt variability which increases as the melt season

Fig. 4. Shadows cast by tress in an open woodland in Wolf Creek basin showing that in spite of low tree density, elongated shadows can shade a disproportionately large part of the snow surface relative to the canopy area. Note also the increased melt around the shrub that emerges from the snow in the foreground. (Photo: M. Giesbrecht)

advances. For a boreal forest, Faria et al. (2000) obtained a covariance relationship between SWE and snowmelt rate, whereby snowmelt becomes more rapid as the forest snow cover diminishes. This is probably due to decreasing albedo as the snow cover thins and increasing heating of the emerging shrubs which, in turn, hastens the snowmelt (Pomeroy et al. 2006).

In the tundra where the snow is unevenly distributed, the snow cover becomes fragmented soon after the commencement of the melt period. This fosters an advection of heat between snow-free and snow-covered patches. Previous works on such heat advection utilize detailed boundary-layer models (Liston 1995), simplified approaches based on Weismann (1977), or parameterization based on the proportion of the area that is snow covered (Neumann and Marsh 1998). While these approaches have met with success in improving the representation of advection in melt enhancement over small areas, a major issue remains in scaling up the results to large areas of the Arctic and the subarctic that regularly experience fragmentation of the snow cover soon after the commencement of the melt period. One difficulty is to properly account for the effects of patch size and geometry on advection (Marsh et al. 1999). To overcome such a problem, Granger et al. (2002) formulated a set of simple equations that reduces the advection term to a power function of the snow patch size. This method requires standard meteorological variables, plus surface temperature of the snow and of the snow-free patches, and the snow patch length.

2.3 Patterns of Snowmelt

For calculating water balances and runoff generation, it is useful to depict the spatial pattern of snow distribution during the melt period. Where sufficient meteorological data are available and where these data are transferable over an area, models have been developed to extend point measurements over small basins to permit computation of snow accumulation, redistribution and melt. An example is from the Trail Valley Creek basin where the accumulation and melt in the tundra environment were successfully simulated (Marsh et al. 2007; Pohl and Marsh 2006). On the site scale, Shook and Gray (1996) applied the log-normal density function to characterize the distribution of SWE, and examined the change in the parameters during melt. Faria et al. (2000) simulated the snow cover area depletion for boreal forests as a function of the distribution of SWE and the covariance between SWE and melt rate (Pomeroy et al. 2007).

Energy fluxes were studied through the use of airborne flux measurements over a large area in the lower MRB during the melt season as part of the Mackenzie GEWEX Study Enhanced Observation Period of the 1998–99 water-year. The maps shown in Fig. 5 illustrate the large spatial variability in sensible and latent heat fluxes over a tundra that was partially snow covered (Brown-Mitic et al. 2001). The results also showed large differences between tundra and boreal forest sites. The tundra landscape converts more than 80% of the net radiation to non-turbulent fluxes (primarily for snow melt and warming of the soil), whereas over the forested region, more than 50% was utilized in latent and sensible heat fluxes.

3 Frozen Ground

The entire MRB experiences winter freezing so that all locations in the Basin are subject to the influence of seasonal frost. Furthermore, permafrost underlies about two-thirds of the Basin. The thermal process of ground freeze-thaw has large effects on water movement as hydraulic conductivity often drops by orders of magnitude when the soil freezes (Burt and Williams 1976). Conversely, freeze-thaw rates are greatly influenced by soil moisture content which affects the soil thermal conductivity. There is thus a strong association between the soil thermal and hydrologic regimes in the cold region.

3.1 Ground Frost

A lack of data in most cold regions necessitates the use of simple schemes with low data requirement to compute ground freeze-thaw. One suitable approach is to model freeze-thaw by heat conduction using the Stefan algorithm (Woo et al. 2007). The equation was originally formulated for lake ice melt but adapted for ground freeze-thaw calculations (Jumikis 1977) and has been applied to Alaska and the Tibetan Plateau (Fox 1992; Li and Koike 2003). It is a robust method grounded on the assumptions that conduction is the primary mechanism of heat flow and all the available heat is used for melting of ground ice. The one-directional freeze-thaw (from the top) scheme was improved by Woo et al. (2004) to take account of both downward and upward freeze and thaw. The algorithm can be incorporated into a land surface scheme (in this case, the Community Land Model 3) to improve the simulation of ground freeze-thaw in northern environments (Yi et al. 2006).

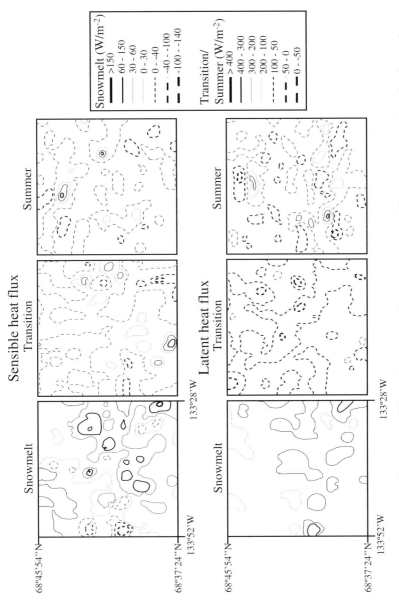

Fig. 5. Spatial variability in sensible and latent heat fluxes over a partially snow covered tundra in northern Mackenzie Basin. (Modified after Brown-Mitic et al. 2001)

The Stefan algorithm was used to simulate ground thaw for the purpose of investigating climatic influence on northern infrastructures, including pipeline (Woo et al. 2007). Simulations using Stefan's model demonstrate the sensitivity of active layer thaw to (1) soil materials, due to their differential thermal properties, (2) moisture content, which largely controls the latent heat requirement for phase change, and (3) inter-annual variations in near-surface temperature and duration of thaw season, which are governed by the climate. The presence of an organic layer on top of mineral soils, a common occurrence in northern areas, greatly limits annual ground thaw because of the insulating properties of porous peat and for saturated peat, the large amount of latent heat needed for phase change.

The Stefan algorithm was also used in conjunction with field observations to study the winter freezing and the thermal regime of four slopes in the Wolf Creek basin, Yukon Territories (60°31.7′ N; 135°07.1′ W); for details of the basin see Martz et al. (2007). On slopes with dry soil but no permafrost, frost penetration reached 1.2–1.6 m (Carey and Woo 2005). For slopes with permafrost, two-sided freezing occurred whereby freezing proceeded downward from the ground surface and upward from the permafrost table. This rapidly closed the seasonal thaw zone which was only 0.6–0.8 m thick. The principal determinants of ground freezing include moisture content of the soil, hence the amount of heat required for phase change, and properties of the soil (organic materials with lower thermal conductivity than mineral soil). Conduction is the dominant mechanism of heat transfer but lateral drainage that occurs on some slopes can convect heat to retard descent of the freezing front.

3.2 Frozen Soil Infiltration

Frozen soils are generally considered to be impervious to meltwater infiltration, but this depends greatly on the amount of ground ice that seals the soil pores. Gray et al. (2001) distinguished frozen soils into three infiltration categories: restricted (water entry impeded by surface conditions), limited (capillary flow dominates), and unlimited (gravity flow allows most meltwater to infiltrate). For limited infiltrability which is common for most mineral soils, cumulative infiltration is expressed as a function of moisture content at the soil surface, average soil saturation (including both water and ice), average temperature of the top 40 cm at the start of infiltration, and a factor known as infiltration opportunity. More detail on frozen soil infiltration is provided in Pomeroy et al. (2007).

For organic materials which commonly overlie mineral soils in permafrost areas, infiltration is often facilitated by the large porosity. With the exception of wetlands, infiltration into the frozen organic cover is usually unimpeded since the soil pores are seldom fully ice-filled. The organic soil may become much compacted below the surface layer, however, and this can greatly decrease the porosity to limit percolating into the deeper layers.

4 Organic Terrain

Organic soils are ubiquitous in northern North America and in Siberia (Aylesworth and Kettles 2000; Zhulidov et al. 1997). These soils may include peat and living plants such as lichens and mosses. The peat can be at various stages of development, ranging from scarcely decomposed and little compacted to highly humified and dense. Quinton et al. (2000) noted that in a peat profile, the hydraulic conductivity can change within a depth of 0.3 m by three orders of magnitude, from almost 100 m d^{-1} near the surface to <1 m d^{-1} at depth. Other hydrologic properties such as porosity, specific yield and retention also change with depth. Peat is also easily erodible so that rills and pipes are often found. A combination of organic and mineral soils produces distinct flow mechanisms in permafrost areas, as was revealed by field studies carried out primarily in two areas: a tundra site near the Mackenzie Delta comprising an earth hummock field, and Wolf Creek where some slopes of different orientations and elevations are covered by an organic layer.

4.1 Flow Mechanisms in Organic Terrain

In hummocky permafrost terrain where mineral hummocks are separated by peat-filled inter-hummock channels, large contrasts in hydraulic conductivity give rise to preferential flow in the organic materials, mainly in the acrotelm which is the hydrologically active top horizon with high porosity. The bottom peat layer, known as the catotelm, is humified and compacted. Slope runoff is concentrated in the inter-hummock channels where water movement is particularly effective through matrix flow and pipeflow in the acrotelm, capable of delivering 0.1 to 1.0 m^3 d^{-1} per unit width (Quinton and Marsh 1999). Such subsurface flow discharges are as rapid as surface runoff, and subsurface drainage is the predominant mechanism of water transmission (Quinton et al. 2000). Lateral flow is obstructed by the mineral earth hummocks (Fig. 6) so that the flow paths tend

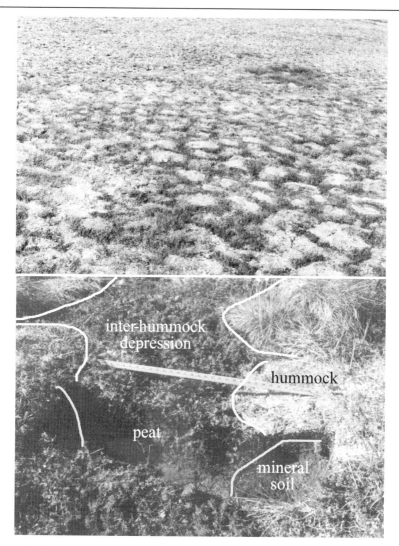

Fig. 6. Earth hummocks in the tundra of Siksik Creek area, Northwest Territories: (top) a hummock field occupying the right and foreground areas, with maximum hummock diameter of about 1 m; to the left of the hummocks is a band of sedge tussocks, (bottom) soil pit excavated down to the frost table, showing a profile that includes mineral soil on either side of a peat-filled inter-hummock depression. (Photos: W.L. Quinton)

to follow a tortuous network that consists of primary drainage oriented in a downslope direction, fed by a number of secondary channels that collect water across the slope. Increasing tortuosity lengthens the time of runoff delivery along the hillslopes (Quinton and Marsh 1998).

On the hillslopes, peat thickness is about 0.3 m on 90 percent of the total area but reaches 0.4–0.7 m near the streambanks. Quinton and Marsh (1999) noted that the hillslopes can be hydrologically distinguished into an upland and a near-stream zone. During the wet period, the saturated area extends to most parts of the slopes. A high water table enables rapid flow delivery in the acrotelm to facilitate hillslope–streamflow connection. However, the duration of high flow contribution from the hillslope to the stream varies, depending on the width of the near-stream saturated area, gradient and length of the slope, the SWE and the melt rate in the spring. In the dry summer season, saturation is confined largely within the near-stream zone and the upland is then hydrologically disconnected from the stream. An expansion of the runoff source area by summer storms will be delayed until the storage capacity of the near-stream zone is satisfied.

In the subarctic Wolf Creek basin, slopes underlain by permafrost have an organic layer of varying thickness overlying mineral soils. During snowmelt, meltwater infiltrates and percolates the frozen organic layer, rendered porous largely through an upward flux of vapor as the ground ice sublimates in the winter (Woo 1982). Percolation is retarded at the organic–mineral interface due to the impermeable nature of the frozen mineral soils, leading to the formation of a perched saturated zone above the organic–mineral interface (Carey and Woo 1999) and the initiation of slope runoff. Both surface flow and subsurface flow in the organic layer can deliver water rapidly downslope and they are collectively termed quick flow. The mode of runoff includes the flow in rill and gullies, along soil pipes and within the matrix of the organic soil (Carey and Woo 2000). Flow in the mineral substrate, if present, tends to be orders of magnitude lower than in the organic layer. This subsurface flow is termed slow flow. Where thick organic soils are distinguishable into acrotelm and catotelm zones, quick flow is restricted to the top layer. Where there is no distinctive catotelm development, the mineral substrate is the slow flow zone. This gives rise to a two-layer flow system in which most runoff is shed through the organic layer as quick flow (Carey and Woo 2001a; Fig. 7).

In the summer, ground thaw continues and the frost table drops into the mineral layer. This is accompanied by a lowering of the saturated zone that sits on the impermeable frozen soil. Slow flow may be produced in the catotelm or in the thawed mineral substrate but its magnitude is meager. Quick flow is re-generated when rain events raise the water table into the acrotelm layer. Typically, storm flow hydrographs respond rapidly to rainfall and runoff is the largest at the sites where antecedent soil moisture is sustained by inflow from upslope.

Fig. 7. Two-layer flow system, Wolf Creek basin, Yukon: (left) peat overlying mineral soil on a subarctic north-facing slope, (right) quick flow in the peat layer demonstrated by an outpour of water when the peat is lifted. (Photos: S. Carey)

4.2 Wetland Hydrology

Many areas of low relief in the discontinuous permafrost zone have developed wetlands with poorly integrated drainage channel networks. Quinton et al. (2003) noted that the wetland-dominated terrain on the Interior Plains can be distinguished into fens, isolated bogs, and peat plateaus underlain and raised by permafrost. Water may shed from the peat plateaus either to their adjacent bogs that serve mainly as a storage, or laterally to the channel fens. Water in the channel fens can drain directly to the basin outlet, but water in the flat bogs has to move to a channel fen before it can flow out of the wetland. Flow connections are intact and extensive during the snowmelt and high rainfall periods. At other times low rainfall, continuous drainage and evaporation will isolate many parts of a wetland from the main channels. This leads to a reduction in streamflow. Quinton and Hayashi (2007) commented that storage capacity of the isolated and flat bogs and the mode of flow should be taken into consideration when developing algorithms for routing water through the peat terrain.

5 Mountain Environment

Mountains constitute the headwater areas of such major rivers as the Liard, the Athabasca, the Arctic Red and the Peel. Field research undertaken in Wolf Creek Basin took advantage of its logistical convenience and data support even though it lies outside of the MRB. Investigations were carried out on four hillslopes and an upland site to emphasize two major landscapes: open woodland slopes and subalpine shrubland. In steep terrain at high latitudes, slope direction plays a major role in affecting the microclimate, as does the vegetation cover. On a long-term basis, vegetation growth, permafrost development and soil formation differ between elevations and among slopes. Shrubs dominate the subalpine zone and below it, spruce and aspen woodlands are common on north- and south-facing slopes, respectively. Many north slopes in the mountainous boreal zone, from central Alaska through Yukon to western Northwest Territories, have permafrost below a layer of peat, while south slopes often have leaf litter and humus on mineral soils without permafrost. The energy and water balances of slopes vary considerably, thus affecting the snow and runoff hydrology.

5.1 Flow Generation

Snowmelt and rainfall are the primary sources of runoff. Runoff production is greatly reduced where infiltration rates are high. Water balance shows that vertical processes prevail at the non-permafrost slope but horizontal water transfer is important on slopes with ice-rich frost. For permafrost slopes, the two-layer flow system (see Sect. 4.1) is the principal mode of lateral water transfer. In the spring, the frost table is shallow and quick flow is common on permafrost slopes. The rate of summer runoff is governed by a rise of the water table to activate quick flow in the organic layer. Evapotranspiration often exceeds rainfall (Carey and Woo 2001a). The potential to evaporate is controlled by radiation balance on slopes, though the actual rates are limited by the phenology of the deciduous forests and by the desiccation of the surface organic layer in late summer when there is a general depletion of soil moisture storage at all the slopes.

Carey and Woo (2001b) found that not all hillslopes in the Wolf Creek basin yield lateral runoff. Many south-slopes with seasonal frost in ice-poor silt permit deep percolation without generating surface flow. Lateral fluxes are produced only from slopes with an organic layer overlying ice-rich mineral substrate that hinders percolation, but the flow becomes negligible when the water table recedes into the mineral soil layer (due to water loss to evaporation and continuous descent of the impermeable frost table). The extent of the flow contributing area is variable, depending on the hillslope wetness and the properties of the organic soil. Thus, during any time of the year, only some parts of the mountainous catchment yield runoff to the streams.

Linkages among various hydrologic units must be established in order to route the flows from one surface to another and from the catchment slopes to the streams. On an upper slope in the subalpine shrub zone in the same Wolf Creek basin, Leenders and Woo (2002) observed surface flow from the shrubland to the woodland downslope to last for only several days during the snowmelt period. With a mineral soil of low ice content, percolation facilitates vertical drainage at the expense of lateral flow. Soon, lateral runoff ceases and flow linkage between the upper shrubland and lower woodland slope is severed. The upper and the lower slope segments essentially operate as isolated hydrologic entities. On the other hand, an east-facing slope site receives inflow mostly through a gully that connects the woodland slope to the shrubland above, suggesting hydrologic connection between these two land units via a channel linkage. These examples demonstrate that there are no straightforward flow connections among various terrain facets. Proper modeling of water delivery in a mountainous setting should take account of flow connectivity between the streams and their catchment areas.

6 Shield Landscape

The Canadian Shield is a mosaic of Precambrian bedrock uplands separated by soil-filled valleys, many of which are occupied by lakes and wetlands (Fig. 1 in Spence and Woo 2007). Crystalline bedrock outcrops in the Shield are not entirely impervious as conventionally considered, but are fissured by joints and faults to varying extent to permit the entry of meltwater and rainwater (Spence and Woo 2002). Frost may have little effect on infiltration if the cracks in the outcrops are not ice-filled to an extent that would inhibit meltwater entry into the bedrocks. Furthermore, sporadic presence of thin soil patches on the rock can absorb snowmelt and rain water, which is then gradually released to augment bedrock seepage or evaporation. Thus, the runoff ratio (ratio of runoff to rainfall or snowmelt) from the Shield upland is highly variable, depending on the density and size of the bedrock cracks, the patchy soil cover and the intensity of water input.

In the spring, the non-saturated, hence not ice-filled, soils within the valleys between the bedrock ridges also permit infiltration (Spence and Woo 2003), though a greater amount of water comes from their adjacent slopes. Slope runoff usually occurs as overland flow or seepage along the bedrock-overburden interface to reach the valleys. The water table at the valley sides are often the first to rise, and the rise then spreads to the center of the valley. Surface flow is initiated in a headwater valley only when the water table rises above ground, otherwise only subsurface flow may occur through the soil matrix and macropores. The valleys perform multiple hydrologic functions. They collect water from rain and snowmelt, and from lateral inflow from adjacent slopes; they alternately store and release storage water; and they transmit water down the valley. Flow along a valley can be intermittent if storage deficit created by evaporation exceeds the inputs. The valley is considered to comprise a number of segments with varying storage conditions. As water is delivered down-valley, each segment has to be filled until its storage threshold is exceeded. Only then will surface or subsurface flow be released to the segments downstream. This mechanism is termed the fill-and-spill runoff system (Spence and Woo 2003, 2007).

Lake–stream systems in the semi-arid Shield region offer instructive examples that illustrate the fill-and-spill concept. Mielko and Woo (2006) showed that rapid and substantial runoff from the local slopes during the snowmelt period leads to a rise of lake levels above their outlet elevations to generate outflow. Continued summer evaporation reduces lake storage below the outflow threshold (represented by the lake outlet elevation). Outflow ceases and the lake together with its surrounding drainage areas upstream are disconnected from the drainage system (Woo and Mielko 2007). To re-establish flow connectivity, the storage deficit has to be filled until lake outflow spills over the threshold. Woo and Mielko noted further that outflow from any lake along a valley is usually independent of the runoff generated at the lakes located above or below it. Each lake contributes runoff at different times, being controlled largely by the amount of runoff received from its direct catchment, and by the antecedent storage status of the lake relative to its outflow threshold.

7 Streamflow and River Ice

7.1 Streamflow Regime

The flow of most rivers in the MRB exhibits a nival regime (Church 1974) in which high flows occur during the snowmelt period. This is followed by flow recession but high flows can be revived in the summer by the occasional large rain events. Winter experiences low flows with streamflow sustained mostly by groundwater discharge. There are modifications to this seasonal flow pattern, notably when the nival regime is affected by lake outflow, glacier discharge or wetland storage (Woo 2000). In areas of high relief, the intensity of spring high flow is accentuated when snow melt occurs simultaneously across a range of elevations. This applies to small catchments such as the Wolf Creek basin (Janowicz et al. 1997) as well as to large areas such as the Liard River basin (Woo and Thorne 2006).

The Mackenzie River, with a drainage area of 1.8×10^6 km^2, also exhibits a nival regime of flow but its discharge is an aggregation of contributions from its tributaries, each of which has a slightly different timing of snowmelt high flow. Rivers from the lowlands and southern latitudes have earlier snowmelt peaks than those from the western mountains or located further north. Figures 8 and 9 show the average monthly flow contributions of the major tributaries, and the mean monthly runoff (in mm) of the Mackenzie River at Arctic Red River. Storage, especially in the great lakes, delays the Mackenzie River flow relative to the tributary discharges. Rivers originating in the Western Cordillera (Athabasca, Peace, Liard and other mountain rivers) provide 60% of the Mackenzie runoff. The Interior Plains and the eastern Shield, covering 40% of the total basin area, provide only about 25% of the Mackenzie discharge (Woo and Thorne 2003). Based on the 1973–2003 record, annual flow of the Mackenzie River at the village of Arctic Red River (before the Mackenzie River enters its delta) averages 8994 m^3 s^{-1}, with a standard deviation of 954 m^3 s^{-1}.

7.2 River Ice and Ice-jam Floods

Over 90% of Canadian rivers acquire an ice cover in winter. Spring floods are exacerbated by the breakup of the ice cover. The processes of river freeze-up and breakup received much attention by MAGS researchers (Hicks and Beltaos 2007). Floods associated with ice jams during spring breakup pose risks to life and property. These floods can be very sudden, with river levels rising several meters within a few minutes. At present, the

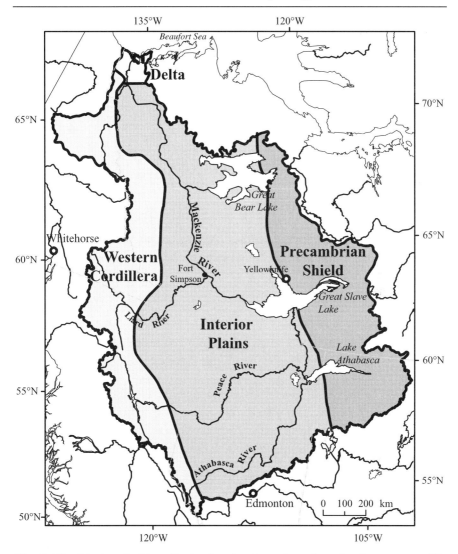

Fig. 8. The Mackenzie River Basin showing its major tributaries that traverse different physiographic regions: Mackenzie Delta, mountainous Cordilleran region, Interior Plains with many wetlands and lakes, and Canadian Shield.

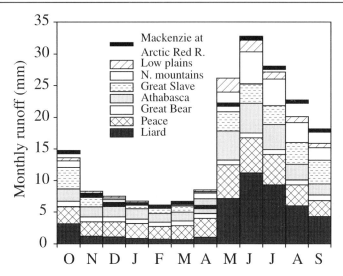

Fig. 9. Mean monthly runoff contribution to the Mackenzie River from its major tributary basins, and the mean monthly runoff of Mackenzie River at Arctic Red River

progression of breakup along inaccessible reaches of rivers is observed by aerial reconnaissance flights. Hicks et al. (2005) used satellite synthetic aperture radar images to characterize river ice during breakup. Although interpretation of the imagery is complicated by such factors as properties of the ice, illumination and geometric considerations, investigations of the ice on the Athabasca River near Fort McMurray demonstrated that this satellite tool has considerable potential for monitoring river ice and ice jams (Fig. 10).

Disintegration of the river ice can be distinguished into thermal and mechanical breakup. Mild weather and low runoff favor thermal breakup in which the ice cover deteriorates gradually, losing much of its strength before the final break up occurs. Mechanical breakup takes place when much of the ice integrity is still retained so that resulting ice jams and their consequent floods are more severe. Beltaos (2003a) developed an empirical function to determine the threshold between mechanical and thermal breakup events in terms of several factors: water surface elevation at which the ice cover starts to move, water surface elevation at which the ice cover formed during the preceding freeze-up, ice cover thickness prior to the start of melt and location on the river.

Historically, it has been virtually impossible to provide more than a few hours of advance flood warning and in many cases such warnings have proven false. A novel approach based on fuzzy logic provides forecasts 4–

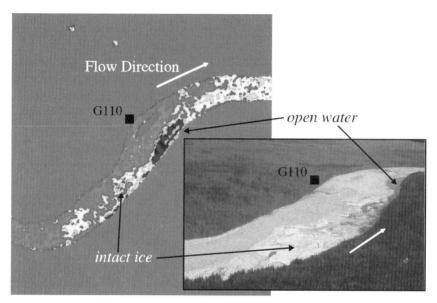

Fig. 10. Downstream end of an ice jam on Athabasca River, April 22, 2003, showing intact ice on the right bank upstream of the toe of the jam, and open water further downstream. (Image source: Hicks et al. 2005)

6 weeks ahead, to facilitate early emergency preparedness planning (Mahabir et al. 2007). Enhanced data acquisition, facilitated by remote sensing and by robust water level sensors developed specifically to survive the floods, permit computational hydraulic models to more accurately forecast the ice jam release surges and the associated floods (Blackburn and Hicks 2002, Hicks and Healy 2003). The interactive effects of meteorologic and hydrologic processes on ice jam occurrence and severity of dynamic breakup were applied to the Athabasca River to aid in ice jam prediction. Additionally, thermal ice process model components have been incorporated into a hydraulic model for application to the Peace River, which has a thermally dominated ice regime due to regulation by the Bennett Dam. A reduction of flood magnitude may have adverse effects on wetlands. On the lower reaches of the Peace River, for example, the rarity of ice-jam floods in the period following construction of the Bennett Dam has resulted in serious habitat degradation and risk to aquatic ecology within large areas of the Peace-Athabasca Delta (Beltaos 2003b). Under climate

warming scenario, the duration of ice season at the Delta is likely to be reduced by 2–4 weeks by the end of the century, while a large part of the Peace River basin may experience frequent and sustained mid-winter thaws (Beltaos et al. 2007). From a model study, Andrishak and Hicks (2007) also found that by the mid-21st century the ice cover duration at the Town of Peace River may be reduced by 28 days.

8 Lake Studies

Lakes are an important component of the MRB, covering about 11% of the total surface. Their size ranges from small ponds to some of the largest freshwater lakes in the world, including Great Bear Lake (65°N), Great Slave Lake (62°N) and Lake Athabasca (59°N). The Shield has the highest concentration of lakes and these are generally much deeper than those in the Interior Plains.

Ice cover plays a prominent role in affecting the thermal and hydrologic behavior of lakes. For small and shallow lakes, ice formation in early winter terminates their evaporation loss. In the spring, lake-ice melt can proceed rapidly with a combination of above-freezing temperatures and strong absorption of solar radiation, allowing vigorous evaporation to proceed early in the thaw season. Heating and cooling of shallow lakes are governed by the temperature of the overlying air masses. Typically, the period of evaporative loss from shallow subarctic lakes is about four months, but becomes shorter in the tundra region. In large deep lakes such as Great Slave and Great Bear lakes there is a strong asymmetry between heating and cooling rates. They take substantial periods to warm, but stay thawed into early winter due to considerable heat storage. Evaporation continues during the autumn ice-free period as the warmer water surface exchanges both heat and mass with the cold overlying air. Evaporation from large lakes operates over a period ranging from 5 to 6 months.

There are large differences in the freeze-up and breakup dates of large lakes. On average Great Slave Lake freezes about ten days later than Great Bear Lake and thaws about 21 days sooner (Walker et al. 2000). This may be related to water throughflow characteristics of the two lakes, with a considerably larger quantity of river discharge into and out of Great Slave Lake. The Canadian Lake Ice Model gives a reasonably accurate representation of the ice cover phenology on northern lakes (Duguay et al. 2003). For the MRB, model results indicate a decrease in the length of the ice cover period in the latter part of the 20th century (Duguay et al. 2003).

Blanken et al. (2000) documented large evaporation and sensible heat fluxes in the late fall and early winter from Great Slave Lake. Evaporation from this lake is episodic in nature, enhanced largely by the entrainment of dry air from above the surface inversion layer in early and mid-summer (Blanken et al. 2007). The interannual and seasonal variability of the surface energy balance and temperature of central Great Slave Lake indicates large lake-to-land temperature differences and their reversals between summer and fall and early-winter (Rouse et al. 2003). Schertzer et al. (2003) conducted an in-depth study of the spatial and seasonal variability in the heat content of Great Slave Lake and found good agreement between heat storage calculated calorimetrically from vertical temperature gradients and those obtained as a residual from energy balance measurements made above the lake. Measurements of the energy budget of a small 6 m deep Canadian Shield lake indicate that almost all of the energy exchange was concentrated in the top layer with little exchange in the bottom waters (Spence et al. 2003).

On the landscape-scale, the number and size of lakes greatly influence the magnitude and timing of large-scale evaporative and sensible heat fluxes to the atmosphere. Nagarajan et al. (2004) indicated substantial effects of small water bodies on the atmospheric heat and water budgets over the MRB. Oswald and Rouse (2004) detailed the energy balance of various-sized Canadian Shield Lakes. In a regional synthesis Rouse et al. (2005) illustrated the important role played by northern lakes in the energy balance in the central Basin. They show that lakes are the most efficient evaporators of any of the high latitude surfaces and that the larger the lake, the more efficient is this process. Large lakes in particular introduce large seasonal lags into the regional energy and water balances (Rouse et al. 2007).

Large lakes dictate, in part, their own climates. This arises from their very considerable heat capacities. Their lake ice cover is slow to melt in spring and final melt occurs near or after the summer solstice. Because the lake waters remain cold during the summer there is a very stable atmospheric regime (temperature inversion) in the lower atmosphere. This inhibits cloud formation (Fig. 11) so that substantially larger amounts of solar radiation are absorbed than over adjacent terrestrial areas (Rouse et al. 2003). The absorbed solar radiation goes mainly into heating the lake during summer (Schertzer et al. 2007). The large stored heat during the summer is utilized in fall and early winter (into December–January) for evaporation and atmospheric heating. Acting as reservoirs of water and heat, therefore, lakes interact with their overlying atmosphere through moisture (evaporation and precipitation) and sensible heat exchanges.

Fig. 11. Lower Great Slave Lake on a summer day showing cloud-free condition over the lake but cloud formation over the adjacent land. (Photo: M.K. Woo)

9 Conclusions

Most hydrologic and land surface processes studies under MAGS have a strong field orientation. In addition to the gathering of valuable data in a remote region and sometimes under hazardous conditions, field experimentation enables the derivation of equations and the quantification of input parameters. The subsequent stage of model development, be it conceptual or numeric, permits generalization of findings. In terms of hydrologic contributions, MAGS can claim achievements in the discovery of new processes or improved quantification of hydrologic functions, conceptualization and model development, and application of knowledge for scientific and practical purposes.

MAGS research has led to the revision of conventional interpretation of several hydrologic phenomena. Examples include the important role of sublimation in winter snow loss, the response of lake evaporation to synoptic-scale rather than diurnal energy fluxes, the highly variable and sometimes pervious nature of bedrock surfaces that allows infiltration, and the extreme contrasts between north and south slopes in terms of runoff production in a subarctic mountainous catchment. New concepts were formulated to explain flow mechanisms (e.g., the two-layer flow system in organic soils) and the connectivity of flow in a drainage network (e.g., the fill-and-spill concept) and in wetlands. Understanding of the physical processes was translated into development of algorithms, including those for frozen-soil infiltration, river ice breakup, and for snowmelt as influenced by advection or by the forest and tundra vegetation. New knowledge enabled the improvement of models for the simulation of snow redistribution and melt, investigation of thermal variability of a large lake, and the macro-scale studies of catchment hydrology (Kerkhoven and Gan 2007; Thorne et al. 2007) and regional water balance (Soulis and Seglenieks 2007). Information obtained and models developed were used for prediction and management purposes. Examples include the forecast of floods associated with river ice breakup, the prediction of climate change impacts on ice thermal regime and ice-jam floods, and the assessment of ground thaw in permafrost terrain.

Chapters in this Volume present aspects of hydrologic research conducted under MAGS. While the study area was the Mackenzie Basin and its surroundings, most of the processes are common to the circumpolar regions and therefore the knowledge gained from our studies is entirely relevant to the cold environment in general.

References

Andrishak R, Hicks F (2007) Impact of climate change on the Peace River thermal ice regime. (Vol. II, this book)

Aylsworth JM, Kettles IM (2000) Distribution of peatlands. In: Dyke LD, Brooks GR (eds) The physical environment of the Mackenzie Valley, Northwest Territories: a base line for the assessment of environmental change. Geol Surv Can B 547:49–55

Beltaos S (2003a) Threshold between mechanical and thermal breakup of river ice cover. Cold Reg Sci Technol 37:1–13

Beltaos S (2003b) Numerical modelling of ice-jam flooding on the Peace-Athabasca Delta. Hydrol Process 17:3685–3702

Beltaos S, Prowse T, Bonsal B, Carter T, MacKay R, Romolo L, Pietroniro A, Toth B (2007) Climate impacts on ice-jam floods in a regulated northern river. (Vol. II, this book)

Blackburn J, Hicks F (2002) Combined flood routing and flood level forecasting. Can J Civil Eng 29:64–75

Blanken PD, Rouse WR, Culf AD, Spence C, Boudreau LD, Jasper JN, Kochtubajda B, Schertzer WM, Marsh P, Verseghy D (2000) Eddy covariance measurements of evaporation from Great Slave Lake, Northwest Territories, Canada. Water Resour Res 36:1069–1077

Blanken PD, Rouse WR, Schertzer WM (2007) The time scales of evaporation from Great Slave Lake. (Vol. II, this book)

Brown-Mitic G, MacPherson IJ, Schuepp PH, Nagarajan B, Yau MK, Bales R (2001) Aircraft observations of surface-atmospheric exchange during and after snowmelt of different arctic environments: MAGS 1999. Hydrol Process 15:3585–3602

Burt TP, Williams PJ (1976) Hydraulic conductivity in frozen soils. Earth Surf Processes 1:349–360

Carey SK, Woo MK (1999) Hydrology of two slopes in subarctic Yukon, Canada. Hydrol Process 13: 2549–2562.

Carey SK, Woo MK (2000) The role of soil pipes as a slope runoff mechanism, subarctic Yukon, Canada. J Hydrol 233:206–222

Carey SK, Woo MK (2001a) Slope runoff processes and flow generation in a subarctic, subalpine catchment. J Hydrol 253:110–129

Carey SK, Woo MK (2001b) Spatial variability of hillslope water balance, Wolf Creek basin, subarctic Yukon. Hydrol Process 15:3113–3132

Carey SK, Woo MK (2005) Freezing of subarctic hillslopes, Wolf Creek basin, Yukon, Canada. Arct Antarct Alp Res 37:1–10

Church MA (1974) Hydrology and permafrost with special reference to northern North America. In: Proc Workshop on Permafrost Hydrology. Canadian National Committee for the IHD, Ottawa, pp 7–20

Déry SJ, Yau MK (2001) Simulation of blowing snow in the Canadian Arctic using a double-moment model. Bound-Lay Meteorol 99:287–316

Déry SJ, Yau MK (2002) Large-scale mass balance effects of blowing snow and surface sublimation. J Geophys Res 107(D23): 4679, doi: 10.1029/2001JD001251

Déry SJ, Yau MK (2007) Recent studies on the climatology and modeling of blowing snow in the Mackenzie River Basin. (Vol. I, this book)

Duguay CR, Flato CK, Jefferies MO, Ménard P, Morris K, Rouse WR (2003) Ice cover variability on shallow lakes at high latitudes: model simulations and observations. Hydrol Process 17:3465–3483

Essery R, Li L, Pomeroy JW (1999) Blowing snow fluxes over complex terrain. Hydrol Process 13:2423–2438

Gray DM, Toth B, Zhao L, Pomeroy JW, Granger RJ (2001) Estimating areal snowmelt infiltration into frozen soils. Hydrol Process 15:3095–3111

Faria DA, Pomeroy JW, Essery RLH (2000) Effect of covariance between ablation and snow water equivalent on depletion of snow-covered area in a forest. Hydrol Process 14:2683–2695

Fox D (1992) Incorporating freeze-thaw calculations into a water balance model. Water Resour Res 28:2229–2244

Giesbrecht MA, Woo MK (2000) Simulation of snowmelt in a subarctic spruce woodland: 2. Open woodland model. Water Resour Res 36:2287–2295

Goodison B (1978) Accuracy of Canadian snow gauge measurements. J Appl Meteorol 17:1541–1548

Granger RJ, Pomeroy JW, Parvianien J (2002) Boundary-layer integration approach to advection of sensible heat to a patchy snow cover. Hydrol Process 16:3559–3569

Hicks F, Beltaos S (2007) River ice. (Vol. II, this book)

Hicks F, Healy D (2003) Determining winter discharge by hydraulic modeling. Can J Civil Eng 30:101–112

Hicks F, Pelletier K, van der Sanden JJ (2005) Characterizing river ice using satellite synthetic aperture radar. Proc. 11th Annual Scientific Meeting, Mackenzie GEWEX Study, Ottawa, Ontario, Canada, pp 490-508

Hydrological Atlas of Canada (1978) Ministry of Supply and Services, Ottawa, Canada

Janowicz JR, Gray DM, Pomeroy JW (1997) Snowmelt and runoff in a subarctic mountain basin. In: Milburn D (ed) Proc of the Hydro-ecology Workshop on the Arctic Environmental Strategy, Symposium No. 16, National Hydrology Research Institute, Saskatoon, pp 303–320

Jumikis AR (1977) Thermal Geotechnics. Rutgers University Press, New Brunswick, NJ.

Kerkhoven E, Gan TY (2007) Development of a hydrologic scheme for use in land surface models and its application to climate change in the Athabasca River Basin. (Vol. II, this book)

Leenders EE, Woo MK (2002) Modeling a two-layer flow system at the subarctic, subalpine treeline during snowmelt. Water Resour Res 38(10):1202, doi:10.1029/2001WR000375

Li X, Koike T (2003) Frozen soil parameterization in SiB2 and its validation with GAME-Tibet observations. Cold Reg Sci Technol 36:165–182

Liston GE (1995) Local advection of momentum, heat, and moisture during melt of patchy snow covers. J Appl Meteorol 34:1705–1715

Liston GE, Sturm M (2004) The role of winter sublimation in the Arctic moisture budget. Nord Hydrol 35:325–334

Mahabir C, Robichaud C, Hicks F, Robinson-Fayek F (2007) Regression and logic-based ice jam flood forecasting. (Vol. II, this book)

Marsh P, Neumann N, Essery R, Pomeroy JW (1999) Model estimates of local scale advection of sensible heat over a patchy snow cover. In: Interactions between the cryosphere, climate and greenhouse gases. IAHS Publication No. 256, pp 103–110

Marsh P, Onclin C, Neumann N (2002) Water and energy fluxes in the lower Mackenzie Valley, 1994-95. Atmos Ocean 4:245–256

Marsh P, Pomeroy JW, Pohl S, Quinton W, Onclin C, Russell M, Neumann N, Pietroniro A, Davison B, McCartney S (2007) Snowmelt processes and runoff at the arctic treeline: ten years of MAGS research. (Vol. II, this book)

Martz LW, Pietroniro A, Shaw DA, Armstrong RN, Laing B, Lacroix M (2007) Re-scaling river flow direction data from local to continental scales. (Vol. II, this book)

Mielko C, Woo MK (2006) Snowmelt runoff processes in a headwater lake and its catchment, subarctic Canadian Shield. Hydrol Process 20:987–1000

Nagarajan B, Yau MK, Schuepp PH (2004) The effects of small water bodies on the atmospheric heat and water budgets over the Mackenzie River Basin. Hydrol Process 28:913–938

Neumann N, Marsh P (1998) Local advection in the snowmelt landscape of arctic tundra. Hydrol Process 12:1547–1560

Oswald CM, Rouse WR (2004) Thermal characteristics and energy balance of various-size Canadian Shield lakes in the Mackenzie River Basin. J Hydrometeorol 5:129–144

Parvianinen J, Pomeroy JW (2000) Multiple-scale modeling of forest snow sublimation: initial findings. Hydrol Process 14:2669–2681

Pohl S, Marsh P (2006) Modelling the spatial-temporal variability of spring snowmelt in an arctic catchment. Hydrol Process 20:1773–1792

Pomeroy JW, Bewley DS, Essery RLH, Hedstrom NR, Link T, Granger RJ, Sicart JE, Ellis CR, Janowicz JR (2006) Shrub tundra snowmelt. Hydrol Process 20:923–941

Pomeroy JW, Parviainen J, Hedstrom N, Gray DM (1998) Coupled modelling of forest snow interception and sublimation. Hydrol Process 12:2317–2337

Pomeroy JW, Gray DM (1995) Snow accumulation, relocation and management. NHRI Science Report No. 7, Environment Canada

Pomeroy JW, Gray DM, Hedstrom NR, Janowicz JR (2002) Prediction of seasonal snow accumulation in cold climate forests. Hydrol Process 16:3543–3558

Pomeroy JW, Gray DM, Marsh P (2007) Studies on snow redistribution by wind and forest, snow-covered area depletion and frozen soil infiltration in northern and western Canada. (Vol. II, this book)

Quinton WL, Hayashi M (2007) Recent advances toward physically-based runoff modeling of the wetland-dominated central Mackenzie River Basin. (Vol. II, this book)

Quinton WL, Marsh P (1998) The influence of mineral earth hummock on subsurface drainage in the continuous permafrost zone. Permafrost Periglac 9:213–228

Quinton WL, Marsh P (1999) A conceptual framework for runoff generation in a permafrost environment. Hydrol Process 13:2563–2581

Quinton WL, Gray DM, Marsh P (2000) Subsurface drainage from hummock-covered hillslopes in the Arctic tundra. J Hydrol 237:113–125

Quinton WL, Hayashi M, Pietroniro A (2003) Connectivity and storage functions of channel flows and flat bogs in northern basins. Hydrol Process 17:3665–3684

Rouse WR, Oswald CJ, Binyamin J, Blanken PD, Schertzer WM, Spence C (2003) Interannual and seasonal variability of the surface energy balance and temperature of Central Great Slave Lake. J Hydrometeorol 4:720–730

Rouse WR, Oswald CJ, Binyamin J, Spence C, Schertzer WM, Blanken PD, Bussières N, Duguay C (2005) The role of northern lakes in a regional energy balance. J Hydrometeorol 6:291–305

Rouse WR, Oswald CJ, Binyamin J, Spence C, Schertzer WM, Blanken PD, Bussières N, Duguay C (2007) The influence of lakes on the regional energy and water balance of the central Mackenzie River Basin. (Vol. II, this book)

Schertzer WM, Rouse WR, Blanken PD, Walker AE (2003) Over-lake meteorology, thermal response, heat content and estimate of bulk heat exchange of Great Slave Lake during CAGES (1998-1999). J Hydrometeorol 4:649–659

Schertzer WM, Rouse WR, Blanken P, Walker AE, Bussières N, Lam DCL, León L, McCrimmon RC, Rowsell RD (2007) Interannual variability of the thermal components and bulk heat exchange of Great Slave Lake. (Vol. II, this book)

Shook K, Gray DM (1996) Small-scale spatial structure of shallow snowcovers. Hydrol Process 10:1283–1292

Soulis ED, Seglenieks FR (2007) The MAGS integrated modeling system. (Vol. II, this book)

Spence C, Rouse WR, Worth D, Oswald CJ (2003) Energy budget processes of a small northern lake. J Hydrometeorol 4:694–701

Spence C, Woo MK (2002) Hydrology of subarctic Canadian Shield: bedrock upland. J Hydrol 262:111–127

Spence C, Woo MK (2003) Hydrology of subarctic Canadian Shield: soil-filled valleys. J Hydrol 279:151–166

Spence C, Woo MK (2007) Hydrology of the northwestern subarctic Canadian Shield. (Vol. II, this book)

Taylor PA, Wilson J (2003) Modelling and parameterization of blowing snow and limited fetch evaporation. In: Proc of the 9th Annual Scientific Meeting of the Mackenzie GEWEX Study, 12-14 November 2003, Montreal, pp 67–75

Thorne R, Armstrong RN, Woo MK, Martz LW (2007) Lessons from macroscale hydrologic modeling: experience with the hydrologic model SLURP in the Mackenzie Basin. (Vol. II, this book)

Walker A, Silis A, Metcalf JR, Davey MR, Brown RD, Goodison BE (2000) Snow cover and lake ice determination in the MAGS region using passive microwave satellite and conventional data. Proc 5th Scientific Workshop, Mackenzie GEWEX Study, Edmonton, Alberta, Canada, pp 39–42

Weismann RN (1977) Snowmelt: a two-dimensional turbulent diffusion model. Water Resour Res 13:337–342

Woo MK (1982) Upward flux of vapor from frozen materials in the High Arctic. Cold Reg Sci Technol 5:269–274

Woo MK (2000) Permafrost and hydrology. In: Nuttall M, Callaghan TV (eds) The arctic: environment, people, policy. Harwood Academic Publishers, Amsterdam, the Netherlands, pp 57–96

Woo MK, Arain MA, Mollinga M, Yi S (2004) A two-directional freeze and thaw algorithm for hydrologic and land surface modeling. Geophys Res Lett 31:L12501, doi:10.1029/2004GL019475

Woo MK, Giesbrecht MA (2000) Simulation of snowmelt in a subarctic spruce woodland: 1. Tree model. Water Resour Res 36:2275–2285

Woo MK, Mielko C (2007) Flow connectivity of a lake–stream system in a semi-arid Precambrian Shield environment. (Vol. II, this book)

Woo MK, Mollinga M, Smith S (2007) Modeling maximum active layer thaw in boreal and tundra environments using limited data. (Vol. II, this book)

Woo MK, Thorne R (2003) Streamflow in the Mackenzie Basin, Canada. Arctic 56:328–340

Woo MK, Thorne R (2006) Snowmelt contribution to discharge from a large mountainous catchment in subarctic Canada. Hydrol Process 20:2129–2139

Yang D, Ishida S, Goodison B, Gunther T (1999) Bias correction of precipitation data for Greenland. J Geophys Res–Atmos 104(D6):6171–6181

Yi S, Arain MA, Woo MK (2006) Modifications of a land surface scheme for improved simulation of ground freeze-thaw in northern environments. Geophys Res Lett 33:L13501, doi:10.1029/2006GL026340

Zhulidov AV, Headley JV, Robarts RD, Nikanovrov AM, Ischenko ASA (1997) Atlas of Russian Wetlands. Environment Canada, Saskatoon

Chapter 3

Analysis and Application of 1-km Resolution Visible and Infrared Satellite Data over the Mackenzie River Basin

Normand Bussières

Abstract National Oceanographic and Atmospheric Administration (NOAA) Advanced Very High Resolution Radiometer (AVHRR) visible and infrared satellite data were collected at 1-km resolution over the Mackenzie Basin specifically for the Mackenzie GEWEX Study (MAGS) water year and the Canadian GEWEX Enhanced Study period. Through the use of satellite data, clear sky products (land surface temperature, evapotranspiration, water temperature time series over lakes of size ≥ 100 km^2) and cloud top temperatures were generated to provide information for the analysis of the water and energy budgets. The satellite database was augmented with 2002–04 Moderate Resolution Imaging Spectroradiometer (MODIS) land surface temperature data. This chapter summarizes results of MAGS studies based on these satellite products, and documents improved methods to compare the satellite-derived products with ground observations from research sites and with regional climate model outputs of low spatial resolution.

1 Introduction

Research was conducted during the study period of the Mackenzie GEWEX Study (MAGS) to improve methods for combining the detailed but intermittent satellite-derived products with regular and accurate observations from local research sites and with lower resolution model outputs. The project objective evolved throughout 1994–2005. The main goal was to document and study the spatial variability of some of the water and energy budget components over the Mackenzie River Basin (MRB) through the acquisition, processing, and analysis of 1-km resolution satellite data for the MAGS water year (1994/95) and for the Canadian GEWEX Enhanced Study (CAGES) period from June 1998 to September 1999. The following products were derived from the National Oceanographic and Atmospheric Administration (NOAA) Advanced Very High Resolution Radiometer (AVHRR): land surface temperature (LST), evapotranspiration, cloud mask, and cloud top temperatures. The AVHRR data enabled elucidation of seasonal variation of surface temperatures for large (≥100 km^2) water bodies. The results facilitated joint studies on the interaction of water bodies with the atmospheric energy and water cycles (Rouse et al. 2007). A method was developed also for comparisons of remotely sensed land surface temperature estimates with local field observations and with outputs from coupled regional climate model and land surface scheme. At the end of MAGS, the satellite database was augmented with 2002–04 MODIS LST data which permits further evaluation of the temperature outputs from the Canadian Regional Climate Model (CRCM). This chapter reports on the application of 1-km resolution clear sky satellite-derived products together with other MAGS information towards understanding and quantifying the water and energy budgets.

2 Study Area

The study area encompasses the entire MRB. At 1-km resolution using AVHRR data, there are 1.8×10^6 observation pixels over the Basin in each satellite scene. The Canada Centre for Remote Sensing (CCRS) land cover data published in 1995 (Beaubien et al. 1997) is available in a 31-category classification. To generalize the land cover distribution over the Basin, these categories are grouped into four broad classes that include water (8.4%), forest (51.5%), agricultural and rangeland (3.9%), and low vegetation, very low vegetation, and barren surfaces (36.2%). Elevation of MRB ranges from 0 m (sea level) to 2200 m (average values at 1 km resolution). Except for a few high altitude observing stations installed just southeast of the Basin for the duration of the FOPEX experiment (Smith 2007), there is no surface observation in the 1200–2200 m elevation range which occupies 14% of the MRB. Satellite offers useful data in such higher elevation regions where ground-based measurements do not exist. In this chapter, the time is reported in UTC units. Near the center of the MRB at 120°W, for example, an observation made at 22:00 UTC corresponds to local solar time of 14:00.

3 Data and Methods

3.1 AVHRR Data Collection and Basic Processing

The NOAA 12 and 14 AVHRR sensor consists of five spectral bands in the optical and infrared regions which observe, at a 1 km nominal resolution, the radiances reflected and emitted from clouds, the atmosphere, land, water, and cryospheric surfaces. The AVHRR radiances permit the derivation of several major variables for MAGS, including clouds, land/water surface temperatures, and solar radiation budget components.

AVHRR binary format data were collected from Environment Canada receiving stations in western Canada. There were 60 scenes collected for the MAGS water year (1994/95) and approximately 2170 scenes for the CAGES period (June 1998 to September 1999). For each scene, the AVHRR pixels were geo-located automatically, using a three dimensional model of the earth–satellite system, to the row and column locations of the map projections defined for this study (1-km resolution Albers conic equal area centered at 118°01'W for the 1994/95 data and 1-km polar stereographic centered at 118°35'W for CAGES). Geo-location of each scene was fine-tuned manually to ensure the best possible spatial match from scene to scene. The AVHRR satellite raw counts were converted to percent albedo (used here to detect clouds) and radiative temperatures following the calibration procedures described in the NOAA Technical Memorandum NESS 107 (Planet 1988). In accordance with these procedures, time-dependent gain and offset coefficients were obtained from NOAA and applied to the linear calibration of the visible channels from raw count to percent albedo. In-orbit time-varying gain and offset coefficients were applied to the linear calibration of the thermal channel data. The thermal data were corrected for non-linearity and converted to temperatures using Planck's function. Figure 1 shows a portion of the study area in the northernmost sector of MRB on July 14, 1999 at 22:11 UTC (approximately 13h15 local solar time at 134°W near the center of the image). The cool (20–22°C) Mackenzie Delta contrasts strongly with the surrounding warmer land surfaces. Temperature of some areas in the southernmost part of the image is above 35°C. The CAGES AVHRR-derived visible data served in several projects are described in other publications (Feng et al. 2002; Guo et al. 2007; Leconte et al. 2007).

Visible and infrared thresholds were used to discriminate whether a pixel was cloudy or clear. Cloud masks created with this simple method have cloud/no-cloud edges that often presented contaminated pixels that may produce large outliers in the clear sky land surface temperatures. Subsequently, a more stringent cloud mask was created for the analysis of clear-sky data, in which the "clear" perimeter pixels (within 1-km) at each cloud edge as found in the first cloud masks were assigned a "cloudy" status.

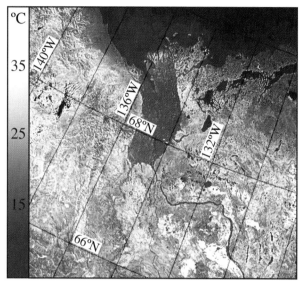

Fig. 1. AVHRR land surface temperatures (LST) on July 14, 1999 at 12:11 UTC, depicting the contrast between the cooler Mackenzie Delta and it surrounding land surfaces

3.2 Land Surface Temperature Information

At each clear-sky pixel location, radiative temperatures were converted (Bussières 2002) to land surface temperature values through the use of a split window algorithm. Various LST products were prepared from the pool of AVHRR scenes. These included the generation of clear-sky time series values at a selected (longitude, latitude) pair and over (longitude, latitude) windows, the generation of 1-km resolution and of spatially averaged instantaneous LST maps, and of time averaged LST maps. Maximum LST maps for the warm season (Bussières 2002) were obtained by extracting the warmest value at each clear sky pixel from the summer time series.

3.3 Evapotranspiration Estimates

The satellite evapotranspiration algorithm developed by Granger (Granger and Bussières 2005) for northern climates and land features was selected for this experiment. This algorithm represents an extension of the Penman equation to the non-saturated case. It was developed for the Boreal Ecosystem-Atmosphere Study (BOREAS) research sites where the algorithm gave an accuracy of ± 0.3 mm day^{-1}. Verification of the algorithm was also performed at the Wolf Creek research site, Yukon Territories (Granger et al. 2000). Granger's method is based on a feedback relationship between the LST and the vapor pressure deficit in the overlying air. In this satellite algorithm, an estimate of the LST is obtained near the time of the afternoon maximum. This estimate is converted to a mean daily LST through an empirical relationship. The water vapor deficit VP_{def} is then computed empirically at each data grid point as follows:

$$VP_{def} = -0.278 - 0.015 T_{ltm} + 0.668 VP*_{Ts} \qquad (1)$$

in which $VP*_{Ts}$ is the saturation vapor pressure at the mean daily surface temperature and T_{ltm} is the long-term mean air temperature to account for seasonal and latitudinal effects at a given location. The air temperature normals required for the algorithm are determined from the objective analysis (spatial interpolation) of climate station data.

Under clear sky conditions, daily net radiation at each grid point is estimated (Bussières et al. 1997) from a linear relationship with instantaneous net radiation obtained at the same time as the LST value. The components for instantaneous net radiation are estimated from outgoing long-wave radiation observed with AVHRR (thermal infra-red), satellite albedo, and air temperature. Estimated net radiation and vapor pressure deficit are then used to drive Granger's model at each grid point.

3.4 Water Temperature Time Series over Water Bodies

When buoy data collected from Great Slave Lake were compared with the MAGS AVHRR dataset over open water, the uncertainty of the LST values was found to be $\pm 2.0°C$ (Bussières and Schertzer 2003). For the processing of water temperatures, the location of water pixels was obtained from the 1-km resolution 1995 CCRS land cover classification (Beaubien et al. 1997). To prevent contamination of water pixels by land, all "water" pixels within 1 km from the shore were discarded. This was accomplished by creating, from the land cover data, a lake/land mask corresponding to the outline of the water bodies, and setting these outlines as non-water pixels. To avoid data limitation associated with the smaller water bodies, only water bodies with area of ≥ 10 km^2 were sampled. This procedure reduces the computational area over all water bodies and completely eliminates many of the small ones. Since only the overall behavior of each water body was sought, the computed temperature is a spatial average of all the water pixels over the water body.

The time series were filtered to include only observations for which 95% or more of the surface of the water body is classified as non-cloudy. To filter out the remaining outliers, the time series values differing by more than 7°C in absolute value from the fitted quadratic curve were rejected. A new quadratic fit was then performed on the portion of the data beginning at the time when above zero temperatures occurred. This quadratic equation (Bussières and Schertzer 2003) takes the form

$$T_w^\mu = At^2 + Bt + C \qquad (2)$$

where T_w^μ (in °C) is the water surface temperature fitted from the water temperature (T_w) observations, t is the number of days since January 1 (00 UTC) of a given year, and A, B, and C are empirical coefficients obtained by least squares fitting. There is no simple empirical equation yet known that describes well the seasonal variation of water temperature. We find that Eq. (2) generates more stable results than higher order polynomials. The third order polynomial fit, for example, cannot reproduce as well as the quadratic the rapid rise in water temperatures in the spring after the water becomes completely ice-free. The quadratic equation is useful to extend the seasonal data back to the beginning of the ice-free period in spring and also to standardize description of the seasonal trend so that various water bodies can be compared. For each water body, the following variables can be computed from Eq. (2): the time (t_0) when the fitted temperature rises above freezing; the time (t_4) when temperature starts to rise above 4°C (i.e., temperature of maximum density for freshwater); and the time (t_{max}) of maximum fitted temperature (T_{w_max}).

3.5 Cloud Top Temperatures at Selected Locations

Clouds and their associated precipitation are fundamental aspects of the MRB climate system (Hudak et al. 2007). It is assumed that over most wavelengths in the 8–14 mm band, the majority of clouds approach a black body status (Carleton 1991). Apparent cloud top temperatures (not using atmospheric correction as in the case of land surface temperatures) were extracted over a 10 km x 10 km area over Fort Simpson. Cloud top temperatures combined with information from MOLTS sounding provided a means to estimate the height of cloud tops in a joint study (Hudak et al. 2004).

3.6 Canadian Regional Climate Model (CRCM) Outputs

The coupling of the Canadian Regional Climate Model (CRCM) with the Canadian Land Surface Scheme (CLASS) was an important element of MAGS (MacKay et al. 2003, 2007). For model verification, a method was developed to compare model outputs with satellite-derived LST values. The model outputs are available on a polar stereographic grid, at 50-km resolution, centered at 124°45'W. The most suitable model variable for comparison with satellite LST is the GT which is a combination of the soil and vegetation canopy skin temperatures for each grid cell.

Initial comparison (referred to as the MAGS-1 method) of model GT with the maximum AVHRR LST map published in Bussières (2002) showed large differences due to differences between times of LST observations and of the standard model grid outputs available in this study. Figure 2 shows a histogram of the number of AVHRR scenes as function of the time of day as well as the corresponding numbers for the archived CCRM gridded outputs. The AVHRR LST values observable in the 21–23

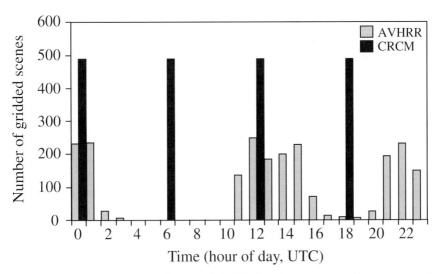

Fig. 2. Comparison of the timing of AVHRR observations with the 6-hourly CRCM outputs

UTC range corresponded to times of higher surface heating than the nearest CRCM grid output presented for 18 UTC or 00 UTC.

The intercomparison method was subsequently improved to avoid this time mismatch. In this method, referred to as the MAGS-2 method, only the AVHRR LST values within one hour of CRCM/CLASS grid output times are used. The MAGS-2 method also includes an improved algorithm for the spatial aggregation of the 1-km AVHRR data to the 50-km model resolution. This algorithm assigns a unique number to each CRCM grid cell location to produce a map of grid cell numbers over the modeling domain. The map is then projected with standard USGS cartographic transformation code into the 1-km resolution AVHRR map projection. All LST values at AVHRR locations with the same CRCM grid number are aggregated to provide the desired comparison points. Comparisons were made under clear sky since AVHRR LST values are not available under cloudy conditions. Only aggregated AVHRR grid values with more than 60% clear sky pixels were used. To further ensure that cloudy CRCM values are not compared with clear sky AVHRR data, grid point values with more than 20% total model integrated cloud were discarded. Thus, when aggregating the clear sky 1-km AVHRR data values, the representativeness of the aggregated grid value at 50-km depends on how many clear sky values are used (maximum of about 2500 values). Since the land surface scheme component is not involved in the generation of clouds, this method of examining CRCM/CLASS outputs under clear sky conditions presents an opportunity to evaluate CLASS within the CRCM.

3.7 Moderate Resolution Imaging Spectroradiometer (MODIS) LST Data

After the NASA TERA and AQUA satellites were launched, additional LST data derived from the MODIS instruments for 2002–04 were ordered from the NASA EOS DACC for comparisons with CRCM GT values. Automated satellite geolocation, calibration, cloud masking, and generalized split window temperature computations were performed at source by NASA teams (Wan et al. 2004). The same authors suggest an uncertainty of ±1 K with the MODIS LST values. As with the AVHRR data, intercomparison with CRCM data was performed using the MAGS-2 method.

4 Results and Discussion

4.1 Land Surface Temperature and Model GT

The computation of maximum LST for the MRB (Bussières 2002) was intended as a tool to explore and develop a better understanding of the complex energy and water processes leading to variations in surface temperatures. The maximum LST patterns of MRB for 1994 can be subdivided in three land zones (>35°C, 33–34°C, and 27–32°C) and a water dominated zone (20.5°C on average). The highest maximum temperatures zone (>35°C) corresponds to a combination of minimal vegetation and low terrain. Maximum LST for summer 1994 decreased with increasing vegetation density, and with elevation at a rate of -4.5°C km^{-1}. This study recognized the role of water surfaces in the cooling of the Mackenzie Delta and over the Canadian Shield east of Great Bear and Great Slave Lakes. Attempts to explain the subdivision of the three land zones with the amount of moisture in the soil and vegetation were made, but these were inferred mainly on the basis of the type of vegetation, as no information is available to ascertain whether the areas of higher LST corresponded with areas of low water content.

Using the MAGS-2 method, the satellite LST values were compared with the CRCM/CLASS GT values. Figure 3 shows histograms of the number of maximum LST values observed by satellite and the modeled values for May–September 1999. The histogram labeled as "ALL AVHRR LST" is the maximum AVHRR LST for all available AVHRR data including those from 20 to 23 UTC done with the MAGS-1 method. The histogram "AVHRR_LST_at_RCM_TIMES", obtained by the MAG-2 method (with a lower peak temperature of 29°C versus 36°C according to MAGS-1 method), is a much better match to the CRCM GT values.

Since climate models are designed to simulate the mean climate rather than the extremes, comparing the mean values should be more meaningful than comparing maximum values. With the MAGS-2 method, the spatial variation of mean temperature for May–September 1999 yielded the same standard deviation (4.4°C) for AVHRR LST as for model GT. Both model GT values and LST exhibit higher temperatures in the central Basin and cooler in the mountains. Both GT and LST values are cooler in the areas east of the three largest lakes (Canadian Shield region) with the model GT values being lower by 1.5°C on average.

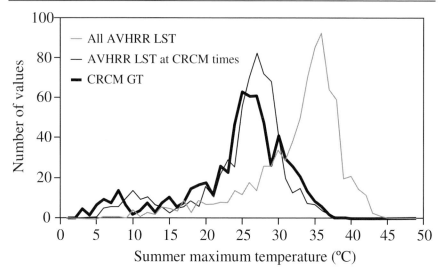

Fig. 3. Histograms of the number of maximum LST values observed by satellite and of the corresponding maximum CRCM/CLASS GT values for May to September, 1999

The temporal variation of LST and GT can be compared using the clear-sky time series at one CRCM model grid point in an area north of Inuvik (Fig. 4). This grid was chosen to include infrared temperature (IRT) measurements made during a Twin Otter flight over the area in 1999 (Brown-Mitic et al. 2001; Nagarajan et al. 2004). This example also illustrates the scaling of 1-km satellite data to 50-km model resolution. The spatial averages of the high-resolution Twin Otter IRT observations are shown for the tundra (centered near 68°40'N, 133°40'W) and the Mackenzie Delta (approximately at 68°40'N, 133°25'W) flight lines. During the spring and until day 163 (June 12), differences between Delta and tundra AVHRR LST values can reach 10°C, as shown by the large size of the grey zone between the two time series. During summer, the differences are reduced, as indicated by the small vertical extent of the grey zone. In general, the AVHRR LST values are within ±2.6°C of the IRT observations. The AVHRR LST aggregated at 50-km resolution and the RCM/CLASS GT values are in good to excellent agreement, except for the Delta during the period of snow melt (Schuepp et al. 2000) on day 145.

There is a strong correlation ($r^2 = 0.98$) between monthly AVHRR LST and CRCM values under clear sky conditions, averaged over the MRB for the period May–September 1999 (Fig. 5). The warmer (>20°C) AVHRR LST values are from the NOAA12 satellite (daytime passes) while the

Application of Visible and Infrared Satellite Data 49

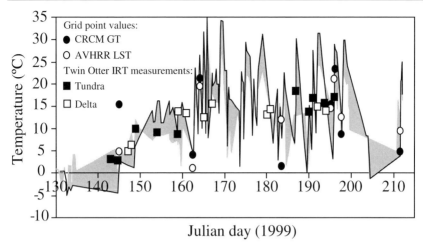

Fig. 4. Clear sky temperatures at a model grid point north of Inuvik. Large black circle are CRCM/CLASS GT values at the grid point and large white circles are the AVHRR LST values obtained for the same grid point and at the same time. Shaded area is bounded by two series of sub-grid scale LST observations, with one series (darker edge) taken over a small area in the Mackenzie Delta and the other series (light edge) over a small area in the tundra. The Delta and the tundra are representative of the two major land cover types found within this model grid square. These two areas are made to coincide spatially with Twin-Otter infrared thermal (IRT) observations shown here as black square (tundra) and as open square (Delta)

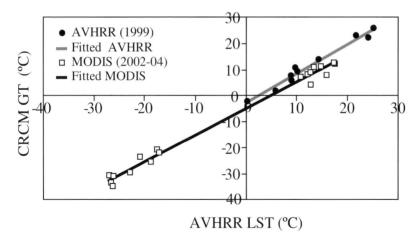

Fig. 5. Relationship between AVHRR LST and CRCM/CLASS GT for May–September 1999 and between MODIS LST and CRCM/CLASS GT for summer and winter 2002–2004

cooler values (0 to 15°C range) are from the NOAA14 satellite (nighttime passes). On a monthly basis, model GT is 3.0°C lower than the LST. Comparison between land surface temperatures and model data can also be made under various constraints (elevation or land cover, for example). Figure 6a indicates that for May–September 1999, the bias (GT minus LST) is negative and increases with elevation. It is noted that most of the data pertain to the 200 to 800 m elevation range.

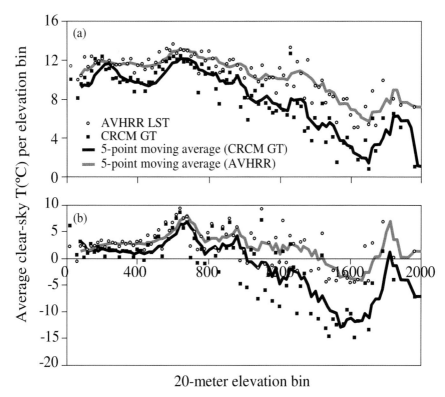

Fig. 6. Elevation effect in AVHRR LST and CRCM GT (°C) over the Mackenzie Basin, for (a) the period from May to September 1999, and (b) for May 1999. Individual points are averages per elevation bin. Lines are 5-point moving averages (equivalent to 100-meter data bins)

Clear sky validation of the CRCM/CLASS GT (skin temperatures) in relation to the timing of snowmelt (Szeto et al. 2007) suggests that the regional climate model has a strong cold bias that leads to delayed snowmelt. The timing of the spring melt can be further investigated using satellite data. For example, if the model retains snow at high altitudes when the actual surface is already bare, a bias should be noticeable when GT is compared with the AVHRR LST. Such a comparison for May 1999 is provided in Fig. 6b, between clear-sky AVHRR LST and CRCM GT as function of elevation. Both datasets showed a general decrease of temperature with elevation. However, departures from this general trend (e.g., peak near 600 m elevation) indicate that elevation is only one of the controlling factors. The average latitude for all observed values below 600 m is 61.5°N, while above 600 m it is 54.6°N. Multiple regression analysis suggests that latitude has a similar effect on temperature as elevation. During the spring, snow is likely to be present when surface temperature remains below 0°C. In the May CRCM values, below zero GT temperatures occur above 1000 m, while below zero LST temperatures occur only above 1400 m in the AVHRR data. The bias (GT minus LST) is about 2°C below 1000 m and more than 5°C above this level. In June to September, RCM GT and AVHRR LST are more closely related than in May. These results support the hypothesis that snowmelt is delayed in the model, especially as GT and LST differences are greater at lower temperatures.

With the MODIS LST data, we choose two periods for study: one being the warmest months (June–August) and the other the coldest (December–January). There is a high correlation ($r^2 = 0.99$) between the monthly MODIS LST and CRCM GT values over MRB (Fig. 5). At this monthly scale, the model GT is 5.0°C cooler than the LST estimated from MODIS, a bias that is stronger than the 3.0°C difference between GT and the AVHRR LST. Differences in the LST derived from AVHRR and from MODIS may be attributed to certain parameters, such as atmospheric correction for temperature, pixel resolution, and cloud detection skill in the satellite algorithms. The June–August clear sky temperatures averaged 13.8°C in the 1999 AVHRR data, which is about 1°C within the range of average values obtained with the MODIS data (12.8, 14.7, and 15.0°C for 2002, 2003, and 2004, respectively).

The collected MODIS LST data allowed comparison with model values. Figure 7 shows the difference (CRCM GT minus MODIS LST) as function of elevation for July–August and for December–February. Again, the difference is less at low elevation than at high elevation, and is also less in winter than in summer. Since a snow cover is expected over large parts of MRB during most of December–January, the CRCM GT seems to reproduce better the remotely sensed estimates of temperatures under snow-covered than snow-free conditions.

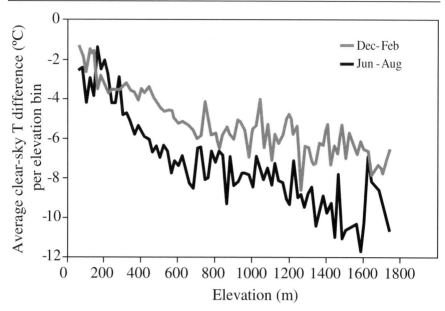

Fig. 7. Difference between MODIS LST and CRCM GT (°C) for summer and winter, at various elevations in the Mackenzie Basin, based on data years 2002 to 2004

4.2 Clear Sky Evapotranspiration Computed with Granger's Algorithm

A series of 30 AVHRR scenes from 1994 data were processed with Granger's feedback algorithm. Computation was performed for the entire MRB, but excluding areas over water bodies. Evapotranspiration varied from 1.4 to 2.2 mm day^{-1}, with the AVHRR pattern suggesting higher rates over the south than over the northern half of the Basin. Although the uncertainty is high and the results are not strictly applicable to the mountainous areas, the algorithm suggests higher evapotranspiration over wetlands and denser vegetation areas. Monthly water balance computations for summer 1994 (Soulis and Seglenieks 2007) suggest monthly amounts of 63.0 mm in June, 66.9 mm in July and 47.2 mm in August. These values give evaporation estimates in the range of 1.6 to 2.2 mm day^{-1}.

4.3 Temperatures of Water Bodies

Although the sampling of surface features with AVHRR data is limited in time by the number of orbits and the presence of clouds, Bussières and Schertzer (2003) obtained sufficient data over water bodies to generate seasonal time series of water temperature. A sample of the results (1999 dataset shown in Table 1) indicates that latitudinal influence is much more important than lake size in affecting the shape and amplitude of the temperature curves. Compared to Great Slave Lake, the duration of open water for Lake Athabasca is longer and begins about 16 days earlier; for Great Bear Lake the open water season is shorter and begins about 45 days later. Fitted maximum temperatures are 15.5°C for Lake Athabasca, 13.7°C for Great Slave Lake, and 6.8°C for Great Bear Lake. Field data collected in 2004 by Rouse's team yielded an observed maximum water temperature of 5.8°C in mid-August. Rouse et al. (2005, 2007) combined the information from AVHRR data with the field data from lakes of various sizes and described the role of high latitude lakes in the regional energy balance context. With the development of a lake component in regional models, satellite-derived water temperatures will provide useful data for the verification of model output.

On the basis of Eq. (2), it is possible to derive the date where the fitted seasonal temperature curve (T_w^μ) reaches 4°C (t_4 shown in Table 1), a time when the lake should be free of ice. With an uncertainty of ±2°C for the estimated temperature T_w^μ, the corresponding uncertainty for t_4 is ±6 days. Figure 8 provides a comparison of the spring breakup date for 1999 over Great Slave Lake, ODAS site, as detected by the AVHRR method and by the passive microwave (SSM/I) method (Walker 2001 pers comm, and as reported in Schertzer et al. 2002). SSM/I technique indicates that breakup began on May 26 or day 146, a date that can be accepted with high confidence, and the microwave data illustrates the transition towards the lake being clear of ice (Fig. 8a). AVHRR method (squares and line fitted Eq. (2), in Fig. 8b) suggests that the area was completely free of ice with T_w^μ = 4°C on June 17 (day 168). The dates estimated by both techniques agree closely with lake observations.

Table 1. Coefficients of water temperature time series during the 1999 ice-free season for various water bodies; $T=At^2+Bt+C$ where t = number of days since January 1, 1999, 00 UTC

Water body name	Time		Coefficients		
	t_0[d][a]	t_4[d][b]	$A*10^{-3}$[°C d^{-2}]	B[°C d^{-1}]	C[°C]
Great Bear	197.9	212.2	-4.25625	2.0247	-233.969
Great Slave	152.5	163.9	-2.62976	1.1809	-118.926
Athabasca	135.9	147.1	-2.37420	1.0284	-95.894
La Martre	143.1	152.7	-3.12388	1.3417	-128.029
Williston	125.8	139.5	-1.73876	0.7543	-67.378
Claire	115.4	124.7	-2.60311	1.0548	-87.049
Cree	121.3	132.9	-2.09680	0.8796	-75.852
Lesser Slave	116.2	128.5	-1.73780	0.7510	-63.800
Conjuror Bay	135.3	147.4	-2.42984	1.0151	-92.840
Tazin East	128.8	139.7	-2.43858	1.0215	-91.101
Cree SW part	119.7	132.3	-1.89427	0.7946	-67.966
Travaillant	142.4	152.3	-3.21195	1.3525	-127.480
Duncan	146.8	157.1	-2.96906	1.2901	-125.400
Prelude	128.9	138.7	-2.82118	1.1657	-103.391
Francois	126.9	137.8	-2.48045	1.0245	-90.058

[a] t_0 = value of t when fitted temperature starts to rise above 0°C.
[b] t_4 = value of t when fitted temperature reaches 4°C.

Water body name	Peak values		Area		
	t_{max}[d][c]	T_{max}[°C]	W[km^2]	E[km^2]	Latitude[°N]
Great Bear	237.9	6.8	30764	27287	65.95
Great Slave	224.5	13.7	27048	21323	61.50
Athabasca	216.6	15.5	7849	6861	59.11
La Martre	214.8	16.0	1687	1236	63.32
Williston	216.9	14.4	1657	567	56.25
Claire	202.6	19.8	1410	916	58.55
Cree	209.7	16.4	1228	624	57.51
Lesser Slave	216.1	17.3	1167	919	55.40
Conjuror Bay	208.9	13.2	100	100	65.73
Tazin East	209.5	15.9	100	64	59.82
Cree SW part	209.7	15.4	77	77	57.32
Travaillant	210.6	14.9	67	67	67.70
Duncan	217.3	14.8	16	16	62.86
Prelude	206.6	17.0	18	18	62.57
Francois	206.5	15.7	13	13	62.44

[c] t_{max} = value of t when fitted temperature reaches maximum value (T_{max}).
W = western section of lake
E = eastern section of lake

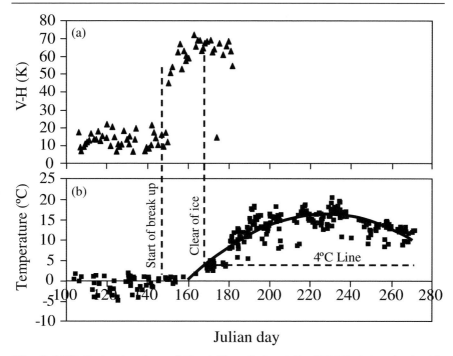

Fig. 8. 1999 Spring breakup of Great Slave Lake at the ODAS, determined with (a) passive microwave technique, 37 Ghz V-H polarization difference, and (b) AVHRR water temperature technique

5 Conclusion

At the beginning of this project, we underestimated the difficulties of undertaking joint studies of the Mackenzie Basin using a combination of satellite data, site observations, and model outputs. As problems were identified and solutions evolved, the goal was attained and two new methods were developed using 1-km resolution satellite thermal infrared data:
- for generating land surface temperature (LST) data and for applying such data in model evaluation;
- for generation of lake water temperature to examine its seasonality, relationship with buoy data, and application to lake evaporation.

The project also explained the temporal and spatial variations of the LST in the context of the MRB, and explored relationships between LST and other types of data and with model outputs. For example, the large number of small water bodies (<1 km^2) in eastern MRB was found to have a collective influence on the observed LST, and the relative influence of small to large water bodies in basin evaporation has been assessed (Rouse et al. 2007). These and other applications of satellite information contribute to understanding the energy and water cycles of the Basin.

An investigation on the relationship between LST and CRCM GT has confirmed that the model GT is 3–5°C cooler than the satellite-derived LST. The bias is larger at higher elevation than at lower elevation and stronger in summer than in winter. Since clouds in CRCM/CLASS are specified and not generated by the model, results of simulation under cloudy conditions do not depend completely on CRCM/CLASS. Comparison of satellite-derived and modeled temperature under clear sky conditions is a useful approach to identify effects that are related to CRCM/CLASS. This project has produced data for 2002–04 that show a smaller bias in winter than in summer. Future investigation is warranted to study this effect which is likely related to the presence or absence of snow and other land cover aspects.

Acknowledgements

Thanks to Ron Goodson of the MSC Branch, Prairie and Northern region, Environment Canada for access to Edmonton station's AVHRR data. The MODIS WEB data access was provided through the courtesy of the NASA EOS program and the NSIDC DAAC data gateway. The Canada Centre for Remote Sensing provided access to their land cover data products. Special thanks to Raoul Granger for his contribution to the evapotranspiration component and to Murray MacKay for help and advice regarding the RCM model and this chapter in general.

References

Beaubien J, Cihlar J, Xiao Q, Chen J, Fung K, Hurlburt P (1997) A new, nationally consistent, satellite-derived land cover of Canada: A comparison of two methodologies. Proc International symposium on geomatics in the era of Radarsat, GER '97, May 25–30, 1997, Ottawa, Ontario, Canada, Department of National Defense, Directorate of Geographic Operations, CD-ROM

Brown-Mitic CM, MacPherson JI, Schuepp PH, Nagrajan B, Yau PMK, Bales R (2001) Aircraft observations of surface-atmosphere exchange during and after snowmelt for different arctic environments: MAGS 1999. Hydrol Process 15:3585–3602

Bussières N (2002) Thermal features of the Mackenzie Basin from multiple NOAA AVHRR sensor observations for summer 1994. Atmos Ocean 40:233–244

Bussières N, Granger RJ, Strong GS (1997) Estimates of regional evapotranspiration using GOES-7 Satellite data: Saskatchewan case study, July 1991. Can J Remote Sens 23:3–14

Bussières N, Schertzer WJ (2003) The Evolution of AVHRR-derived water temperatures over lakes in the Mackenzie Basin and hydrometeorological applications. J Hydrometeorol 4:660–672

Carleton AM (1991) Satellite remote sensing in climatology. Studies in climatology series, vol 1, Belhaven, London

Feng J, Leighton HG, MacKay MD, Bussières N, Hollmann R, Stuhlmann R (2002) A comparison of solar radiation budgets in the Mackenzie River Basin from satellite measurements and a regional climate model. Atmos Ocean 40:221–232

Granger RJ, Pomeroy J, Bussières N, Janowicz R (2000) Parameterization of evapotranspiration using remotely-sensed data. Proc 6th Annual scientific workshop for the Mackenzie GEWEX Study, November 15–17, 2000, Saskatoon, Saskatchewan, pp 120–129

Granger RJ, Bussières N (2005) Evaporation/evapotranspiration estimates with remote sensing. In: Duguay CR, Pietroniro A (eds) Remote sensing in northern hydrology. Am Geophys Un Mono 163:143–154

Guo S, Leighton HG, Feng J, MacKay M (2007) Comparison of solar radiation budgets in the Mackenzie River Basin from satellite observations and a regional climate model. (Vol. I, this book)

Hudak D, Currie B, Stewart R, Rodriguez P, Burford J, Bussières N, Kochtubajda B (2004) Weather systems occurring over Fort Simpson, Northwest Territories, Canada during three seasons of 1998–1999. Part 1: cloud features. J Geophys Res 109, D22108, doi:10.1029/2004/D004876

Hudak D, Stewart R, Rodriguez P, Kochtubajda B (2007) On the cloud and precipitating systems over the Mackenzie Basin. (Vol. I, this book)

MacKay MD, Szeto K, Verseghy D, Chan E, Bussières N (2003) Mesoscale circulations and surface energy balance during snowmelt in a Regional Climate Model. Nordic Hydrol 34:91–106

MacKay MD, Bartlett P, Chan E, Verseghy D, Soulis ED, Seglenieks F (2007) The MAGS regional climate modeling system: CRCM-MAGS. (Vol. I, this book)

Leconte R, Temimi M, Chaouch N, Brissette F, Toussaint T (2007) On the use of satellite passive microwave data for estimating surface soil wetness in the Mackenzie River Basin. (Vol. II, this book)

Nagarajan B, Yau MK, Schuepp PH (2004) The effects of small water bodies on the atmospheric heat and water budgets over the Mackenzie River Basin. Hydrol Process 28:913–938

Planet WG (1988) Data extraction and calibration of TIROS-N/NOAA radiometers, NOAA technical memorandum NESS 107, Rev 1, October 1998

Rouse WR, Binyamin J, Blanken PD, Bussières N, Duguay CR, Oswald CJ, Schertzer WM, Spence C (2007) The influence of lakes on the regional energy and water balance of the central Mackenzie River Basin. (Vol. I, this book)

Rouse WR, Oswald CM, Binyamin J, Spence C, Schertzer WM, Blanken PD, Bussières N, Duguay C (2005). The role of northern lakes in a regional energy balance. J Hydrometeorol 6:291–305

Schertzer WM, Rouse WR, Walker AE, Bussières N, Leighton HG, Blanken P (2002) Heat exchange, evaporation and thermal response of high latitude lakes. Proc 2nd joint GAME-MAGS international workshop, October 8–9, 2001, Institute of Low Temperature Science, Sapporo, Hokkaido, Japan, ISSN:1480–5308, pp 43–51

Soulis ED, Seglenieks FR (2007) The MAGS integrated modeling system. (Vol. II, this book)

Schuepp PH, MacPherson IJ, Brown-Mitic C, Nagarajan B, Yau P (2000) Airborne observations of surface-atmosphere energy exchange over the northern Mackenzie Basin. Proc 6th Annual scientific workshop for the Mackenzie GEWEX Study, November 15–17, 2000, Saskatoon, Saskatchewan, pp 27–37

Smith CD (2007) The relationship between monthly precipitation and elevation in the Alberta foothills during the foothills orographic precipitation experiment. (Vol I, this book)

Szeto K, Tran H, MacKay MD, Crawford R, Stewart RE (2007) Assessing water and energy budgets for the Mackenzie River Basin. (Vol. I, this book)

Wan A, Zhang Y, Zhang Q, Li ZL (2004) Quality assesment and validation of the MODIS global land surface temperature. Int J Remote Sens 25:261–274

Chapter 4

On the Use of Satellite Passive Microwave Data for Estimating Surface Soil Wetness in the Mackenzie River Basin

Robert Leconte, Marouane Temimi, Naira Chaouch,
François Brissette and Thibault Toussaint

Abstract A method is presented to obtain dynamic estimates of basin wetness (*BWI*) and fractional water surface (*FWS*) indices at the scale of the Mackenzie River Basin using SSM/I remotely sensed brightness temperature measurements. The approach accounts for the seasonal evolution of the vegetation state and the basin surface heterogeneity. Results demonstrate that the approach can filter out the vegetation effect and produce reasonable estimates of *FWS* at the basin scale. However, the low resolution of SSM/I and other passive microwave sensors precludes the use of this approach for monitoring soil wetness at a smaller scale. Also, the *FWS* cannot distinguish moisture effects from open water bodies from that of soil surface. A methodology based on the combined use of passive microwave and visible data, along with topographic information, has been developed to separate the open water from the soil surface component in estimating the *BWI*, and to downscale the index at the digital elevation models scale using a topographic index (*TI*) which is continuously adjusted to account for vegetation growth. When applied to the Peace-Athabasca-Delta area, this method improved the correlation between soil wetness and precipitation measured at a meteorological station, compared to an approach based on a time invariant *TI*.

1 Introduction

Soil wetness, also referred as to soil moisture, is a key variable for climate studies, as it controls the exchange of water and heat energy between land surface and the atmosphere through evaporation and transpiration, and therefore affects the development of weather patterns and the production of precipitation. It is also important to those concerned with flood forecasts, flood control, reservoir management, soil erosion and water quality through its prime role in partitioning rainfall into infiltration and surface runoff. Yet, soil wetness is poorly monitored over much of the globe, mainly because of the difficulties of accurately measuring that state variable and to extrapolate the measurements over a range of scales, from local to regional to global.

Soil wetness is defined here as the reservoir component of the land surface hydrologic cycle. On an annual basis and at the watershed scale it is often assumed that changes in soil water storage are zero. However, the year-to-year changes in storage in the Mackenzie River Basin (MRB) can be as much as 25% of the mean annual runoff because long term fluctuations in groundwater, lake, and glacial storage can occur. Since this is within the range of the closure error in many hydrologic models, independent measures of basin storage are required to verify model values.

An ability to better calibrate hydrologic models will undoubtedly translate into more reliable flood forecast estimates. Models that explicitly make use of soil moisture-derived information to generate flood forecasts usually fall in the category of 'physically based' models, though the degree of conceptualisation and parameterisation can be significant. Nevertheless many of these models often do not lend themselves to direct and easy use, and the water resource practitioner may want to look for simpler tools to forecast flows. Soil wetness information coupled with discharge data offer some potential for improving operational flood forecasting as the magnitude and persistence of high or low river flows is related to soil moisture status in the basin.

The main objectives of this research are two-fold: first, to develop an approach to evaluate the temporal distribution of soil wetness applicable to the MRB and sub-basins and, second, to use soil wetness estimates for flood forecasting. Because of the extreme heterogeneity of landscape, soil and vegetation of the MRB, it is not feasible to obtain direct measurements of soil wetness at the watershed scale. Passive microwave remote sensing offers a promising avenue to retrieve soil moisture at the watershed scale because of the high sensitivity of the microwave frequencies to the amount of water at the soil surface (Jackson 1993; Lakshmi 1996; Njoku et al. 2003; Wigneron et al. 2003), and also because satellite imagery offers the possibility of basin wide daily coverage. Brightness temperatures measured at low frequencies (L band) were recommended for soil moisture retrieval since they minimize atmospheric and vegetation effects. Although there are currently no satellite sensors operating at L band, missions devoted specifically to soil moisture retrieval, such as the Soil Moisture and Ocean Salinity (SMOS) mission, are planned for launch in the near future. Note that remote sensing can only provide information pertaining to soil moisture at the near surface, i.e., up to a few cm, depending on sensor electromagnetic frequency. Moreover, the low spatial resolution of passive microwave sensors (typically in the order of 20–40 km) complicates validation of the soil moisture derived products. Whether surface moisture gives a reasonable estimation of soil wetness depends on the watershed physiographic features. For example, a decoupling between surface and deeper soil moisture is frequently observed for low vegetated clayey soils. Watersheds that do not present strong vertical soil moisture gradients offer better perspectives for inferring soil wetness from surface soil moisture information. The MRB falls in the latter category.

2 A Dynamic Basin Wetness Index for the Mackenzie River Basin

2.1 The Basin Wetness Index

The Basin Wetness Index (*BWI*) suggested by Basist et al. (1998) is based on the correlation between the decrease in emissivity and the presence of water at or near the soil surface which affects the brightness temperature differences measured by SSM/I (Special Sensor Microwave Imager) at 19, 37, and 85 GHz. This index accounts for surface heterogeneity due to vegetation as well as seasonal variations related to vegetation growth, and it allows the calculation of the Free Water Surface coverage (*FWS*) which is a potential indicator of water storage within the upper soil layer. The moisture in the soil reduces its emissivity and affects the differences between emissivities estimated at different frequencies. The reduction in emissivity can be computed as:

$$\Delta\varepsilon = \beta_0[\varepsilon(v_1) - \varepsilon(v_2)] + \beta_1[\varepsilon(v_3) - \varepsilon(v_2)] \quad (1)$$

where β_0 and β_1 are two empirical parameters, and ε is the emissivity at the frequencies v_1 to v_3. This relationship between emissivity reduction and emissivity differences can be expressed as a linear combination of brightness temperature differences that defines the *BWI*. Thus, the *BWI* is written as:

$$BWI = \beta_0[T_b(v_2) - T_b(v_1)] + \beta_1[T_b(v_3) - T_b(v_2)] \quad (2)$$

where $T_b(v_1)$, $T_b(v_2)$, and $T_b(v_3)$ are brightness temperatures remotely sensed by the SSM/I at 19, 37, and 85 GHz, respectively. These correspond to the vertically polarized channels in order to reduce the emissivity variation caused by the presence of wet surfaces. The empirical parameters β_0 and β_1 account for the non-linear reduction of emissivity with liquid water at high frequencies. In addition, these parameters account for the mismatching field of views (FOV) of the different channels of the satellite sensor, which vary from 60 km at 19 GHz to 15 km at 85 GHz (Basist et al. 2001; Tanaka et al. 2002).

Wetness reduces the emissivity and therefore the sensed brightness temperature. To accurately estimate the surface temperature, the wetness effect should be taken into account by adding the *BWI* to the brightness temperature. The *BWI* can therefore be considered as compensation for the vertically polarized brightness temperature reduction at 19 GHz due to soil wetness (Basist et al. 1998). The surface temperature therefore can be estimated as:

$$T_s = \left(T_{b(19V)} + BWI\right)\varepsilon_0^{-1} \tag{3}$$

where ε_0 is the dry land emissivity assumed to be equal to 0.95 and approximately frequency independent. Fily et al. (2003) used a value of 0.97, which resulted in a 3% relative difference in the final *FWS* estimate.

A Free Water Surface coverage (*FWS*) index has been defined in several studies (e.g., Choudhury 1991; Fily et al. 2003; Prigent 1997; Tanaka et al. 2000). It is the fraction of free-water surface within a pixel. Generally, the *FWS* is estimated by an emissivity difference ratio:

$$FWS = \left(e_{draysurface} - e_{soil}\right)/\left(e_{draysurface} - e_{wetsurface}\right) \tag{4}$$

In the Basist et al. (1998) study, this parameter is written as:

$$FWS = \left[1 - \frac{T_{b(19V)}}{\left(T_{b(19V)} + BWI\right)/\varepsilon_0}\right]/0.33 \tag{5}$$

where $T_{b(19V)}$ is the vertical polarization brightness temperature at 19 GHz. The value of 0.33 is the average (empirically determined) difference between wet and dry soil surface emissivities.

The *FWS* and *BWI* are affected by liquid water at or near the surface within the satellite FOV. In addition, both are sensitive to water intercepted by vegetation and flooded or wet open areas between plants. Their sensitivity to water on the ground depends on the state of vegetation. A dense vegetation cover should attenuate the underlying soil signal and reduce the index sensitivity, especially at high microwave frequencies. Temporal evolution of vegetation cover is therefore critical for the determination of *BWI* and *FWS*. The dynamic effect should be filtered out in order to increase the *BWI* sensitivity to the water at the soil surface.

Basist et al. (1998) used constant β_0 and β_1 values. However, the temporal evolution of the vegetation and the surface conditions should have an effect on the emission process and consequently on the relationship between the *BWI* and the brightness temperature differences. It is expected that both β_0 and β_1 follow a seasonal pattern that should be repeatable from year to year. β_0 and β_1 are also expected to be spatially variable due to differences in vegetation types and density. Thus, the relationship between the emissivity reduction due to wetness, and the brightness temperature differences that define the *BWI*, needs to be readjusted over time and geographical locations.

To overcome the classification that would otherwise be required to define vegetation type related β_0 and β_1 values, and to account for the temporal evolution of the vegetation cover, it was decided to generate the parameters of the *BWI* for each satellite image within summer–fall seasons and on a pixel by pixel basis (Temimi et al. 2003). Representative time series of β_0 and β_1 values are then produced. This dynamic readjustment should improve the *BWI* sensitivity to the liquid water on the soil surface. A mobile window was programmed to scan each satellite image, leading to the generation of β_0 and β_1 versus time functions. The window width was fixed at 5 pixels. The value of the estimated empirical parameters is expected to vary with the size of the mobile window. Despite this variation, the parameters should maintain the same trend. Considering that it is the trend of the parameters that will be used to estimate *BWI* and *FWS*, the size of the mobile window is unlikely to be a crucial consideration as long as it is not so big as to encompass large terrain heterogeneities.

2.1 Application

The proposed method was applied to the MRB. The physical characteristics of this Basin are described in Woo et al. (2007). SSM/I data were extracted from the National Snow and Ice Data Center (NSIDC) database. The resolution of the SSM/I imported images is approximately 55, 33, and 14 km for 19, 37, and 85 GHz, which are mapped in polar azimuthally equal-area grids (ascending mode). Air temperatures were computed by the Global Environmental Multiscale (GEM) model, a meteorological forecast model developed by the Canadian Meteorological Centre (Côté et al. 1998). Air temperatures were used as an approximation for soil surface temperatures.

Brightness temperatures and air temperatures measured during the summers of 1998 and 1999 were used to calibrate the *BWI* model, i.e., to estimate the empirical parameters of the *BWI*, daily and on a pixel by pixel basis. This was done first by solving Eq. (3) to obtain *BWI* estimates, and then Eq. (2) to retrieve β_0 and β_1 values by linear regression from computed *BWI* and corresponding observed T_b values inside the mobile window. The resulting time varying curves were filtered to extract temporal trends of the parameters, and then averaged for the two calibration seasons. Variable parameters thus obtained from the fitted averaged curves were subsequently used to compute the *BWI* and therefore the *FWS* for the other years.

Figure 1 illustrates the variability of the empirical parameter β_0 during the summer seasons. Results shown are for one pixel in the MRB Delta (69°01'N; 134°08'W). The figure shows that this parameter is time variant. Parameter β_1 (not shown) exhibits a similar seasonal trend. It is hypothesised that the temporal variability of β_0 and β_2 and its repeatability over time are mainly caused by the evolution of the vegetation cover. Fluctuations around the fitted curve are a combination of the changing atmospheric conditions, surface state evolution and the heterogeneity effect, which all affect the determination of empirical parameters using the moving window. The repeatability of the seasonal patterns of the β_0 and β_1 values was confirmed in all of the MRB except in the mountainous areas in the west, with correlation coefficients above 0.7. A topography effect that has not been considered in this research is probably responsible for the observed lack of correlation in mountainous landscape.

Figure 2 is an example of average monthly *FWS* estimates computed from Eq. (5) using SSM/I data measured during the summer season of 2000. FWS values vary between 0 (dry land) and 100 (water). The *FWS* decreases progressively from the southern to the northern part of the Basin. The largest lakes, i.e., Great Bear, Great Slave, and Athabasca, gradually become open water surfaces as their lake ice melts. The southwestern part of the Basin has the highest average annual precipitation (Szeto et al. 2007). According to the *FWS* maps, it is generally the wettest part of the Basin. However, there is some uncertainty due to the presence of relatively dense vegetation and the topographic effect not considered in our study. The low area between Great Bear Lake and Great Slave Lake has higher *FWS* values compared to other parts of the Basin.

The variation of the *FWS* over the Basin's outlet is presented in Fig. 3. The *FWS* values were computed for the summers of 1997, 1998, 1999 and 2000. *FWS* values computed for the summer of 1998 are higher than those for other years. The summer season of 1998 was the eighth wettest on record (Kochtubajda et al. 2002), and mean temperature of the Basin was

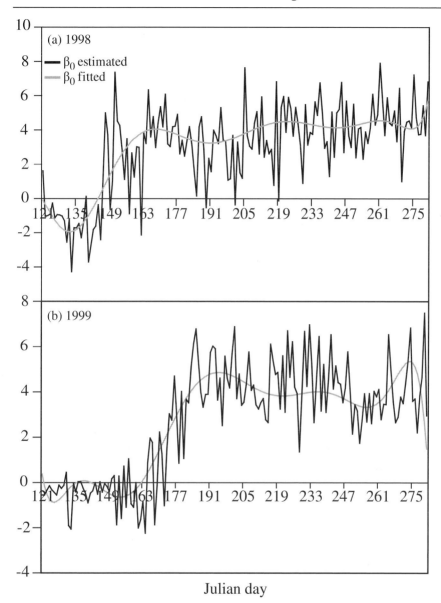

Fig. 1. Variability of β_0 during the summer seasons of (a) 1998 and (b) 1999

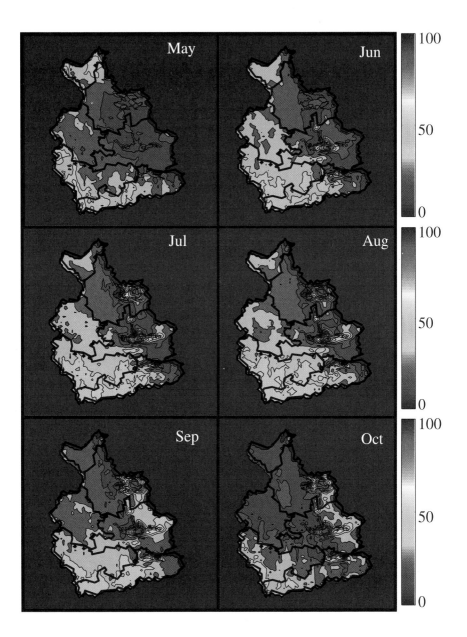

Fig. 2. Spatial variations of the *FWS* over the Mackenzie River Basin during the summer of 2000

above normal. This probably explains the higher values of *FWS* obtained, especially at the beginning of the summer. In addition, in August 1998 the Basin was considered a source of moisture as evaporation exceeded precipitation by nearly 17 mm month^{-1}, compared to 6.92 and 9.33 mm month^{-1} for August 1997 and 1999, respectively, which are considered normal years (Proctor et al. 2000). This corroborates the steeper slope of the *FWS* obtained at the end of the summer of 1998, as compared to the other years.

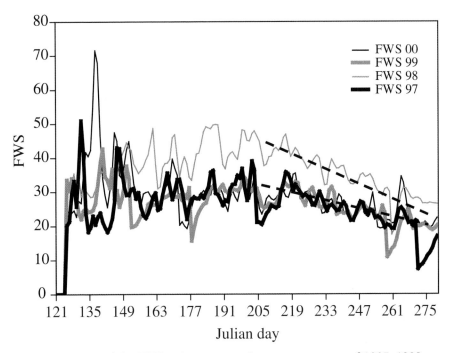

Fig. 3. Comparison of the *FWS* estimates over the summer season of 1997, 1998, 1999 and 2000 in the Basin outlet

A comparison between the *FWS* retrieved using constant and time varying parameters β_0 and β_1 was also performed at various locations within the MRB (not shown here). Using constant parameters, the *FWS* remained approximately constant for the entire summer, before significantly augmentation during the wet fall season. On the other hand, the time varying parameters produced *FWS* values that exhibit a seasonal pattern, characterized by an increase of the index shortly after snowmelt, followed by a gradual decrease as the Basin becomes progressively drier when evaporation exceeds precipitation. Using time varying empirical parameters produced a more 'natural' behavior of the *FWS*, hence of the *BWI*, reflecting their sensitivity to moisture conditions at the soil surface.

Low resolution of the SSM/I sensor precludes application of the proposed approach to monitor regional and local scale surface soil moisture patterns. Furthermore, the approach is based on the simultaneous utilization of 19, 37, and 85 GHz passive microwave data, which is not optimal for quantitative retrieval of surface soil moisture estimates, as vegetation and atmospheric effects may be significant, even after using time varying parameters to calculate the *BWI* and *FWS* indices. The AMSR-E (Advanced Microwave Scanning Radiometer - EOS) sensor onboard NASA's Aqua satellite launched in May 2002 offers a larger spectrum of microwave frequencies/spatial resolutions, and therefore should be more amenable to carry out *BWI* estimates at a regional scale. Moreover, merging low resolution passive microwave data with information from high resolution sensors in the visible range should further improve the spatial resolution of *BWI* estimates. The next section presents such an approach.

3 A Basin Wetness Index at the Regional Scale

Terrain-based wetness indices have been proposed to provide a systematic distribution of soil moisture at the watershed scale (e.g., Beven and Kirkby 1979; Chaplot and Walter 2003; Crave and Gascuel-Odoux 1997; Gomez-Plaza et al. 2000; Romano and Palladino 2002; Wilson et al. 2005). Most of these indices assume that the change in soil moisture is controlled by two main factors, namely the spatial variability of soil properties such as hydraulic conductivity, and topographic attributes such as slope, aspect, and the distance to the stream bank.

Classic topography-based wetness indices are static and cannot capture the temporal variability of the soil water content. However, most of the proposed indices in the literature do not explain more than 50% of the spatial distribution of soil moisture (Western et al. 1999). It is well known that vegetation cover exerts a major influence on soil moisture spatial patterns (Gomez-Plaza et al. 2000, 2001; Qiu et al. 2001; Western et al. 2002). Thus, it is important that topographic indices take into account the vegetation cover in order to improve their prediction of the spatial distribution of the water content in catchments. This requires assessing the vegetation cover development. Ancillary data such as high resolution MODIS (Moderate-Resolution Imaging Spectroradiometer) images from the Terra satellite (250-meter resolution) can be used for a daily assessment of the vegetation cover.

We propose an approach that combines low resolution passive microwave data and high resolution topographic/vegetation attributes to monitor the spatial and temporal variations of soil wetness at a regional, i.e., kilometer, scale (Temimi et al. 2004). It is expected that this combination will improve the low resolution of the passive microwave map and overcome the static behavior of the developed wetness indices. The approach involves retrieval of a basin wetness index (*BWI*) from the AMSR-E 37 GHz vertically polarized data, which will provide the low spatial, but high temporal resolution of surface soil wetness. Achievement of a high resolution *BWI* will be made possible by mapping topographic information, in the form of a topographic index *TI*, onto the *BWI*. The influence of seasonal vegetative growth will also be taken into account through a dynamic, as opposed to a static, adjustment of the *TI*.

3.1 The Topographic/Vegetation Index

The classic wetness index proposed by Beven and Kirkby (1979) is based on two parameters and is written as: $TI = ln\,(a/\tan\beta)$, where *a* is the contributing area and $\tan\beta$ is the local slope of the terrain. It merely represents the wetness state as a function of the ratio of the drained area divided by the local slope, without explicitly considering the vegetation. However, Qiu et al. (2001) concluded that the differences in vegetation resulting from different land use types can dominate the spatial distribution of the soil moisture, and Gomez-Plaza et al (2001) stated that the topographic attributes effect on the wetness state depends largely on the vegetation cover. Therefore, we suggest modifying the static index by weighting its components using the vegetation canopy density which largely influences the spatial distribution of the soil moisture:

$$TI = V \ln(a) + (1-V) \ln(1/\tan\beta) \qquad (6)$$

V is the fractional vegetation cover estimated using the following relationship (Eagleson 1982):

$$V = 1 - \exp(-\mu\, LAI) \qquad (7)$$

where *LAI* is the Leaf Area Index and μ is the extinction coefficient. The value of μ varies depending on the canopy density between 0.35 (low vegetation density) to 0.7 (high vegetation density).

The first component of the index reflects the contribution of the vegetated soil, which is usually wetter, and for which non-local controls such as the contributing area dominate the soil moisture response (Grayson et al. 1997). The second component accounts for bare, or non-vegetated, soils where moisture is more affected by local control conditions such as slope, aspect, and soil profile.

Combining Eqs. (6) and (7) yields:

$$TI = \ln(a) - \ln(a \tan \beta) e^{-\mu * LAI} \qquad (8)$$

in which *TI* is no longer static as the fractions estimated from the *LAI* values are variable with time; and this formulation accounts for the vegetation effect on the spatial distribution of soil moisture. As the *LAI* can be estimated from satellite imagery, such as MODIS, Eq. (8) can be applied on a per pixel basis to provide spatial and temporal maps of *TI*.

3.2 The Basin Wetness Index

An index is proposed that combines information from AMSR-E passive microwave data and MODIS visible data. The proposed methodology first consists in using the 37 GHz vertically and horizontally polarized AMSR-E brightness temperature to extract the fractional water surface (*FWS*) index on a pixel by pixel basis by applying a mixing model:

$$PR_{obs} = ow \cdot PR_{ow} + f \cdot PR_f + nf \cdot PR_{nf} + sm \cdot PR_{sm} \qquad (9)$$

where
$PR = (T_{bv} - T_{bh}) / (T_{bv} + T_{bh})$, is the polarization ratio
obs = observed
nf = non-flooded, i.e., dry, soil fraction
f = flooded soil (bare and vegetated) fraction
ow = fraction of the permanent open water surface
sm = fraction of the upper soil moisture contribution

The 37 GHz frequency of the AMSR-E sensor was selected because of its higher spatial resolution, 8–14 km, as compared to SSM/I at 28–37 km. Lower frequencies available with AMSR-E, for example 6.9 and 10.7 GHz, are less affected by atmospheric and vegetation effects. They are not considered here due to their much lower spatial resolution (at 56 and 38 km respectively). The amount of water at or near the surface as sensed by AMSR-E, denoted here as *FWS*(AMSR), includes the open water, flooded area, and upper soil moisture. Typical values of *PR* for the open water, flooded and non-flooded soil surfaces were calibrated from selected pixels on the AMSR-E imagery. The fraction *ow* was retrieved from a water body extent vector imported from a USGS database. To further reduce the number of unknowns in Eq. (9), the surface soil moisture contribution was explicitly removed and its contribution indirectly taken into account through the flooded soil surface component term, an assumption equivalent to treating surface soil moisture as 'free' water. Then, $ow + f + nf = 1$, with $FWS(\text{AMSR}) = ow + f$. Equation (9) can be solved for the unknown values of *f* and *nf*.

In a second step, MODIS images were used to obtain $FWS(\text{MODIS})$. Because MODIS is a sensor operating in the visible spectrum, the fractional water surface provided by this sensor, $FWS(\text{MODIS})$, is the sum of the contributions of permanently open water and of flooded bare soil areas. This suggests that information about the soil wetness (including flooded soil fraction) can be inferred from the difference between the *FWS* provided by the AMSR-E and MODIS. This difference can be related to a Basin Wetness Index, computed as:

$$BWI = [FWS(\text{AMSR}) - FWS(\text{MODIS})] / [1 - FWS(\text{MODIS})] \qquad (10)$$

A major difficulty is that the MODIS sensor is incapable of 'seeing' through clouds, thus hampering the use of Eq. (10) to provide daily estimates of *FWS*. To circumvent this difficulty, it is suggested to relate the *FWS* to river flow through a flow rating curve model. Several studies have shown that there is a strong correlation between hydrologic variables (discharge or stage) and flooded areas (Frazier et al. 2003; Mosley 1983; Smith et al. 1996; Vörösmarty et al. 1996). Based on these findings, the following model is suggested to relate the *FWS* to river flow (*Q*):

$$FWS(t) = aQ(t)^b \qquad (11)$$

Equation (11), when calibrated, can be used as a surrogate to MODIS to obtain *FWS* when clouds prevent a direct estimation of the index. Combining Eqs. (10) and (11) results in:

$$BWI = [FWS(\text{AMSR}) - aQ^b(t)] / [1 - aQ^b(t)] \qquad (12)$$

The *BWI* in Eq. (12) reflects the fraction of saturated soils within the entire non-flooded area. On the other hand, the *TI* provides a map as a prediction of a systematic spatial organisation of soil moisture over the same non-flooded area. Note that the *BWI* derived from passive microwave data and MODIS images retrieves only the water content at the surface and the upper soil layers (a few millimeters). However, soil moisture in the deeper soil layers may, under some situations, be inferred from near surface soil moisture. For example, Ragab (1995) obtained a linear relationship between surface (0–10 cm) and root zone (0–50 cm) soil moisture for vegetated surfaces where transpiration is considered the dominant mechanism by which water is depleted. The proposed approach could therefore be useful for assessing root zone soil moisture, valuable information for hydrologic simulations and forecasts.

3.3 Application

The model was applied to the Peace-Athabasca-Delta (PAD) area in the south-eastern part of the MRB (58°00'–59°00'N; 111°00'–112°00W). With an area of about 4000 km^2, it is one of the largest inland deltas in the world. It is located at the western end of Lake Athabasca, and was created by the confluence of the Peace, Athabasca and Birch rivers. The Peace River has the largest discharge, with a mean annual flow of 2100 m^3 s^{-1} at Peace Point. The flow is generally northward but this direction can be reversed when the water level in the Peace River exceeds that of Lake Athabasca (Leconte et al. 2001). Several wetland basins surround the delta (Toyra et al. 2002), and its central part is composed mainly of large and shallow lakes connected to Lake Athabasca by several channels.

A digital elevation model (DEM) of the PAD at a 1 km resolution area has been imported from the USGS database, from which the topographical attributes required for solving Eq. (6) were retrieved and a 'classic' topographic index map produced. MODIS-Terra images were used to derive representative seasonal patterns of the LAI over the PAD area on a pixel-by-pixel basis by fitting the retrieved LAI estimates to polynomial relationships. These relationships were used to develop daily maps of the vegetation coverage. It was found that the variation of the LAI over the PAD area is significant during a summer season, with values ranging from 1 to 3.6. Correspondingly, computed TI values displayed seasonal variability. The geographic distribution of the TI changed as vegetation gradually shifted the mechanism that influences the spatial distribution of soil moisture from a local control process dominated by slope effects to a more non-local control process in which the contributing area plays an increasing role.

Application of Eq. (12) to obtain daily BWI maps required the calibration of the flow rating curve (Fig. 4). Discharge measurements of the Slave River at Fitzgerald (59°52'N; 111°35'W) were used. MODIS images were classified to estimate the open water extent. The spatial resolution of the MODIS sensor (250 m) permits an accurate estimation of the water surface area.

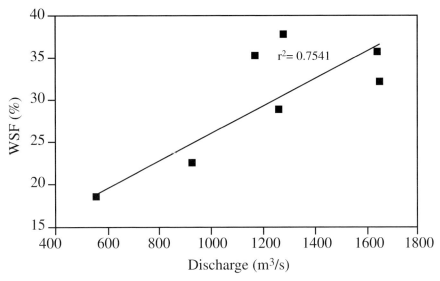

Fig. 4. Rating curve developed for the region of the Peace-Athabasca Delta using 2002 and 2003 summer season MODIS images and discharge data (after Temimi et al. 2005)

To account for the variability of the *BWI*, this index, averaged over the PAD, was compared to the precipitation and temperature measurements from Fort Chipewyan meteorological station (58°46'N; 111°07'W). Figure 5 shows the results for summer–fall 2003 in which weekly mean temperature and weekly total precipitation are used. The seasonal variability of the *BWI* is coherent with corresponding variations of air temperature and precipitation. As the temperature increases during the spring and early summer, snow melt produces large amounts of liquid water and the *BWI* increases sharply. The *BWI* remains relatively stable during June and July, and is minimal during this period, as evaporation exceeds precipitation. The *BWI* starts to increase in late summer and fall as precipitation exceeds evaporation. The decline of the *BWI* by mid-November is probably related to soil freezing, as the mean daily air temperature recorded at Fort Smith (60°01'N; 111°57'W) fell below the freezing point on October 27 and dropped to an average of -8.5°C in the first two weeks of November. On the other hand, the sole effect of precipitation on the *BWI* is more difficult to ascertain. Although Fig. 5 shows high frequency variations of the *BWI*, it is unclear whether such fluctuations are attributable to the precipitation events. Aggregating the meteorological data on a weekly basis may have masked the precipitation-*BWI* relationship. The use of satellite-derived precipitation information, such as AMSR-E precipitation products, would help in identifying possible relationships between precipitation occurrence/intensity and spatially aggregated *BWI* values. Other diagnostic tests such as exploring possible statistical relationships between 36 GHz brightness temperature and *BWI* would help to better identify the source of fluctuations of the wetness index. Note that an analysis with daily values is difficult to perform because the retrieved *BWI* information is noisy, as the original brightness temperature values measured by the AMSR-E and used to compute the index are affected by factors such as atmospheric effects which were not, or only partially, accounted for.

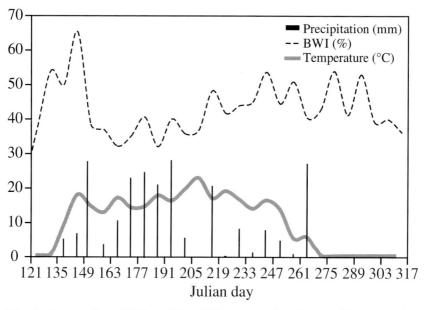

Fig. 5. Temporal variability of the *BWI* compared to the precipitation and temperature values observed during the summer of 2003 (after Temimi et al. 2005)

The *BWI* and *TI* index maps were combined to generate downscaled maps of wetness indices. This was done by assuming that the pixels with the highest *TI* would be wetted up or dried down first. In other words, the *TI* map indicates the priority of each pixel to either be wetted up or dried down depending on the *BWI* fluctuations. In addition, if two pixels present an equal *TI*, priority is given to the one with the highest/lowest elevation when wetting up/drying down the basin. Figure 6 shows downscaled *BWI* maps of the PAD. The fine scale surface moisture variability (1 km resolution) is showing up on these maps. In particular, the area between Lakes Athabasca and Claire and surrounding Mamawi Lake is generally wetter (light blue colour) than the areas north and south of these lakes. This is in good agreement with results from previous studies (e.g., Toyra et al. 2002) in which high resolution remote sensing data (Radarsat, Spot) were able to identify flooded and non-flooded land. It is also in accordance with the topography of the area (Peters et al. 2006). Observe that the PAD is overall drier in August (darker blue colour) than in June, which is in accord with the hydrologic regime characterized by evaporation exceeding precipitation during the summer season.

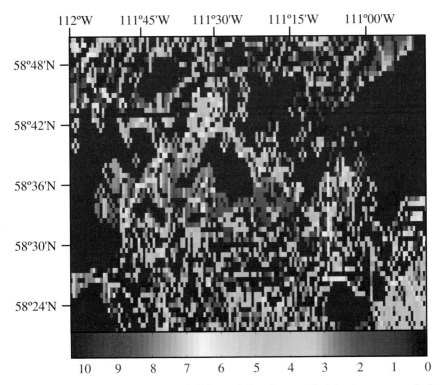

Fig.6. Seasonal and spatial variability of the downscaled basin wetness index (*BWI*) over the Peace-Athabasca Delta area

The results obtained from the downscaled *BWI* were compared to the observed precipitation at the Fort Chipewyan station. The downscaled *BWI* values were also averaged over a window of 3x3 pixels surrounding the station (or 750m x 750 m) to account for possible geo-location errors of the satellite imagery. Correlation coefficients were determined between the observed precipitation and the average of the *BWI* values over the window for the summers of 2002 and 2003. Using the time and space varying *TI* in the downscaling process resulted in a correlation coefficient of 0.50, as compared with a value of 0.37 when downscaling the *BWI* using the classic static index approach. Although preliminary, the results are a significant improvement over the static case approach. It would be worthwhile to test the potential of this approach using a higher resolution DEM, especially given the very gentle topography surrounding the PAD, as well as calculating *BWI* values from 6.9 and 10.7 GHz brightness temperatures, which are less affected by vegetation and water vapor.

4 Discussion and Conclusion

This research has demonstrated that satellite passive microwave data offer significant potential for monitoring surface soil wetness over very large and heterogeneous watersheds such as the MRB. Using data from both SSM/I and SMMR sensors, it is now possible to construct time series of surface soil wetness back to 1978. The analysis of long time series allows investigating the cyclic behavior of surface wetness at the watershed scale, and to identify various 'anomalies' in the cycle that could possibly be linked to climatic shifts. Also, the time series could be compared against soil moisture maps produced by distributed hydrologic models to qualitatively evaluate how well the models perform. As validating remotely sensed and model derived products remains a daunting issue over large and remote watersheds such as the MRB, results from this research should be useful to achieve better calibration of hydrologic models. In turn, improved models will lead to enhanced understanding of the hydrologic cycle in northern environments, as it will provide new insights on the interaction between the climate and terrestrial water.

One must bear in mind that the tools developed in this research were based on using passive microwave sensors working at frequencies that are less than optimal for accurate soil moisture retrieval. The retrieved remotely sensed information is related to basin storage, but much remains to be done to extract 'true' soil moisture information per se, over soil depths even as shallow as a few centimeters. However, new satellites such as the Soil Moisture and Ocean Salinity Mission (SMOS) have been specifically designed for monitoring soil moisture at the global scale. The configuration of the passive and active microwave sensors onboard these satellites will enable the sensing of wetness over depths ranging from several cm to close to 10 cm, depending on the moisture status in the basins. Moreover, vegetation and atmospheric effects that significantly affect sensors such as

SSM/I, would be considerably reduced, allowing an improved estimation of soil moisture. This information should prove invaluable to use with hydrologic models such as WATCLASS (Soulis and Seglenieks 2007), since these models conceptualize the vertical soil moisture transport at a scale compatible with the penetration depths of SMOS.

5 References

Basist A, Grody NC, Peterson TC, Williams CN (1998) Using the special sensor microwave/imager to monitor land surface temperatures, wetness, and snow cover. J Appl Meteorol 37:888–911

Basist A, Williams C, Grody N, Ross TF, Shen S, Chang ATC, Ferraro R, Menne MJ (2001) Using the special sensor microwave imager to monitor surface wetness. J Hydrometeorol 2:297–308

Beven KJ, Kirkby MJ (1979) A physically based, variable contributing area model of basin hydrology. Hydrol Sci B 24:43–69

Chaplot V, Walter C (2003) Subsurface topography to enhance the prediction of the spatial distribution of soil wetness. Hydrol Process 17:2567–2580

Côté J, Gravel S, Méthot A, Patoine A, Roch M, Staniforth A (1998). The operational CMC-MRB global environmental multiscale (GEM) model. Part I: design considerations and formulation. Mon Weather Rev 126:1397–1418

Crave A, Gascuel-Odoux C (1997) Influence of topography on time and space distribution of soil surface water content. Hydrol Process 11:203–210

Choudhury BJ (1991). Passive microwave remote sensing contribution to hydrological variables. Surv Geophys 12:63–84

Eagleson PS (1982) Ecological optimality in water-limited natural soil-vegetation systems 1. Theory and hypothesis. Water Resour Res 18:325–340

Fily M, Royer A, Goita K, Prigent C (2003) A simple retrieval method for land surface temperature and fraction of water surface determination from satellite microwave brightness temperatures in sub-arctic areas. Remote Sens Environ 85:328–338

Frazier P, Page K, Louis J, Briggs S, Robertson AI (2003) Relating wetland inundation to river flow using Landsat TM data. Int J Remote Sens 24:3755–3770

Gomez-Plaza A, Alvarez-Rogel J, Albaladejo J, Castillo VM (2000) Spatial patterns and temporal stability of soil moisture across a range of scales in a semi-arid environment. Hydrol Process 14:1261–1277

Gomez-Plaza A, Martinez-Mena M, Albaladejo J, Castillo VM (2001) Factors regulating spatial distribution of soil water content in small semi-arid catchments. J Hydrol 253:211–226

Grayson RB, Western AW, Chiew FHS (1997) Preferred states in spatial soil moisture patterns: local and non-local controls. Water Resour Manag 33:2897–2908

Jackson TJ (1993) Measuring surface soil moisture using passive microwave remote sensing. Hydrological Processes 7: 139-152.

Kochtubajda B, Stewart RE, Gyakum JR, Flannigan MD (2002) Summer convection and lightning over the Mackenzie River Basin and their impacts during 1994 and 1995. Atmos Ocean 40:199–220

Lakshmi V (1996) Use of special sensor microwave imager data for soil moisture estimation. Ph.D. Dissertation, Princeton University

Leconte R, Pietroniro A, Peters DL, Prowse TD (2001) Effects of flow regulation on hydrologic patterns of a large inland delta. Regul River 17:51–65

Mosley MP (1983) Response of braided rivers to changing discharge. J Hydrol 22:18–67

Njoku EG, Jackson TJ, Lakshmi V, Chan TK, Nghiem SV (2003) Soil moisture retrieval from AMSR-E. IEEE T Geosci Remote Sens 41:215–229

Peters DL, Prowse TD, Pietroniro A, Leconte R (2006) Flood hydrology of the Peace-Athabasca Delta, northern Canada. Hydrol Process 20:4073–4096

Prigent C (1997) Microwave land surface emissivities estimated from SSM/I observations. J Geophys Res 102(D18):21,867–21,890

Proctor BA, Strong GS, Smith CD, Wang M, Soulis ED (2000) Atmospheric moisture budgets for the Mackenzie GEWEX project water years 1994-1995 through 1998-1999. 6th Scientific Workshop for MAGS: Nov. 15–17, 2000, pp 164–177

Qiu Y, Fu BJ, Wang J, Chen LD (2001) Soil moisture variation in relation to topography and land use in a hill slope catchment of the Loess Plateau, China. J Hydrol 240:243–263

Ragab R (1995) Towards a continuous operational system to estimate the root-zone soil moisture from intermittent remotely sensed surface moisture. J Hydrol 173:1–25

Romano N, Palladino M (2002) Prediction of soil water retention using soil physical data and terrain attributes. J Hydrol 265:56–75

Smith LC, Isacks BL, Bloom AL (1996) Estimation of discharge from three braided rivers using synthetic aperture radar satellite imagery: potential application to ungauged basins. Water Resour Res 32:2021–2034

Soulis ED, Seglenieks FR (2007) The MAGS integrated modeling system. (Vol. II, this book)

Szeto KK, Stewart RE, Yau MK, Gyakum J (2007) The Mackenzie climate system: a synthesis of MAGS atmospheric research. (Vol. I, this book)

Tanaka M, Adjadeh TA, Tanaka S, Sugimura T (2002) Water surface area measurement of Lake Volta using SSM/I 37-GHz polarization difference in rainy season. Adv Space Res 30:2501–2504

Tanaka M, Sugimura T, Tanaka S (2000) Monitoring water surface ratio in the Chinese floods of summer 1998 by DMSP-SSM/I. Int J Remote Sens 21:1561–1569

Temimi M, Leconte R, Brissette FP, Toussaint T (2003) A dynamic estimate of a soil wetness index for the Mackenzie River Basin from SSM/I measurements, IGARSS 2003, July 21–25, 2003, Toulouse, France, vol 2, pp 920–922

Temimi M, Leconte R, Brissette F, Chaouch N, Magagi R (2004) Near real time flood monitoring over the Mackenzie River Basin using passive microwave data. IGARSS 2004, Sept. 20–24, 2004, Anchorage, Alaska, USA, vol 3, pp 1862–1865

Temimi M, Leconte R, Brissette F, Chaouch N, Magagi R (2005) Flood and soil wetness monitoring over the Mackenzie River Basin using AMSR-E 37-Ghz brightness temperature. IGARSS 2005, July 25–29, 2005, Seoul, Korea, vol 1, pp 59–62

Toyra J, Pietroniro A, Martz L, Prowse TD (2002) A multi-sensor approach to wetland flood monitoring. Hydrol Process 16:1569–1581

Vörösmarty CJ, Willmott CJ, Choudhury BJ, Schloss AL, Steams TK, Robeson SM, Dorman TJ (1996) Analyzing the discharge regime of a large tropical river through remote sensing, ground-based climatic data, and modeling. Water Resour Res 32:3137–3150

Western AW, Grayson RB, Bloschl G (2002) Scaling of soil moisture: a hydrologic perspective. Ann Rev Earth Pl Sci 30:149–180

Western AW, Grayson RB, Bloschl G, Willgoose GR, McMahon TA (1999) Observed spatial organization of soil moisture and its relation to terrain indices. Water Resour Res 35:797–810

Wigneron JP, Calvet JC, Pellarin T, Van de Griend AA, Berger M, Ferrazzoli P (2003) Retrieving near-surface soil moisture from microwave radiometric observations: current status and future plans. Remote Sens Environ 85:489–506

Wilson DJ, Western AW, Grayson RB (2005) A terrain and data-based method for generating the spatial distribution of soil moisture. Adv Water Resour 28:43–54

Woo MK, Rouse WR, Stewart RE, Stone J (2007). The Mackenzie GEWEX Study: a contribution to cold region atmospheric and hydrologic sciences. (Vol. I, this book)

Chapter 5

Studies on Snow Redistribution by Wind and Forest, Snow-covered Area Depletion, and Frozen Soil Infiltration in Northern and Western Canada

John W. Pomeroy, Donald M. Gray and Phil Marsh

This chapter is dedicated to the memory of Don Gray, one of the original investigators in MAGS and a scientist of many contributions to cold regions hydrology, who passed away in January 2005, after a brief illness.

Abstract Important advances in our understanding of snow and frozen soil processes have been made, especially in regard to the transport and sublimation of blowing snow, interception and sublimation of snow in forest canopies, snow spatial distributions in complex environments, snowmelt in open environments and under forest canopies, advection of energy from bare ground to snow, snowcover depletion during melt, and heat and mass transfer during infiltration to unsaturated frozen mineral soils. These studies, conducted at the Division of Hydrology at the University of Saskatchewan, covered a range of northern environments including the tundra–taiga transition, the cordilleran sub-arctic, the southern boreal forest, and the northern prairie. Results from field research have led to the development and improvement of algorithms related to snow and infiltration processes, which have contributed to hydrologic and atmospheric models in the Mackenzie GEWEX Study.

1 Introduction

Storage, melt, infiltration, and runoff related to the seasonal snowcover are primary hydrologic events in most cold regions including the Mackenzie River Basin (MRB). Building on the 30-year tradition of cold regions hydrology research at the Division of Hydrology at the University of Saskatchewan, a research program was implemented to improve the understanding of the physical processes governing snow accumulation, ablation, and infiltration. These processes became the subject of extensive investigations in the Mackenzie GEWEX Study (MAGS). Led by Gray and Pomeroy from 1992 to 1999, field studies were conducted using a transect of instrumented research basins in different cold region environments, viz., the arctic–taiga transition, the cordilleran sub-arctic, the southern boreal forest, and the northern prairie. Physically-based algorithms were devised that describe these processes, and many of these algorithms have subsequently been incorporated in a range of hydrologic and atmospheric models.

This chapter presents the major findings pertaining to the investigation and modeling of several processes important to the water and energy cycles of cold environments, including interception and sublimation of forest snow, blowing snow transport and sublimation, snowmelt energetics and depletion of snow-covered area, and the infiltration of meltwater into frozen soils.

2 Methodology

Both field investigation and modeling were used in the study of snow and infiltration processes.

Intensive research on snow accumulation, ablation, and frozen soil infiltration were conducted at four locations, each representing a major northern environment:

1. Inuvik, Northwest Territories (arctic–taiga transition), e.g., Pomeroy et al. 1995
2. Wolf Creek, Yukon Territory (cordilleran subarctic), e.g., Pomeroy and Granger 1999
3. Waskesiu, Saskatchewan (southern boreal forest), e.g., Pomeroy et al. 1997a
4. Kernen Farm, Saskatchewan (northern prairie), e.g., Shook and Gray 1996)

Field measurements employed standard meteorological observations as well as direct observations of sensible and latent heat fluxes over snow surfaces using eddy correlation systems, radiation, soil heat flux, blowing snow flux from an optoelectronic particle detector, intercepted snow mass measurements with a suspended, weighed full-size coniferous tree, and infiltration to frozen soils using twin-probe gamma attenuation devices, soil thermocouples, and time domain reflectometry.

Algorithms were developed for the following cold regions hydrologic processes. These algorithms were examined with respect to their performance and they have undergone enhancements where appropriate:
1. Canopy snow interception, unloading, and sublimation (Essery et al. 2003; Hedstrom and Pomeroy 1998; Parviainen and Pomeroy 2000; Pomeroy and Gray 1995; Pomeroy and Schmidt 1993; Pomeroy et al. 1998b)
2. Blowing snow model – threshold condition, scaling, and sublimation (Li and Pomeroy 1997a, b; Pomeroy and Li 2000; Pomeroy et al. 1998a)
3. Complex terrain blowing snow, redistribution, and sublimation (Essery 2001; Essery et al. 1999; Pomeroy et al. 1997b)
4. Snow ablation, open environment snowmelt, snowcover depletion, and small-scale advection to patchy snow (Faria et al. 2000; Pomeroy and Granger 1997; Pomeroy et al. 1998a, 2003; Shook and Gray 1996)
5. Infiltration to frozen soils – fully coupled mass and energy balance and operational algorithm (Gray et al. 2001; Zhao and Gray 1997; Zhao et al. 1997).

3 RESULTS

3.1 Interception, Unloading and Sublimation of Canopy Snow

Field results from several winters of observation of snow accumulation under forest canopies, snowfall, and the load of snow on a series of cut, weighed, suspended trees show that leaf area, canopy closure, species type, time since snowfall, snowfall amount, and existing snow load control the efficiency of snow interception (Hedstrom and Pomeroy 1998; Pomeroy and Gray 1995). A physically-based algorithm describing these results was developed from observations using a weighed suspended tree (see Fig. 2 in Woo and Rouse 2007). The sensitivity of this algorithm to winter leaf area index, air temperature, snowfall and initial canopy snow load is shown in Fig. 1.

Physically-based equations describing snow interception (Hedstrom and Pomeroy 1998) and sublimation processes (Pomeroy et al. 1998b) were coupled and applied to calculate canopy intercepted snow load using a fractal scaling technique (Pomeroy and Schmidt 1993) to provide a snow-covered forest boundary condition for the one-dimensional Canadian Land Surface Scheme, CLASS (Verseghy et al. 1993). Parviainen and Pomeroy

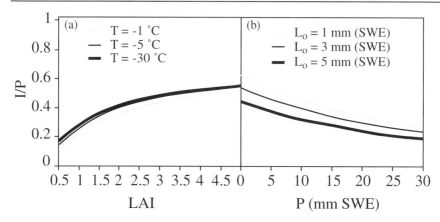

Fig. 1. Modeled interception efficiency (snow interception/snowfall) as a function of (a) winter leaf area index and air temperature, and (b) snowfall and initial canopy snow load (L_o)

(2000) found that CLASS could be modified to provide an appropriate treatment of turbulent transfer and within-canopy ambient humidity to accommodate this nested control volume approach. Tests in late winter in a southern boreal forest against measured sublimation found that the coupled model provides good approximations of sublimation losses on half-hourly and event basis (Fig. 2). A complete model was tested in Russia and confirmed with long term data (Gelfan et al. 2004). The model has been incorporated into the Hadley Centre land surface scheme and compared well to a range of field measurements from Canada and the United States (Essery et al. 2003). Pomeroy et al. (2002) developed a parametric form of this model to predict snow accumulation in northern forests as a function of accumulation in small clearings or cumulative snowfall.

3.2 Blowing Snow

Sublimation fluxes during blowing snow have been estimated to return 10–50% of seasonal snowfall to the atmosphere as vapor in North American prairie and arctic environments (Essery et al. 1999; Pomeroy and Gray 1995; Pomeroy et al. 1997b). These fluxes are calculated as part of blowing snow two-phase transport models with provision for phase change based upon a particle-scale energy balance driven by measured wind speed, air temperature, and humidity (Pomeroy and Li 2000; Pomeroy et al. 1993) and a snow transport threshold algorithm that uses air temperature to index snow cohesion (Li and Pomeroy 1997a). The probability of

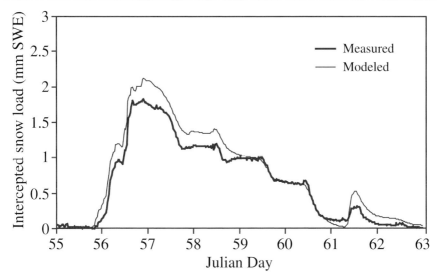

Fig. 2. Measured intercepted snow load (from suspended weighed pine tree), and sublimation model calculated intercepted load (after Pomeroy et al. 1998)

occurrence of blowing snow over time or space (for uniform terrain) follows a cumulative normal distribution that is controlled by snow temperature, snow age, vegetation exposure, and occurrence of melt or rain (Li and Pomeroy 1997b). An algorithm describing blowing snow probability provides a means to scale blowing snow fluxes from point to large areal averages in a computationally simple manner (Pomeroy and Li 2000). An example of the model operation for tundra surfaces at Trail Valley Creek is shown in Fig. 2. The model was tested extensively in the prairie environment as well and found to redistribute snow appropriately between land cover classes (Pomeroy et al. 1998a).

Landscape classifications from a LANDSAT image, a complex terrain windflow model, snowmelt, and blowing snow process routines can be used to determine blowing snow fluxes over irregular arctic land surfaces. The resulting runs with a distributed blowing snow model represented the distribution of SWE in test basins and matched basin snow accumulation within 6% (Pomeroy et al. 1997b). Sublimation losses were small for the subarctic basin, about 21% over the arctic basin and 30% from tundra surfaces (Pomeroy and Marsh 1997). Subsequent tests with a more physically-based distributed blowing snow model show that the arctic tundra comprises a variety of blowing snow flow zones that are largely controlled by vegetation cover and by topographically induced zones of convergence and divergence of windflow. Results that included sublimation as modeled by Pomeroy and Li (2000) produced spatial distributions of snow accumulation that were close to the values obtained by extensive snow surveys (Essery et al. 1999). However, results with suppressed blowing snow sublimation redistributed too much snow to the treeline, where modeled accumulations were much greater than observed. This provides a regional mass balance confirmation of significant sublimation from blowing snow on arctic tundra and calls into question models that produce minimal sublimation from blowing snow in this environment. Later work developed scaling relationships from this model (Essery 2001) and examined the influence of shrub tundra coverage on blowing snow fluxes (Essery and Pomeroy 2004). Figure 3 provides an example of the mapped SWE distribution simulated with the blowing snow model, for an arctic domain centered about Trail Valley Creek.

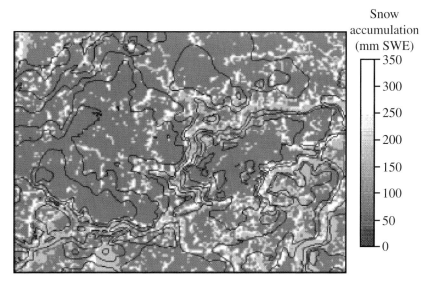

Fig. 3. Mapped distribution of late winter snow accumulation (mm SWE) in the Trail Valley Creek domain; simulation produced with the Distributed Blowing Snow Model (DBSM), a version of PBSM coupled to the MS3DJH/3R complex terrain boundary-layer model (after Essery et al. 1999)

Blowing snow models such as the Prairie Blowing Snow Model (PBSM) have normally been evaluated based upon their ability to reproduce diagnostic mass flux gradient measurements and regional-scale snow redistribution patterns and snow mass. Direct evidence was obtained in MAGS that large latent heat fluxes (40–60 W m^{-2}) that result in sublimation rates of 0.05–0.075 mm-SWE hour^{-1} are associated with mid-winter, high-latitude blowing snow events (Fig. 4). For events with wind speeds above the threshold level for snow transport, these fluxes are within the range of those predicted by Pomeroy and Li (2000). The fluxes are well in excess of those which can be predicted by standard bulk aerodynamic transfer equations, suggesting that blowing snow physics needs to be addressed by land surface schemes and hydrologic models in order to properly represent snow surface mass and energy exchange in open environments (Pomeroy and Essery 1999).

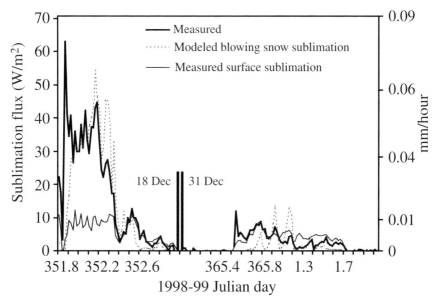

Fig. 4. Measured sublimation flux, modeled blowing snow sublimation (PBSM) and modeled surface sublimation (bulk transfer) measured at over a level Prairie surface (after Pomeroy and Essery 1999)

3.3 Ablation of Seasonal Snow-covers

Advection of energy from bare ground to patchy snow was recognized to have an important role in the energy equation for snowmelt in open environments. Shook (1995) applied the advection expressions of Weisman (1977) in small-scale gridded model to a synthetic snowcover; grid cells as they became snow-free became sources of advected energy to the remaining snow-covered cells. The synthetic snowcover was generated using the

Fractal Sum of Pulses Method and had the same statistical properties as real snowcovers (Shook and Gray 1997a). Shook and Gray (1997b) showed that advection was important for late spring snowmelt under strong insolation as is typical in the northern prairies and arctic. This detailed model is the basis behind the simplified advection efficiency scheme of Marsh and Pomeroy (1996) and Marsh et al. (1997).

Because snow depth is self-similar and fractal, the frequency distribution of snow water equivalent (SWE) can be simulated using a 2-parameter log-normal density function (Shook and Gray 1996). By applying an even melt rate to this distribution the snow covered area (SCA) can be calculated during snow depletion (Shook and Gray 1997c). The more variable the initial SWE, the more rapid the SCA depletion, other factors held constant (Fig. 5).

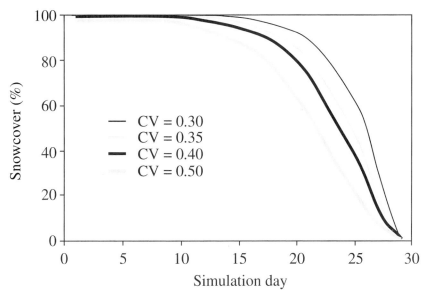

Fig. 5. Snow-covered area depletion curves modeled by applying a constant, uniform melt rate to snow covers with SWE that have a log-normal distribution. Fractional snow cover is plotted against time from an initial SWE of 130 mm for various coefficients of variation (CV) of SWE

Pomeroy et al. (1998a) provided the statistical parameters needed to calculate snow depletion for various environments and showed how this might be incorporated in melt models. When this approach was tested in complex terrain, it was shown that variable melt energy needs be addressed, even in open environments, but this could be done by landscape stratification (Pomeroy et al. 2003).

The influence of forest canopy cover and variable melt energetics on depletion of snowcover in the boreal forest was then investigated. The results can be distinguished between variability within the forest stand and that between forest stands. Between forest stands, Pomeroy and Granger (1997) found that melt energy was much lower as stand density increased. The radiation model developed by Pomeroy and Dion (1996) accounts for the shortwave component of these melt differences. An examination of the shrub tundra environment led to extension of these concepts to shrub terrain that covers much of the sub-arctic (Pomeroy et al. 2006). Within stands, Faria et al. (2000) found the frequency distribution of SWE under boreal canopies fit a log-normal distribution, with the most dense stand displaying the most variable SWE prior to melt. Within stands, snowmelt energy below the canopies was found to be spatially heterogeneous and inversely correlated to SWE (Fig. 6).

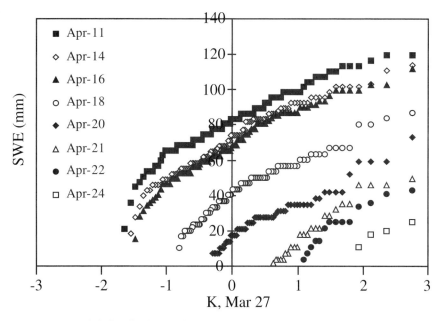

Fig. 6. Sequential distributions of snow water equivalent (SWE) during melt in a Pine Stand. K is the frequency factor for the log-normal distribution of SWE, the K for SWE = 0 reflects snow covered area. Note that melt is greater for smaller SWE (after Faria et al. 2000)

The variability of melt energy within a stand decreased with overall stand density. Within-stand covariance between the spatial distributions of SWE and melt energy promoted an earlier depletion of snowcover than if melt energy were uniform (Pomeroy et al. 2001). This covariance was largest for the most heterogeneous stands (usually medium density). Stand scale variability in mean SWE and mean melt energy resulted in the most rapid SCA depletion for stands with lower leaf area. Because of the heterogeneity in the spatial distributions of SWE and melt energy in forest environments, it is necessary that these variations be included in calculations of SCA depletion (Faria et al. 2000). Figure 7 shows an example calculation where initial SWE and mean melt energy at the stand scale are used to drive SCA depletion calculations which rely on the initial sub-stand distribution of SWE and the covariance between SWE and melt. Comparisons of the measured depletion with simulated depletion showed improved fit for simulations that included covariance over those that neglect this feature (Faria et al. 2000), and the between-stand variation is consistent with the findings of Pomeroy and Granger (1997).

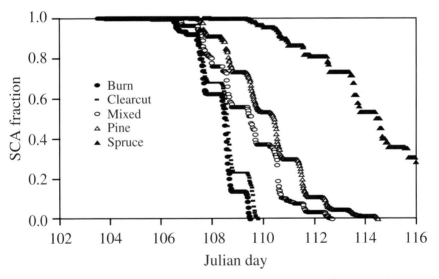

Fig. 7. Simulated snowcover depletion curves using measured mean melt rate at each site to calculate change in snow covered area with time as a function of the distribution of SWE and the covariance between SWE and melt rate (after Faria et al. 2000)

3.4 Infiltration to Frozen Soils

Infiltration into frozen ground involves simultaneous coupled heat and mass transfers with phase changes, such that the infiltrating water conveys heat transfer into the ground and modifies the soil temperature regime. Field measurements and model simulations (Zhao et al. 1997) demonstrate that both the infiltration rate and the surface heat transfer rate (conduction) in a frozen soil decrease with time following the application of meltwater to the surface (Fig. 8). Zhao et al. (1997) separate these variations into a transient regime and a quasi-steady-state regime. The transient regime follows immediately the application of water on the surface and during this period the infiltration rate and the heat transfer rate decrease rapidly. The quasi-steady-state regime occurs when the changes in the infiltration rate and the heat transfer rate with time are relatively small. The duration of the transient period is usually short (a few hours) and the energy used to increase the soil temperature is largely supplied by heat conduction at the surface (i.e., high heat transfer rate at the surface). In the quasi-steady-state regime, the energy used to increase the soil temperature at depth is supplied by latent heat released by the refreezing of percolating meltwater in the soil layers above (i.e., low heat transfer rate at the surface). Zhao et al.

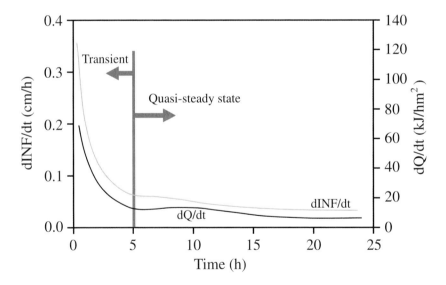

Fig. 8. Variations in infiltration rate (dINF/dt) and surface heat flux rate (dQ/dt) with time during snowmelt infiltration into a frozen silty clay soil (after Zhao et al. 1997)

1997) estimate that as much as 90% of the latent heat released by the refreezing of meltwater is conducted deeper in the soil where it used for melting and increasing the soil temperature

Zhao and Gray (1997, 1999) and Gray et al. (2001) reported the development and testing of a general parametric correlation for estimating snowmelt infiltration into frozen soils. Cumulative infiltration (*INF*) is expressed through the following relationship with the soil surface saturation (S_o) during melting, the total soil moisture saturation (water and ice) (S_I), the temperature (T_I, in K) at the start of snow ablation, and the infiltration opportunity time (*t*) (i.e., the time that meltwater is available at the soil surface for infiltration):

$$INF = C S_o^{2.92} (1-S_I)^{1.64} \left[\frac{273.15 - T_I}{273.15} \right]^{-0.45} t^{0.44} \qquad (1)$$

in which *C* is a bulk coefficient that characterizes the effects on infiltration of differences between model and natural systems.

Figure 9 compares modeled and measured profiles of soil moisture (water and ice) at different infiltration opportunity times in the boreal forest. There is reasonable agreement (likely within measurement accuracy) between measured and modeled values in the sandy loam soil at Waskesiu, and similarly good agreement among measured and modeled cumulative infiltration with a maximum difference of about 3.5 mm, indicating that Eq. (1) gives reasonable estimates of snowmelt infiltration.

The coefficient *C* in Eq. (1) characterizes the effects on infiltration of differences between model and natural systems. For example, the expression assumes that surface saturation is constant, the soil is uniform and homogeneous, and the soil moisture and temperature throughout the soil profile at the start of infiltration are constant. These conditions are rarely found in nature. Zhao and Gray (1999) suggest representative values of *C* = 1.0–1.3 for frozen sandy soils in a boreal forest and *C* = 2.05 for various fine-textured (sandy loam, loam, silty clay, and clay) frozen Prairie soils.

4 Conclusions

The results reported demonstrate that cold regions hydrologic processes can have profound and previously undocumented impacts on the calculation of surface water and energy fluxes of the land surface environments found in the MRB. Progress has been made in describing many of the pro-

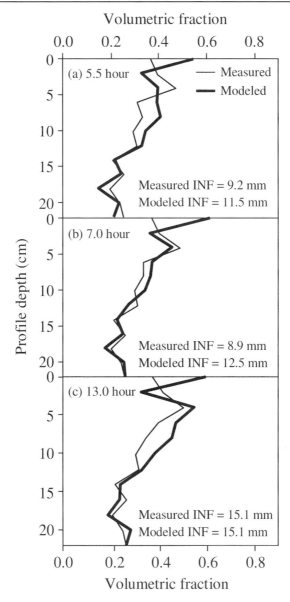

Fig. 9. Comparison of modeled and measured profiles of soil moisture (water + ice) into a frozen sandy loam soil in Prince Albert National Park after (a) 5.5 h, (b) 7.0 h, and (c) 13.0 h of snowmelt infiltration. Simulation initiated at 1200 h March 25, 1999 and compared to measurements of soil moisture (water + ice) at: (a) 1730 h March 25, (b) 0900 h March 26, and (c) 1800 h March 26

cesses in a physical manner, evaluating the process descriptions, and in developing operational algorithms for some of the processes. Some coupling and/or comparison of process algorithms with standard land surface scheme calculations has been demonstrated. The observed multi-scale operation and horizontal interaction of some of these processes means that phenomena operating at very small scales can affect large-scale water and energy balances. The relative success in transposing hydrologic process descriptions from one environment to another can be attributed to the strong physical basis of the descriptions employed.

References

Essery RLH (2001) Spatial statistics of windflow and blowing snow fluxes over complex topography. Bound-Lay Meteorol 100:131–147

Essery RLH, Li L, Pomeroy JW (1999) Blowing snow fluxes over complex terrain. Hydrol Process 13:2423–2438

Essery RLH, Pomeroy JW (2004) Vegetation and topographic control of windblown snow distributions in distributed and aggregated simulations for an Arctic tundra basin. J Hydrometeorol 5:734–744

Essery RLH, Pomeroy JW, Parviainen J, Storck P (2003) Sublimation of snow from boreal forests in a climate model. J Climate 16:1855–1864

Faria DA, Pomeroy JW, Essery RLH (2000) Effect of covariance between ablation and snow water equivalent on depletion of snow-covered area in a forest. Hydrol Process 14: 2683–2695

Gelfan A, Pomeroy JW, Kuchment L (2004) Modelling forest cover influences on snow accumulation, sublimation and melt. J Hydrometeorol 5:785–803

Gray DM, Toth B, Pomeroy JW, Zhao L, Granger RJ (2001) Estimating areal snowmelt infiltration into frozen soils. Hydrol Process 15:3095–3111

Hedstrom NR, Pomeroy JW (1998) Measurements and modelling of snow interception in the boreal forest. Hydrol Process 12:1611–1625

Li L, Pomeroy JW (1997a) Estimates of threshold wind speeds for snow transport using meteorological data. J Appl Meteorol 36:205–213

Li L, Pomeroy JW (1997b) Probability of blowing snow occurrence by wind. J Geophys Res 102(D18):21,955–21,964

Marsh P, Pomeroy JW (1996) Meltwater fluxes at an arctic forest-tundra site. Hydrol Process 10:1383–1400

Marsh P, Pomeroy JW, Neumann N (1997) Sensible heat flux and local advection over a heterogeneous at an arctic tundra site during snowmelt. Ann Glaciol 25:132–136

Parviainen J, Pomeroy JW (2000) Multiple-scale modelling of forest snow sublimation: initial findings. Hydrol Process 14:2669–2681

Pomeroy JW, Bewley DS, Essery RLH, Hedstrom NR, Link T, Granger RJ, Sicart JE, Ellis CR, Janowicz JR (2006) Shrub tundra snowmelt. Hydrol Process 20:923–941

Pomeroy JW, Dion K (1996) Winter radiation extinction and reflection in a boreal pine canopy: measurements and modelling. Hydrol Process 10:1591–1608

Pomeroy JW, Essery R (1999) Turbulent fluxes during blowing snow: field tests of model sublimation predictions. Hydrol Process 13:2963–2975

Pomeroy JW, Granger RJ (1997) Sustainability of the western Canadian boreal forest under changing hydrological conditions – I: snow accumulation and ablation. In: Rosjberg D, Boutayeb N, Gustard A, Kundzewicz Z, Rasmussen P (eds) Sustainability of water resources under increasing uncertainty. International Association Hydrological Sciences Publ no 240. IAHS Press, Wallingford, UK. pp 237–242

Pomeroy J, Granger R (1999) Wolf Creek research basin: hydrology, ecology, environment. National Water Research Institute. Ministry of Environment, Saskatoon, pp 15–30

Pomeroy JW, Granger RJ, Pietroniro A, Elliott JE, Toth B, Hedstrom N (1997a) Hydrological pathways in the Prince Albert model forest: final report. NHRI Contrib Series no CS-97007

Pomeroy JW, Gray DM (1995) Snow accumulation, relocation and management. National Hydrology Research Institute Science Report No. 7. Environment Canada, Saskatoon

Pomeroy JW, Gray DM, Hedstrom NR, Janowicz JR (2002) Prediction of seasonal snow accumulation in cold climate forests. Hydrol Process 16:3543–3558

Pomeroy JW, Gray DM, Landine PG (1993) The prairie blowing snow model: characteristics, validation, operation. J Hydrol 144:165–192

Pomeroy JW, Gray DM, Shook KR, Toth B, Essery RLH, Pietroniro A, Hedstrom N (1998a) An evaluation of snow accumulation and ablation processes for land surface modelling. Hydrol Process 12:2339–2367

Pomeroy JW, Hanson S, Faria D (2001) Small-scale variation in snowmelt energy in a boreal forest: an additional factor controlling depletion of snow cover? Proc 58th Eastern Snow Conference, pp 85–96

Pomeroy JW, Li L (2000) Prairie and Arctic areal snow cover mass balance using a blowing snow model. J Geophys Res 105(D21):26619–26634

Pomeroy JW, Marsh P (1997) The application of remote sensing and a blowing snow model to determine snow water equivalent over northern basins. In: Kite GW, Pietroniro A, Pultz TJ (eds) Applications of remote sensing in hydrology. NHRI Symposium No. 17, National Hydrology Research Institute, Saskatoon, SK, pp 253–270

Pomeroy JW, Marsh P, Gray DM (1997b) Application of a distributed blowing snow model to the Arctic. Hydrol Process 11:1451–1464

Pomeroy JW, Marsh P, Jones HG, Davies TD (1995) Spatial distribution of snow chemical load at the tundra-taiga transition. In: Tonnessen KA, Williams

MW, Tranter M (eds) Biogeochemistry of seasonally snow-covered catchments. IAHS publ no 228. IAHS Press, Wallingford, UK. pp 191–206

Pomeroy JW, Parviainen J, Hedstrom N, Gray DM (1998b) Coupled modelling of forest snow interception and sublimation. Hydrol Process 12:2317–2337

Pomeroy JW, Schmidt RA (1993) The use of fractal geometry in modelling intercepted snow accumulation and sublimation. Proc. 50th Eastern Snow Conference, pp 1–10

Pomeroy JW, Toth B, Granger RJ, Hedstrom NR, Essery RLH (2003) Variation in surface energetics during snowmelt in complex terrain. J Hydrometeorol 4(4):702–716

Shook K (1995) Simulation of the ablation of prairie snowcovers. Ph.D. thesis, University of Saskatchewan, Saskatoon

Shook K, Gray DM (1996) Small-scale spatial structure of shallow snowcovers. Hydrol Process 10:1283–1292

Shook K, Gray DM (1997a) Synthesizing shallow seasonal snowcovers. Water Resour Res 33:419–426

Shook K, Gray DM (1997b) Ablation of shallow seasonal snowcovers. In: Iskandar IK, Wright EA, Sharratt BS, Groenvelt PH, Hinzman L (eds) International symposium on physics, chemistry and ecology of seasonally-frozen soils. Fairbanks, Alaska, June 10–12, 1997. Special report 97-10, US Army Cold Regions Research and Engineering Lab, Hanoever, NH, pp 280–286

Shook K, Gray DM (1997c) The role of advection in melting shallow snowcovers. Hydrol Process 11:1725–1736

Verseghy DL, McFarlane NA, Lazare M (1993) CLASS: a Canadian land surface scheme for GCMs II. Vegetation model and coupled runs. Int J Climatol 13:347–370

Weisman RW (1977) Snowmelt. A two-dimensional turbulent diffusion model. Water Resour Res 13:337–342

Woo MK, Rouse WR (2007) MAGS contribution to hydrologic and surface process research. (Vol. II, this book)

Zhao L, Gray DM (1997) A parametric expression for estimating infiltration into frozen soils. Hydrol Process 11:1761–1775

Zhao L, Gray DM (1999) Estimating snowmelt infiltration into frozen soils. Hydrol Process 13:1827–1842

Zhao L, Gray DM, Male DH (1997) Numerical analysis of simultaneous heat and water transfer during infiltration into frozen ground. J Hydrol 200:345–363

Chapter 6

Snowmelt Processes and Runoff at the Arctic Treeline: Ten Years of MAGS Research

Philip Marsh, John Pomeroy, Stefan Pohl, William Quinton,
Cuyler Onclin, Mark Russell, Natasha Neumann,
Alain Pietroniro, Bruce Davison and S. McCartney

Abstract Under the Mackenzie GEWEX Study, extensive snowmelt and runoff research was carried out at the Trail Valley and Havikpak Creek research basins at the tundra-forest transition zone near Inuvik, Northwest Territories. Process based research concentrated on snow accumulation, the spatial variability of energy fluxes controlling melt, local scale advection of sensible heat from snow-free patches to snow patches, percolation of meltwater through the snowpack, storage of meltwater in stream channels, and hillslope runoff. Building on these studies, process based models were improved, as shown by a better ability to model changes in snow-covered area during the melt period. In addition, various land-surface and hydrologic models were tested, demonstrating an enhanced capability to model melt related runoff. Future research is required to accurately model both snow-covered area and runoff at a variety of scales and to incorporate topographic and vegetation effects correctly in the models.

1 Introduction

Prior to the Mackenzie GEWEX Study (MAGS), there had been numerous studies of snowmelt and snowmelt runoff in northern Canada. Examples include: the distribution of the spatially variable end-of-winter snow cover (Woo et al. 1983) caused by blowing snow events (Pomeroy et al. 2007); surface energy balance of snow covered terrain (Dunne et al. 1976); water storage in snow choked channels (Woo and Sauriol 1980); and melt water flow through cold snowpacks (Dunne et al. 1976; Marsh and Woo 1984a, 1984b). However, by the start of MAGS in the mid 1990s, many unknowns remained (Marsh 1999) and there was considerable uncertainty in our ability to model snow accumulation, melt and runoff at a range of scales in northern environments. To help address these issues, the National

Water Research Institute (NWRI) of Environment Canada conducted research in two basins in the boreal forest–tundra transition zone, in an environment representative of the northern portions of the Mackenzie River Basin (MRB). Although the terrain and vegetation types constitute a small portion of the total Basin area, they are typical of a large area of the circumpolar Arctic.

Under the umbrella of this NWRI research, the topics of investigation included:
- blowing snow and snow accumulation (reported in Pomeroy et al. 2007);
- spatial variability of energy fluxes controlling melt, including: point estimates measured from towers; larger scale variability estimated from aircraft; advection of sensible heat from snow-free to snow covered patches; scale considerations for modeling turbulent fluxes and radiation for a typical modeling grid; and modeling changes in snow-covered area during melt;
- percolation of meltwater through the snowpack;
- storage of meltwater in stream channels prior to the initiation of streamflow in the spring;
- improvement of snowmelt and snowmelt runoff models; and
- incorporation of an appropriate variety of hydrologic and land-surface models for use in northern environments.

2 Study Area and Methods

The lower Mackenzie Valley is characterized by a vegetation transition from northern boreal forest to tundra and the occurrence of continuous permafrost with thickness of 100 to 150 m. Hydrologic research has been conducted in two basins in this area since 1992 (Fig. 1). Trail Valley Creek (TVC), located approximately 50 km northeast of Inuvik ($68°17'N$; $133°24'W$), has an elevation range from 48 to 205 m asl, and is dominated by tundra vegetation (Marsh and Pomeroy 1996). The basin area has been reported as 63 km^2 (Marsh and Pomeroy 1996) or 68 km^2, according to the Water Survey of Canada; but recently obtained LiDAR data indicate that an upper part of this estimated basin area drains into an adjacent basin. Using RiverTools to estimate drainage area covered by the LiDAR data provides an estimated basin area of 55.1 km^2. Combined with an additional 2 km^2 outside the LiDAR coverage (estimated from DEM), a best estimate of the TVC basin area is 57 km^2. This example illustrates an often

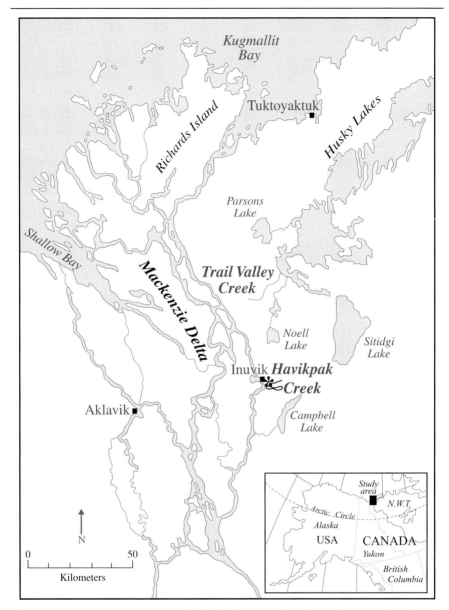

Fig. 1. Location map showing Trail Valley and Havikpak Creeks. The Inuvik MSC (formerly AES) upper air station is indicated by asterisk (*)

overlooked error associated with the drainage networks on topographic maps. Such an error can render it difficult to compare modeled and measured discharge. With an overestimate of basin area by >10%, an even distribution of precipitation and evaporation over the entire basin would cause a similar overestimation of discharge by any hydrologic model.

Havikpak Creek (HPC) situated a few kilometers north of the Inuvik airport (Marsh et al. 2002) is 17 km^2 in area, has an elevation range from 73 to 231 m asl, and is predominantly covered by northern boreal forest, with tundra in the higher elevations. The contrasting vegetation between TVC and HPC has significant effects on aspects of their hydrology, including snow accumulation and melt, evaporation and runoff (Russell 2002).

A Meteorological Service of Canada (MSC) upper air station is located near the outlet of HPC, and a MSC weather station is located within the TVC basin. NWRI has also operated meteorological stations at both research basins since 1992, recording air temperature, rainfall, snowfall, net radiation, incoming and outgoing solar radiation, soil moisture and soil temperature, and turbulent fluxes using eddy correlation during periods of intensive measurements. Incoming long-wave radiation measurement was added during the latter stage of MAGS. Although these stations operated throughout the year, they were unattended for much of the winter, with implications to the accuracy of some of the measurements. Snow surveys, stratified by terrain type, were conducted at both TVC and HPC in most years. Water Survey of Canada (WSC) has operated discharge gauging stations at TVC and HPC since 1979 and 1994, respectively. Additional determination of discharge by current metering from a boat or by wading, and by dye injection (Russell et al. 2004) were made by NWRI during the spring to reduce the large errors typical of discharge estimates in snow/ice clogged channels. Routine meteorological and hydrologic measurements by the MSC and WSC are used to fill in missing research basin measurements. These observations have resulted in a consistent, long term data set that is available for many purposes. For example, Marsh et al. (2004) used these data to consider variations in the annual water balance of TVC and HPC for the period 1992 to 2000. Marsh et al. (2002) analyzed the long term MSC and WSC records, to consider the maximum late winter SWE, timing of the winter to summer transition, the date of the onset of runoff from TVC and HPC, and the time of the peak flow at these basins. This analysis demonstrated that the 1994/95 MAGS study year had the earliest winter to summer transition on record, with early spring melt and its attendant runoff.

3 Spatial Variability of Snowmelt Processes

The TVC and HPC land surface is heterogeneous at different scales and in surface patterns due to their irregular topography and vegetation cover. Differences in the vertical transfers of radiation and sensible and latent heat from different portions of the land surface often result in large spatial differences in snowmelt. Combined with a spatially variable end-of-winter snow distribution (Pomeroy et al. 2007), these processes produce a highly variable and patchy snow cover during the melt period. This patchiness significantly affects the fluxes of energy from the land surface to the atmosphere, as well as runoff.

3.1 Measured Variability in Turbulent and Radiative Fluxes

The snowmelt landscapes of open windswept environments typically have patchy snow covers during the melt period (Marsh 1999), with large differences in albedo, surface roughness, and surface temperature between the snow and snow-free patches. Observations of turbulent fluxes and radiation from both towers and aircraft have confirmed such spatial variability. Examples are provided from tower observations. Figure 2 shows the sensible and latent heat fluxes at TVC over a tundra and a shrub site. Although latent heat was similar at both sites early in the melt period, sensible heat flux was typically larger at the shrub site throughout the melt season, while latent heat became larger at the shrub site in the latter parts of the melt period. Figure 3 compares measurements over a persistent snow patch and over a snow-free patch. As expected, net radiation was much smaller at the snow site, while the sensible, latent, and ground heat fluxes were in different directions at the two sites. Such large variations in fluxes clearly indicate that a single point measurement of turbulent fluxes cannot represent an entire basin.

To obtain a better estimate of the spatial variability of fluxes at both TVC and HPC, the NRC Twin Otter Flux aircraft of the National Research Council was employed during the 1999 Canadian GEWEX Enhanced Study (CAGES) field period in the Inuvik area (Brown-Mitic et al. 2001). The sensible heat, latent heat, and net radiation measured from the aircraft (for a 1 km line centered over the tower) agreed well with the tower data (Fig. 4a), suggesting that the TVC tower is representative of a larger area around it. Figure 4b shows hourly flux data from the TVC tower for the entire melt period, as well as average flux measurements along each of 9 flight lines covering all of TVC. Since the aircraft data were obtained over a short period (approximately a 1 hour period each day), they can be considered as "snapshots" of conditions covering the entire basin. Early in the melt period, when the basin was completely snow covered, the fluxes from the aircraft and the tower were in agreement. As the snow covered area (SCA) declined over the melt season, the peak fluxes measured by the

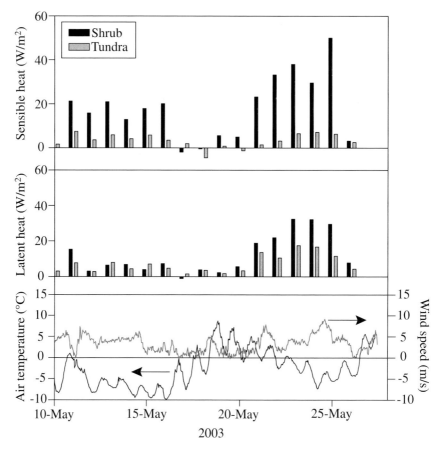

Fig. 2. Sensible and latent heat flux for tundra and shrub sites located in the vicinity of TVC. Also shown are air temperature and wind speed at both sites (after Marsh et al. 2003)

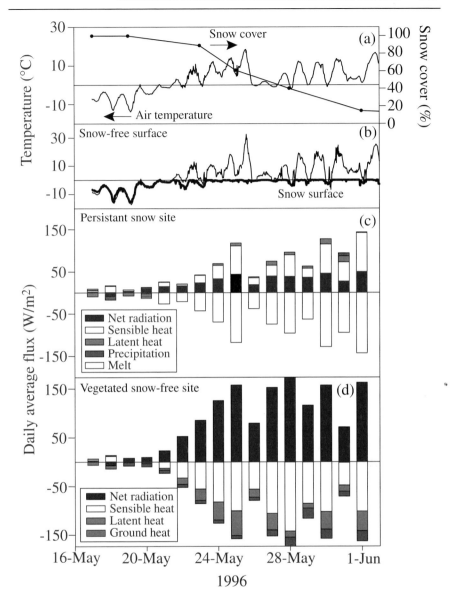

Fig. 3. Trail Valley Creek: (a) air temperature and fractional snow cover, (b) snow surface temperature and snow-free surface temperature, (c) energy balance terms for a persistent snow site and (d) for a snow-free vegetated site. May 17 to June 1 covers the period immediately before the start of snowmelt until the time with only residual snow covering approximately 10% of the basin area. Snowcover data are from SPOT satellite image and all energy balance terms are considered to be spatially representative averages (after Neumann and Marsh 1998).

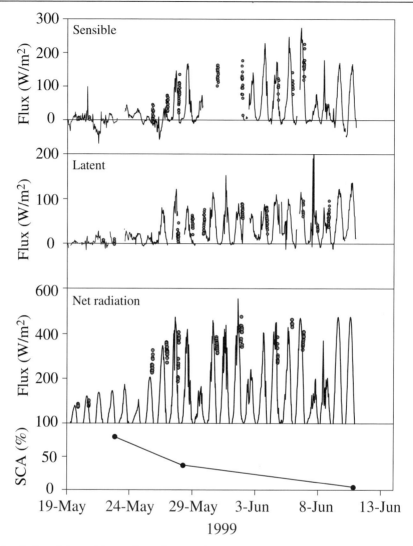

Fig. 4. Sensible and latent heat flux, and net radiation as measured from both the TVC tower (line), and from the NRC aircraft (circle). Also shown is the basin snow covered area. Data are either average values from a single line passing over the tower, or averages for each of the nine flight lines

tower increased, and the range of fluxes measured by the aircraft also increased, clearly showing an increase in the spatial variability in fluxes. It is believed that these observed spatial variations in fluxes at scales of less than 1 km play an important role in the hydrology of northern areas. As a result, the following observations and modeling studies were undertaken in order to better understand the importance of such variations, and to assess the appropriate scale needed for simulating the hydrology of these areas.

3.2 Advection of Sensible Heat from Bare Patches to Snow Patches

The primary sources of energy for snowmelt are sensible heat and radiant energy, while local advection of sensible heat from the bare ground can increase the fluxes at the upwind edge of the snow patches (Shook 1995). Local advection is further complicated by the gradually changing area and size distribution of the patches during the snowmelt period. Attempts were made to incorporate advection into simple melt models. Marsh and Pomeroy (1996) showed that the average advection of sensible heat to snow patches can be related to the available sensible heat from the snow-free areas, and suggested that the ratio between the two, termed the advection efficiency factor (F_S), varied over the snowmelt period. F_S varies between 0 and 1, and its determination requires estimates of the spatially averaged sensible heat flux to the snow patches, the sensible heat flux at the downwind edge of a large snow patch where local advection is negligible, the sensible heat over vegetated snow-free area, and the fraction of the basin that is snow-free and snow-covered.

Marsh et al. (1997, 1999) and Neumann and Marsh (1998) used field observations and a boundary layer model to show that F_S decreases with increasing snow-free area (Fig. 5), but increases with wind speed and decreasing snow patch size. Neumann and Marsh (1998) suggested that the relationship between F_S and the snow-free fraction (P_V) can be described as: $F_S = 1.0 \exp(-3.2 P_V)$. Differences in the relationship between F_S and P_V from field and modeling studies could be attributed to differences in patch size between the observations and model.

Pohl and Marsh (2006) incorporated F_S into a snowmelt model. This improved the accuracy of the estimate of the snow cover area during the melt period when snow coverage decreased from 40% to 5% of the basin area. Since the usage of F_S has only been tested in limited conditions, further work is needed to better understand how the relationship changes between the magnitude of local scale advection and such factors as terrain condition, patch geometry, and patch size.

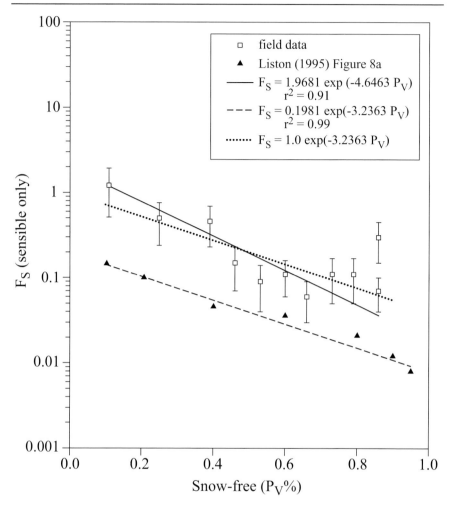

Fig. 5. Advection efficiency term (F_S) vs. snow-free fraction. Dashed line represents the best fit to the model output from Liston (1995) for model runs with a single snow patch of varying size. Solid line is the best fit for F_S estimated from measurements at TVC, with error bars for the case of a +/-40% error in $\hat{Q}_{H_S}$ (i.e., sensible heat flux at 1.8 km downwind of the leading edge of a large snowpack, as estimated from Eq. 3). Dotted line has an intercept of 1 and the same slope as the best fit line through Liston model data (after Neumann and Marsh 1998)

3.3 Variability in Turbulent Fluxes and Incident Solar Radiation

It is well known that incident solar radiation varies with slope angle and aspect in areas of high relief. However, Pohl et al. (2006b) demonstrated that even in relatively low relief areas such as TVC, spatial variability in incident solar radiation is important in controlling snowmelt. They modelled incident solar radiation by calculating clear sky horizontal plane radiation and cloudy sky radiation, and then using standard methods for determining the topographic effects on incident short-wave radiation at a scale of 40 m. Direct beam and diffuse radiation were evaluated by a combination of estimated and measured global short-wave to obtain a cloudiness index. Model runs compared well to observed direct and diffuse radiation. Accumulated net solar radiation over the entire melt period showed large variations over the study area, with notable effects on snowmelt. One indication of the importance is that the difference in absorbed solar radiation was equivalent to about 53 mm of melt potential. Since open tundra areas within TVC typically have a SWE of 50 to 120 mm, such differences in solar radiation have considerable implications for the development of a patchy snow cover.

There can be also large spatial variability of turbulent fluxes due to topographically induced wind speed variations. Pohl et al. (2006a) implemented a simple wind model to simulate topographic effects on the surface wind field at a scale of 40 m. Hourly wind observations were distributed by the model and used to calculate spatially variable sensible and latent turbulent heat fluxes for TVC. Simulations showed that, even though the study area is characterized by relatively low relief, the small-scale sensible and latent heat fluxes varied considerably. Overall, turbulent fluxes within the research area varied by as much as 20% from the mean, leading to differences in potential snowmelt of up to 70 mm SWE over the entire melt period. Such potential variations in snowmelt rates further contribute to the development of a patchy snow cover.

3.4 Variability in Snowmelt and Snow-covered Area

Pohl and Marsh (2006) combined model results of spatial variability in incident solar radiation and turbulent fluxes with estimates of local scale advection and a spatially variable end-of-winter snow cover to demonstrate the relative importance of each. This study also illustrated appropriate methods to model snowcover melt, depletion, and runoff at a variety of scales.

When mapped over the TVC domain, combined radiation, turbulent fluxes and local advection yielded large spatial differences in total accumulated energy balance (Fig. 6), with melt energy varying from +30 to -60 mm SWE over the entire domain. Pohl and Marsh (2006) provided similar maps for the radiation and turbulent flux components comprising the accumulated energy balance distribution in Fig. 6, and described the major forcing factors controlling their distribution. In addition, the melt magnitude is greatly influenced by wind direction (Fig. 7). At this treeline site, spatial variability increases considerably on days with a southerly wind as turbulent fluxes and radiant fluxes are both largest on south-facing slopes. In contrast, days with a north wind have a much more uniform melt, with larger turbulent fluxes on north-facing slopes and the largest radiative fluxes on south-facing slopes.

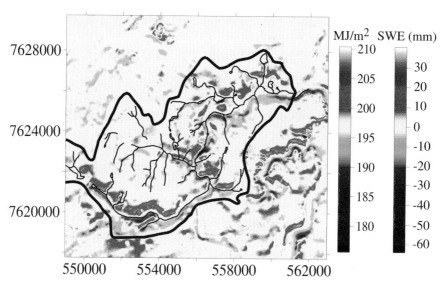

Fig. 6. Total energy balance over entire study period and as potential snowmelt amount (expressed as difference from the mean potential snowmelt) (after Pohl and Marsh 2006)

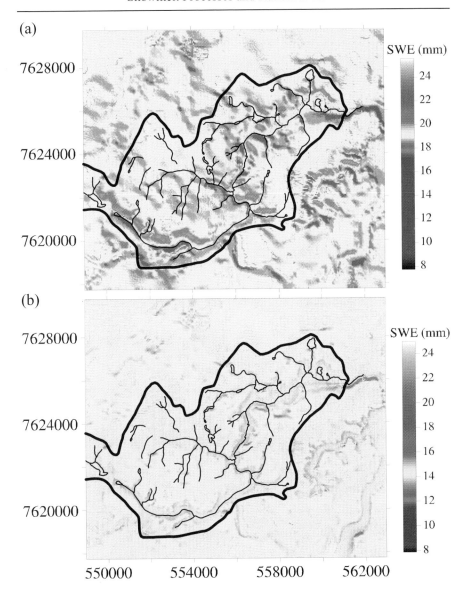

Fig.7. Influence of wind direction on daily melt rates for consecutive days with similar snowcover, air temperature, and wind speed. (a) 23 May (southerly winds, mean melt 19 mm, S=28 MJ m^{-2} or 0.9 mm) and (b) 24 May (northerly winds, mean melt 14 mm, S = 0.11 MJ m^{-2} or 0.3 mm) (after Pohl and Marsh 2006)

The basin average for each component (radiation, turbulent flux, and local advection) over the entire TVC basin is shown in Fig. 8. Note that early in the melt period net radiation tends to dominate melt, but as the snow becomes increasingly fragmented, the relative importance of turbulent fluxes and local advection increases.

Pohl and Marsh (2006) carried out a variety of model runs to compare the relative importance of distributed snow, distributed melt, and advection on modeled changes in snow covered area during melt at TVC (Fig. 9). With uniform snow and uniform melt, the SCA obviously is a step function with the SCA changing instantly from completely snow-covered to completely snow-free. In contrast, distributed snow/distributed melt provides a gradual decrease in SCA that is similar to observed.

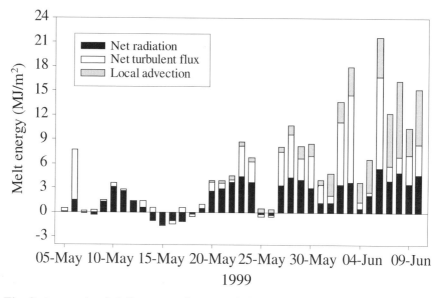

Fig. 8. Accumulated daily snowmelt energy balance averaged over the TVC basin (after Pohl and Marsh, 2006)

Pohl and Marsh (2006) applied the fully distributed melt model to examine the changing pattern of the SCA. The simulated patterns captured the progress of changes in SCA as shown in satellite and aircraft observations (Fig. 10). The first areas to become bare were mainly located on northwest- to southwest-facing slopes due to a combination of eroded snow cover and above-average melt energy fluxes at those locations. Pohl et al. (2006a) showed that west-facing slopes have higher melt rates as a result of receiving maximum solar radiation in the afternoon, coinciding

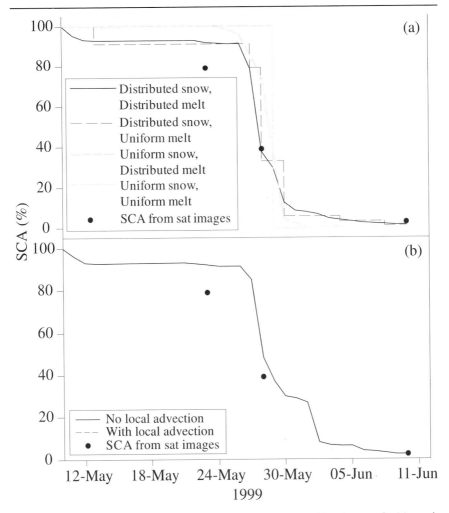

Fig. 9. Snowcovered area (SCA) for various combinations of (a) uniform/distributed snow and melt, and (b) for local scale advection. In all cases, observed SCA is obtained from satellite images (after Pohl and Marsh 2006)

with the highest air temperatures and a ripe snow cover that has recovered from any energy deficit that might have been incurred during the previous night. The model predicted that the bulk of the open upland tundra areas would become snow free between 27 and 29 May, as was verified by aerial photography (Fig. 10). Most of the north-facing open tundra slopes and shrub tundra areas were predicted to become snow-free over the next two days (30–31 May). Satellite images indicate that shrub tundra areas showed a higher variability, and melting earlier than or simultaneously

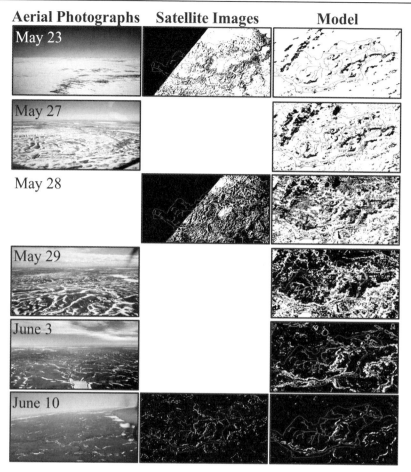

Fig. 10. Comparison of modeled melt patterns with satellite images and aerial photographs (after Pohl and Marsh 2006)

with the open tundra. This discrepancy was not unexpected, since the model was set up for open, vegetation-free areas (especially through assumptions made about albedo and surface roughness) that were not corrected for vegetation influence. By 3 June the simulation showed residual snow only in drift areas along the sides of river valleys, around lake margins, and in the channels themselves. This result was validated by aerial photography for that day. These late-lying snow patches are attributed to a combination of higher than average end-of-winter snow accumulation and below-average snowmelt energy, especially on steep, north-facing slopes and in the valley bottoms.

3.5 Effects of Shrubs on Snowmelt

As suggested previously, shrubs can significantly affect snowmelt in the study area. To consider this effect, Marsh et al. (2003) carried out a study of snow accumulation and melt at a large shrub area in the vicinity of TVC. End-of-winter snow surveys showed that in late winter 2003, the SWE varied from 98 mm for tundra sites, to 141 mm for shrub sites, 499 mm for drift locations, and 155 mm at forested sites, with a basin average of 142 mm. Although the shrub site had a pre-melt SWE approximately 40% higher than at the tundra site, the SWE and the SCA at the shrub site deceased faster than at the tundra site (Fig. 11), suggesting a higher melt rate at the shrub site compared to the tundra site (cf., Pomeroy et al. 2006). Ongoing work will produce results for implementation in the Pohl and Marsh (2006) melt model to improve the ability to simulate snowmelt, changes in SCA, and snowmelt runoff. This will be especially important when considering runoff scenarios, as the shrub areas may increase rapidly in the coming decades due to climate warming.

4 Processes Controlling Lag Between Melt and Runoff

Percolation of surface melt through a snowpack is controlled by both the requirement to wet the dry snow to satisfy its irreducible water content, and to warm the snow to 0°C. Marsh and Pomeroy (1996) applied the Marsh and Woo (1984b) percolation model to consider the effect of meltwater percolation on the timing and volume of water availability for runoff. This model includes a variable flow path and meltwater percolation algorithm, with the melt flux applied to mean snow-cover depth and density in each landscape type in TVC.

Model results indicate that the initial release of meltwater first occurred on the shallow upland tundra sites, with meltwater released nearly two weeks later from the deep snow drifts. The delay between the initiation of melt and arrival of meltwater at the base of the snowpack varied from 6 days for 0.45 m deep snow at tundra sites, to 10 days for 1.85 m deep snow at drift sites. During the beginning of the melt season, not all meltwater is available for runoff. Instead, some snowpacks contribute partially to runoff, and the spatial variation of runoff contribution corresponds to landscape type. At the tundra and drift sites, it was not until 9 days and 16 days respectively that all of the melt water was available for runoff. At TVC, shrub tundra sites were intermediate between the tundra and drift sites in terms of the delay between start of melt and start of runoff.

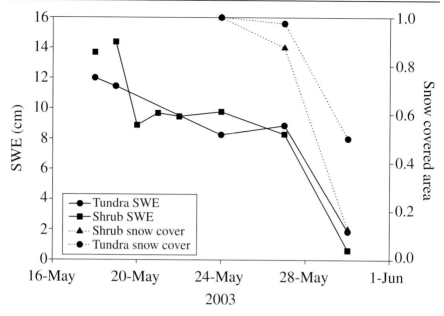

Fig. 11. Change in snow water equivalent and snow covered area during the 2003 melt season at TVC for both tundra and alder shrub sites (after Marsh et al. 2003)

Combining model results with the distribution of landscape types in TVC, Marsh and Pomeroy (1996) were able to map the contributing areas of meltwater runoff on a daily basis. On May 19, 1993 for example, when the tundra areas (70% of the basin) were fully contributing meltwater to runoff, the shrub tundra areas (22% of the basin) were partially contributing meltwater, while the drift areas (8% of the basin) were not contributing any meltwater. Such daily calculations of the contributing areas throughout the melt period could be used to drive a distributed hydrologic model. This large spatial and temporal variability of meltwater release has significant implications for predicting runoff in these environments.

Water balance of TVC and HPC shows that streamflow begins well after the start of melt, indicative of rapid rises in liquid water storage in the snow that reaches a peak near the start of discharge (Fig. 12). The development of large meltwater ponds in the stream channels suggested that the main channels of TVC and HPC may be a major storage area of meltwater prior to discharge commencement. However, field observations of the liquid water stored in TVC stream channels showed that it was only a minor component of total storage, holding only about 1 mm (averaged over the entire basin) of the approximately 100 mm of basin storage at the beginning of melt. This result clearly demonstrates that the largest portion of meltwater storage must include the following: liquid water in the snowpack and in numerous meltwater ponds (not in the stream channels) distributed around the basin, and infiltration into the frozen soils,. Further work is required to assess the relative importance of these stores.

The relative travel times of water through the major components (meltwater percolation through snow, infiltration into the ground, surface and subsurface flows, and stream flow) of the overall flow system, from the surface of melting snowpacks to the basin outlet, as well as the temporal variations in these relative rates were investigated by Quinton and Marsh (1998a). They reported that about 90% of the overall travel time to outlet was spent in the snowmelt percolation pathway. During the middle of the melt period, the hillslope runoff segment dominated (ca. 80%) the overall travel time. Later in the melt-runoff period, the critical factor controlling the overall travel time was whether or not an ephemeral stream was active.

5 Snowmelt Runoff

5.1 Processes

The tundra basins have extensive hummocky permafrost terrain (Fig. 6, Woo and Rouse 2007). Quinton and Marsh (1998b) and Quinton et al. (2000) reported that the combination of mineral earth hummocks and inter-hummock pathways (dominated by peat of about 0.3 m thick) significantly affects slope runoff. Mineral earth hummocks have low permeability while the inter-hummock peat areas have a permeability that is typically three orders of magnitude larger than the hummocks, with the near surface portions of the peat having a permeability up to six orders of magnitude higher than the hummocks. In these inter-hummock zones the physical properties of the peat change abruptly with depth, with bulk density increasing fourfold and the active porosity decreasing from approximately 0.85 near the surface to approximately 0.50 in the basal peat; and the permeability decreasing by two to three orders of magnitude over a 20 cm increase in depth. Mineral earth hummocks influence hillslope runoff by: (1) concentrating flow through the inter-hummock zone, (2) obstructing flow from following a direct path to the stream banks, (3) attenuating flow by interacting with the saturated layer of the inter-hummock zone, and (4) raising the water table in the inter-hummock area by displacement.

Quinton and Marsh (1999) subsequently showed that subsurface flow is the dominant mechanism of runoff to the stream channel, that this flow is

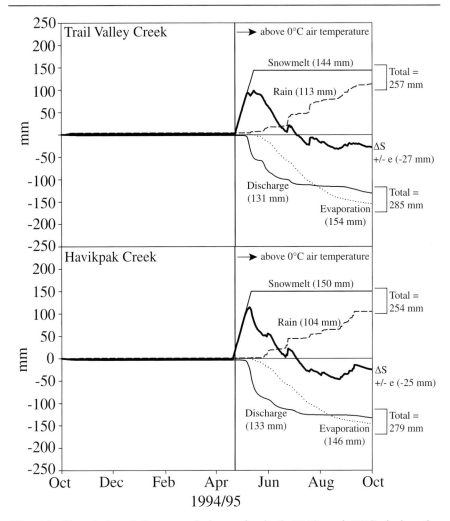

Fig. 12. Cumulative daily water balance for both TVC and HPC during the 1994/95 water year. Note that snowmelt occurs well before the start of discharge from both streams and as a result, the storage term rises rapidly, reaching a maximum at approximately the time when discharge begins (after Marsh et al. 2002)

conveyed predominately through the peat of the inter-hummock areas, and that subsurface flow through highly conductive upper peat layer and soil pipes is as rapid as surface flow. They also suggested that TVC can be distinguished into three hydrologic units (viz., stream channel, near-stream area, and uplands), and that the hydrologic response of each unit varies greatly depending on subsurface water levels. For example, when the water table is close to the ground surface, flow is rapid and the source area for the production of storm flow is relatively large as the source area extends away from the channels to include the upland areas. However, when the water table is in the lower peat layer, the source area is reduced (typically just the near-stream area), and flows are slow due to slow seepage flow in the lower peat layers.

5.2 Modeling

A major limitation in hydrologic modeling at large scales is the availability of appropriate input data including, for example, precipitation, temperature, and radiation (cf., Thorne et al. 2007). Russell (2002) compared field observations at TVC and HPC to the Canadian Meteorological Centre Global Environmental Multiscale Model (GEM) weather prediction output and found that pressure, specific humidity and air temperature compared well, but that GEM precipitation was typically lower than measured. These results demonstrate that numerical weather prediction data are potentially a suitable source of input data for hydrologic models in northern Canada.

Pohl et al. (2005) utilized the fully distributed hydrology land-surface scheme WATCLASS (Soulis and Seglenieks 2007) to model changes in snow covered area and spring snowmelt runoff in TVC. WATCLASS was able to predict satisfactorily the runoff volumes (on average within 15% over five years of modeling) and mean SWE, as well as timing of snowmelt and meltwater runoff for open tundra (Fig. 13). Melt was underestimated in the energetically more complex shrub tundra areas of the basin. Furthermore, the observed high spatial variability of the SCA at a 1-km resolution was not captured well (Fig. 14). Several recommendations are made to improve model performance in Arctic basins, including a more realistic implementation of the gradual deepening of the thawed layer during the spring, and the use of topographic information in the definition of land cover classes for the GRU approach. Davison et al. (2006), for example, attempted to improve the sub-grid variability of the snow cover and the subsequent melt in TVC by including wind-swept tundra and drift classes based on topography rather than the traditionally used vegetation land classes. This approach improved the ability of WATCLASS to simulate the variability in snow covered area during the melt period.

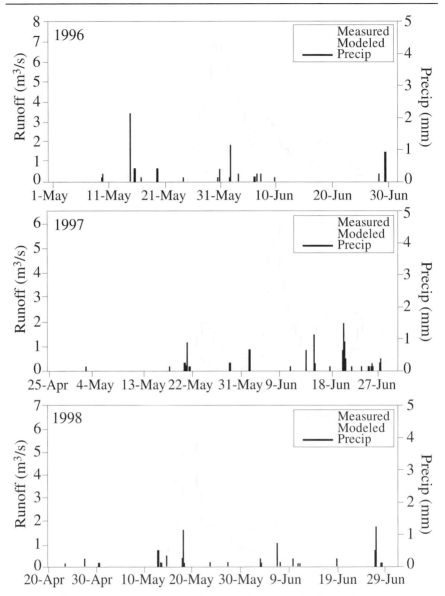

Fig. 13. Observed vs WATCLASS simulated hydrographs for spring melt periods in TVC. 1997 and 1998 were model calibration periods (after Pohl et al. 2006)

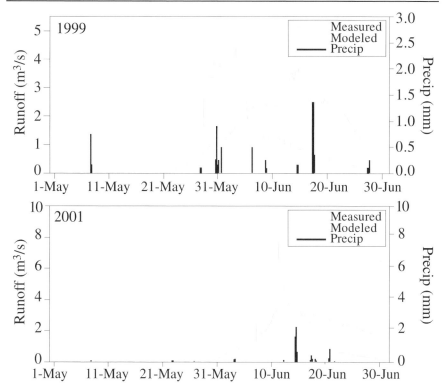

Fig. 13. (cont.)

6 Conclusion

Field investigations in two small basins at the treeline have advanced knowledge of understanding of snowmelt and runoff processes in the northern environment. A combination of tower and aircraft flux measurements have demonstrated conclusively that the fluxes of sensible and latent heat, and radiation vary significantly over typical 10x10 km grids. Combined with a spatially variable snow distribution at the end of winter, this results in a patchy snow cover during the melt period, with implications on the fluxes of energy and water to the atmosphere and for runoff. Implementation of these processes in a model allowed the simulation of spatially variable fluxes, and produced realistic changes in the snow cover pattern over the study area. These studies demonstrated the importance of a variable snow cover, variable energy fluxes, and local scale advection in con-

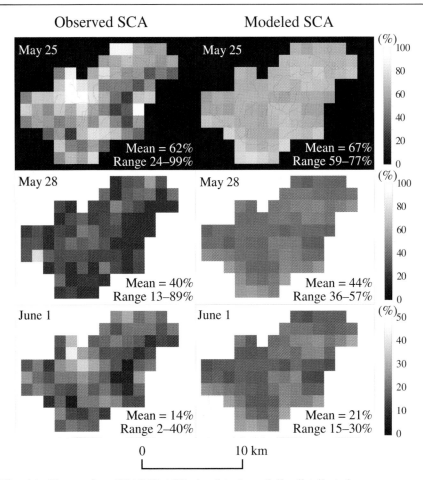

Fig. 14. Observed vs WATCLASS simulated spatially distributed snow covered area for TVC in 1996

trolling the snowmelt process at small scales. In addition, runoff studies showed that meltwater storage in the snowpack and flow pathways of meltwater as influenced by such features as earth hummocks and peat terrain, have major impacts on hillslope runoff and therefore basin discharge. The combined land-surface/hydrology model WATCLASS was tested for the TVC basin. Although this model has produced improvements in combining hydrology with detailed surface energy balance, comparison with detailed datasets has revealed areas where future efforts should focus, including the ability to represent properly the spatial variability in melt. Ongoing studies continue to use detailed observations, process studies, and parameterizations from TVC and HPC to further improve the models to simulate the hydrology of northern Canada at a variety of scales.

Acknowledgements

The work reported here was carried out over a decade, with support of many colleagues at the National Water Research Institute, other Government Departments and Universities. This research was supported by: the National Water Research Institute of Environment Canada; Mackenzie GEWEX Study (MAGS) through funding by Environment Canada; the Climate Change Action Fund; the Polar Continental Shelf Project; the Aurora Science Institute; and Indian and Northern Affairs Canada.

References

Brown-Mitic CM, MacPherson JI, Schuepp PH, Nagarajan B, Yau PMK, Bales R (2001) Aircraft observations of surface-atmosphere exchange during and after snowmelt for different arctic environments: MAGS 1999. Hydrol Process 15:3585–3602

Davison B, Pohl S, Dornes P, Marsh P, Pietroniro A, Mackay M (2006) Characterizing snowmelt variability in Land-Surface-Hydrologic Model. Atmos Ocean 44:271–287

Dunne T, Price AG, Colbeck SC (1976) The generation of runoff from subarctic snowpacks. Water Resour Res 12:677–685

Marsh P (1999) Snowcover formation and melt: recent advances and future prospects. Hydrol Process 13:2117–2134

Marsh P, Neumann NN, Essery RLH, Pomeroy JW (1999) Model estimates of local advection of sensible heat over a patchy snow cover. Interactions between the cryosphere, climate and greenhouse gases. Proc IUGG 99 Symp HS2. Birmingham, July 1999, IAHS Publ 256, pp 103–110

Marsh P, Onclin C, Neumann N (2002) Water and energy fluxes in the lower Mackenzie Valley, 1994/95. Atmos Ocean 40:245–256

Marsh P, Onclin C, Russell M (2004) A multi-year hydrological data set for two research basins in the Mackenzie Delta region, NW Canada. Northern research basins water balance. Proc Workshop held at Victoria, Canada, March 2004, IAHS Publ 290, pp 205–212

Marsh P, Onclin C, Russell R, Pohl S (2003) Effects of shrubs on snow processes in the vicinity of the Arctic treeline in NW Canada. Northern research basins. Proc 14th Int Symposium and Workshop, Kangerlussuaq/Sdr. Strømfjord, Greenland. August 25–29, 2003, pp 113–118

Marsh P, Pomeroy JW (1996) Meltwater fluxes at an Arctic forest-tundra site. Hydrol Process 10:1383–1400

Marsh P, Pomeroy JW, Neumann N (1997) Sensible heat flux and local advection over a heterogeneous landscape at an Arctic tundra site during snowmelt. Ann Glaciol 25:132–136

Marsh P, Woo MK (1984a) Wetting front advance and freezing of meltwater within a snow cover 1. Observations in the Canadian Arctic. Water Resour Res 20:1853–1864

Marsh P, Woo MK (1984b) Wetting front advance and freezing of meltwater within a snow cover 2. A simulation model. Water Resour Res 20:1865–1874

Neumann N, Marsh P (1998) Local advection of sensible heat in the snowmelt landscape of Arctic tundra. Hydrol Process 12:1547–1560

Pohl S, Davison B, Marsh P, Pietroniro A (2005) Modelling spatially distributed snowmelt and meltwater runoff in a small arctic catchment with a hydrology land-surface scheme (WATCLASS). Atmos Ocean 43:193–211

Pohl S, Marsh P (2006) Modelling the spatial-temporal variability of spring snowmelt in an arctic catchment. Hydrol Process 20:1773–1792

Pohl S, Marsh P, Liston G (2006a) Spatial-temporal variability in turbulent fluxes during spring snowmelt. Arct Antarct Alp Res 38:136–146

Pohl S, Marsh P, Pietroniro A (2006b) Spatial-temporal variability in solar radiation during spring snowmelt. Nord Hydrol 37:1–19

Pomeroy JW, Bewley DS, Essery RLH, Hedstrom NR, Link T, Granger RJ, Sicart JE, Ellis CR, Janowicz JR (2006) Shrub tundra snowmelt. Hydrol Process 20:923–941

Pomeroy JW, Gray DM, Marsh P (2007). Studies on snow redistribution by wind and forest, snow-covered area depletion, and frozen soil infiltration in northern and western Canada. (Vol. II, this book)

Shook KR (1995) Simulation of the ablation of prairie snowcovers. Ph. D. thesis, University of Saskatchewan. Saskatoon, Saskatchewan

Quinton WL, Gray DM, Marsh P (2000) Subsurface drainage from hummock-covered hillslopes in the Arctic tundra. J Hydrol 237:113–125

Quinton WL, Marsh P (1998a) The influence of mineral earth hummocks on subsurface drainage in the continuous permafrost zone. Permafrost Periglac Process 9:213–228

Quinton WL, Marsh P (1998b) Melt water fluxes, hillslope runoff and stream flow in an Arctic permafrost basin. Proc 7th Int Conf Permafrost, Yellowknife, Canada. Centre D'etudes Nordiques, Université Laval, Laval, Canada; pp 921–926

Quinton WL, Marsh P (1999) A conceptual framework for runoff generation in a permafrost environment. Hydrol Process 13:2563–2581

Russell MK (2002) Water balances across arctic tree-line and comparisons to a numerical weather prediction model. M.Sc. thesis, University of Saskatchewan, Saskatoon, Saskatchewan

Russell M, Marsh P, Onclin C (2004) A continuous dye injection system for estimating discharge in snow-choked channels. Arct Antarct Alpine Res 36:539–554

Soulis ED, Seglenieks FR (2007) The MAGS integrated modeling system. (Vol. II, this book)

Thorne R, Armstrong RN, Woo MK, Martz LW (2007) Lessons from macroscale hydrological modeling: experience with the hydrological model SLURP in the Mackenzie Basin. (Vol. II, this book)

Woo MK, Heron R, Marsh P, Steer P (1983) Comparison of weather station snowfall with winter snow accumulation in high arctic basins. Atmos Ocean 21:312–325

Woo MK, Rouse WR (2007) MAGS contribution to hydrologic and surface process research. (Vol. II, this book)

Woo MK, Sauriol J (1980) Channel development in snow-filled valleys, Resolute, NWT, Canada. Geografiska Annaler 62A:37–56

Chapter 7

Modeling Maximum Active Layer Thaw in Boreal and Tundra Environments using Limited Data

Ming-ko Woo, Michael Mollinga and Sharon L. Smith

Abstract The variability of maximum active layer thaw in boreal and tundra environments has important implications for hydrologic processes, terrestrial and aquatic ecosystems, and the integrity of northern infrastructure, including oil and gas pipelines. For most planning and management purposes, the long-term probability distribution of active layer thickness is of primary interest. This study presents a robust method for calculating maximum active layer thaw, employing Stefan's equation to compute phase change of the moisture in soils and using air temperature as the sole climatic forcing variable to drive the model. Near-surface ground temperatures, representing the boundary condition for Stefan's algorithm, were estimated based on empirical relationships established for several sites in the Mackenzie Valley. Active layer thaw simulations were performed for typically saturated soils (one with 0.2 m peat overlying mineral substrate and one with 1.0 m peat) in tundra and in boreal forest environments. The results permit an evaluation of the probability distributions of maximum active layer thaw for different locations in permafrost terrain.

1 Introduction

Ice-rich permafrost is prevalent in the northern environment. Thawing of ice-rich ground can lead to major changes in the natural landscape and have important implication on the integrity of engineering structures such as building foundations, bridge footings, railways, and pipelines (e.g., Smith et al. 2001). In the Mackenzie River Valley, an existing buried oil pipeline (Norman Wells to Zama pipeline) and a proposed natural gas pipeline route to transport gas from the Mackenzie Delta traverse both the continuous and discontinuous permafrost zones, crossing many areas that are sensitive to fluctuations in active layer thaw. Characterization of active layer conditions and the ground thermal regime is essential for the evaluation of terrain stability, sound design of northern infrastructure and assessment of environmental impacts associated with northern development. Typically, finite difference schemes have been used to solve for ground temperature variations (e.g., Goodrich 1982; Oelke et al. 2003) but this approach requires detailed soil property information usually not available for remote locations. Several simplified methods using a frost index approach or n-factors (ratio of the air thawing/freezing degree day index to the ground surface thawing/freezing degree day index) have been applied to assess the presence or absence of permafrost or to estimate the temperature at the top of the permafrost (Henry and Smith 2001; Nelson 1986; Romanovsky and Osterkamp 1995; Taylor 2000).

Like most remote northern regions, the Mackenzie Valley suffers from a scarcity of long term thaw depth and ground temperature measurements, and a paucity of weather stations that can offer sufficient climatic data to permit the simulation of ground temperatures using detailed heat balance schemes. On the other hand, we are often concerned with the depth of annual maximum thaw rather than the daily progression of the thaw front and for most planning and management purposes, the long-term probability distribution of active layer thickness is of primary interest. The purpose of this study is to provide a robust method to estimate the probability of maximum active layer thaw.

2 Study Area and Data

The Mackenzie River Valley extends from the Beaufort Sea coast above the Arctic Circle to the temperate grasslands in the south. This region is also one where increased hydrocarbon development, including a proposed gas pipeline, is projected to occur. Along the Valley, tundra and boreal forest are two environments underlain by permafrost. The tundra landscape is largely flat to rolling, with shrubs and herbs, grasses and sedges, lichens and mosses growing on organic soil of various thickness (from centimeters to over a meter), overlying mineral substrates that range from marine and lacustrine clay, to sand, gravels and boulders (Aylsworth et al. 2000; Rampton 1988). Segregated and pore ice, and sometimes massive ice bodies, can be encountered at all depths (Heginbottom 2000). Winters are long and experience considerable radiative losses, resulting in intense coldness with mean January temperatures falling below -25°C. Snow stays on the ground for about 8 months and its uneven distribution is largely caused by the many drifting events resulting from exposure to wind due to lack of vegetation/tree cover. Summers have long daylight hours and moderate temperatures (Fig. 1). The boreal region has a longer thaw season (Fig. 1) and the snow cover lasts for a shorter period (6–7 months). The snow is more evenly distributed, mainly due to the presence of trees (mostly spruce in the north, both spruce and aspen further south) which reduces snow redistribution by wind. As in the tundra areas, the soils often have an organic top layer (Aylsworth and Kettles 2000) and ground ice is prevalent in both the organic and the mineral soils (in particular the lacustrine silty clay). Many tundra and boreal sites in the Valley are occupied by wetlands and their soils are saturated during all or most parts of the thawed season (Zoltai et al. 1998).

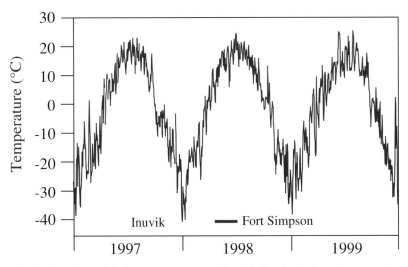

Fig. 1. Samples of daily temperatures at Inuvik (typical of tundra conditions) and Fort Simpson (boreal environment) for 1997–99

Two of Environment Canada's weather stations, Aklavik (68°08'N; 135°00'W) and Inuvik (68°17'N; 133°24'W), reported daily air temperatures for periods greater than 30 years. Aklavik temperature data is available for the periods 1926 through 1961 and 1981 to the present, while Inuvik data spans the years 1957 to the present, with some gaps. A comparison of overlapping data from the two sites showed a high degree of correlation ($r = 0.99$). This allowed us to merge the data from both sites into a single time series of 65 years to represent temperature conditions in a tundra zone. In the boreal environment, Fort Simpson (61°54'N; 121°24'W) provided 39 years of data for this study. In addition, several sites operated by the Geological Survey of Canada (GSC) offer short (<20 years) records of air temperature and near-surface ground temperature measured at 0.05 m below ground. Both air and near-surface ground temperatures are obtained with thermistors connected to single channel data-loggers that record temperatures at 4 hour intervals to an accuracy of ± 0.5°C with a resolution of ± 0.3°C. Where the short-term sites are located within 100 km radius from the long-term weather stations, their mean daily air temperatures are strongly correlated (correlation coefficients of 0.95 or greater), indicating that the Aklavik-Inuvik and Fort Simpson data are representative of their nearby areas.

Several sites in the Mackenzie Valley that are part of the GSC's active layer monitoring network provided maximum annual thaw depth measurements (Nixon et al. 2003; Tarnocai et al. 2004). These measurements were made with frost tubes (Mackay 1973) and are used for comparison with the simulation results.

For this study, we use two composite soil profiles typical of many areas in the North. Both profiles comprise ice-rich silty clay overlain by an organic layer, which in one case is thin (0.2 m) and in the other thick (1.0 m), representing two soil conditions commonly encountered in the Mackenzie Valley. Only saturated soil conditions are investigated but this is highly relevant as it reflects the conditions existing in a large part of the Valley. The soil properties used in this study of active layer thaw are given in Table 1.

Table 1. Properties of the composite soil profile used in the modeling of active layer thaw

Soil type	Bulk density [kg m^{-3}]	Porosity	Organic fraction	Mineral fraction	Minimum fraction of unfrozen water
Organic	60	0.9	0.1	0	0.05
Mineral	1600	0.44	0	0.56	0.1

3 Methods

This study makes use of long-term air temperature data to obtain the probability of maximum active layer thaw depths, using empirical relationships to estimate near-surface ground temperature and the Stefan's equation to calculate ground thaw.

3.1 Commencement of Ground Thaw

Ground thaw commonly begins as soon as the winter snow cover is depleted (Carey and Woo 1998). To determine the date of snow disappearance requires information on winter snow accumulation and spring snow melt, which is generally unavailable for remote locations. It is also noted that a rise of the near-surface ground temperature to the freezing point always lags behind the rise of mean daily air temperature above 0°C (which is indicative of the commencement of snowmelt). This lag time can be considered as a proxy for the duration of snowmelt. Owing to an increase in snowmelt rate as the summer advances, particularly at high latitudes where the days become very long around the solstice, the duration of snowmelt becomes shorter if the melt begins late. This relationship is confirmed by plotting the lag time against the day of year when air temperature rises above 0°C. Plotting data (Fig. 2) from Fort Simpson and Ochre River (boreal environment) and from Reindeer Station and YaYa Lake (tundra), we obtain the empirical relationship

$$LAG = 43.3 - 0.24 \, DOY \qquad (1)$$

where LAG is the delay (in days) of ground thaw commencement after the air temperature reaches 0°C; DOY is the day of year (DOY=1 for January 1). The correlation coefficient is 0.83 and the standard error is 3.6 days.

3.2 Air and Near-surface Temperature Relationship

From an energy balance consideration, air temperature is a manifestation of the sensible heat flux while ground temperature is controlled by ground heat flux. Heating of the air and the ground depends on the partitioning of the energy balance components, and the ratio of the ground heat to the sensible heat may change systematically during the snow-free period. We take the mean daily near-surface temperature (Tg) and mean daily air temperature (Ta) as indicators of these fluxes to form Tg/Ta ratios. The ratios are then plotted against time over the duration of the thaw period.

In order to reduce noise in the plotted data caused by fluctuations in Ta that are too rapid to permit a synchronous response in Tg, the ratios are plotted as 3-day running averages, starting at the second day after thaw commencement. Figure 3 shows the plotted data for Martin River near Fort Simpson (boreal site) and for Lousy Point in the Mackenzie Delta (tundra site). Both graphs suggest that the ratios change non-linearly with time and can be approximated by an inverse exponential relationship until the freeze-back season. The data for each site show some scattering but the relationship takes the form

$$Tg/Ta = b - c \exp(s \, P) \qquad (2)$$

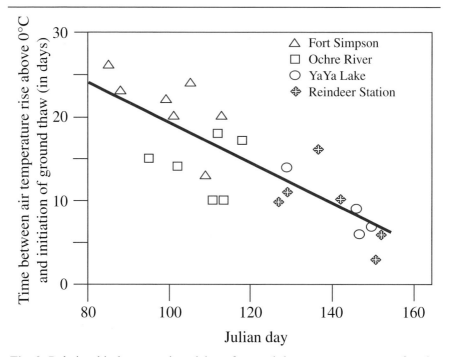

Fig. 2. Relationship between time delay of ground thaw commencement after the air temperature has attained 0°C (LAG, in days) and the date when air temperature rises to 0°C (Day of Year), based on data from two tundra and two boreal forest sites

where P is the day number, plus 1, after the beginning of ground thaw; b, s and c are coefficients. The value of b is the maximum recorded ratio, and the other two coefficients can be obtained by least square fitting. The data are bracketed by an upper enveloping curve that represents strong surface heating and a lower enveloping curve that signifies conservative warming of the ground surface in relation to air temperature. For these enveloping curves, the c and s coefficients are obtained by inspection so that together, they encompass most of the data points. Numerical values of the coefficients for the two study sites are provided in Table 2. The ground surface temperature is estimated by multiplying the T_g/T_a ratio from Eq. (2) by the mean (3-day running average) air temperature of the day. Note that only the period of thaw, and not the freeze-back, is of interest to this study so that there is no need to extend the relationship to the autumn when air temperature falls below 0°C.

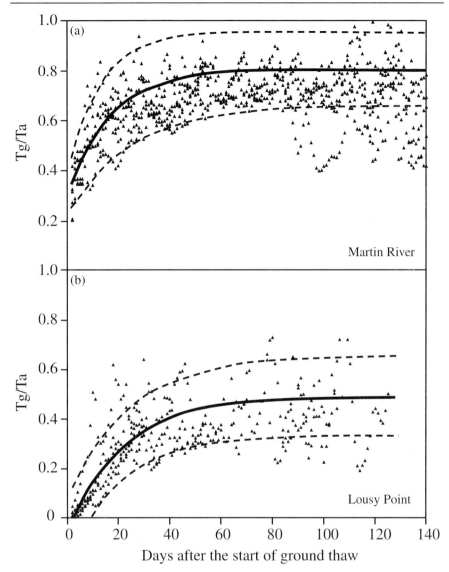

Fig. 3. Change of Tg/Ta ratio (ratio of near-surface ground temperature and air temperature, as 3-day running averages) with time since the beginning of ground thaw for (a) a boreal site at Martin River, and (b) a tundra site at Lousy Point. Also shown are the curves that represent the average conditions and the curves that envelop most of the data points. The coefficients are obtained by least-square fitting (for the average curves) and by inspection (for the enveloping curves)

Table 2. Parameter values for equation relating Tg/Ta ratio to day after start of ground thaw (P): Tg/Ta = b-c exp(s P)

	Parameters		
	b	c	s
Boreal site at Martin River			
Upper curve	0.87	0.55	-0.08
Middle curve	0.74	0.47	-0.06
Lower curve	0.60	0.40	-0.04
Tundra site at Lousy Point			
Upper curve	0.66	0.58	-0.04
Middle curve	0.50	0.54	-0.05
Lower curve	0.33	0.50	-0.06

3.3 Active Layer Thaw Calculation

The Stefan's equation offers a robust yet physically sound approach to calculate ground thaw (Woo et al. 2004). Originally formulated for lake ice melt, it has been adapted for ground freeze-thaw calculations (Jumikis 1977):

$$dz_t/dt = k_t D_t / \lambda \theta_z \qquad (3)$$

where dz_t/dt is the rate of thaw front descent (m s^{-1}), k_t is the thermal conductivity of the thawed soil (J m^{-1} s^{-1} K^{-1}), D_t is the thawing degree days, λ is the volumetric latent heat of fusion of water (J m^{-3}), and θ_z is the volumetric fraction of soil moisture content at depth z. Thawing is initiated at the top of the soil column. The near-surface ground temperature is the forcing that drives the heat to the freeze-thaw front through conduction, which is a function of the thermal properties of the soil. Heat convection is ignored and all available heat is used for phase change, not for warming or cooling of the soil.

The physical properties of the soil, including bulk density, porosity, mineral and organic contents, as well as the moisture content, are estimated based on field data. These properties are used to calculate the thermal conductivity of the soil above the thaw front (Farouki 1981). The advance of the thaw front is calculated daily until freeze-back occurs in the fall. The thaw front position immediately before freeze-back marks the maximum active layer thaw for that year. In permafrost terrain, the entire profile would be frozen back in the winter before ground thaw computation resumes in the subsequent year.

4 Results of Application

Mean daily air temperatures from Aklavik-Inuvik and from Fort Simpson, representing the tundra and the boreal environments respectively, were used to simulate the annual maximum active layer thaw in soils with a thin (0.2 m) and a thick (1.0 m) organic cover.

Year to year variations in the maximum annual thaw depth in response to warm and cold summer conditions are well known (e.g., Young and Woo 2003). The simulated annual thaw depths were tallied at 0.05 m intervals for each simulated data set and converted into probability estimates (divided by the number of years of simulation). Three probability distributions were obtained for each test case, using the upper, middle and lower Tg/Ta curves to derive the near-surface temperature. They represent the maximum annual thaw probabilities under conditions of strong, average and weak surface heating responses to air temperature. As expected, usage of the upper and lower enveloping curves for the Tg/Ta ratio expands the range of possible active layer thickness for each site.

The simulation results are provided in Fig. 4. A comparison was made with some active layer thickness measurements available from two sites in GSC's active layer monitoring network (Lousy Point in tundra and Martin River in boreal forest). Considering that we used only average soil conditions for the simulations, it is unlikely that the simulated thaw depths would encompass every observation made in sites where actual soil parameter values differ from the idealized values. Given these limitations, the model was still able to bracket most of the observed values for saturated soil with both a thick and a thin organic cover. This partial validation adds confidence to subsequent interpretation of the simulated results.

4.1 Effects of Location

The boreal region, being further south than the tundra, has a longer thawed season and warmer summer, as is seen in the air temperature differences between Fort Simpson and Aklavik-Inuvik in the Mackenzie Delta (Fig. 1). Such contrasts give rise to deeper active layer thaw in the boreal than in the tundra zone. Applying the mean Tg/Ta ratio curves (middle curves in Fig. 3) to areas with a thin organic layer overlying mineral substrate (Fig. 4a), the simulated deepest annual thaw in the Mackenzie Delta has a median value of 0.60 m; but deepens to 1.36 m in the Fort Simpson area. For areas with a thick organic layer, the median for simulated maximum thaw depth is 0.36 m for the Delta and 0.65 m for the boreal location (Fig. 4b).

These results reflect the effect of latitude on active layer thickness in typical soils under saturated conditions.

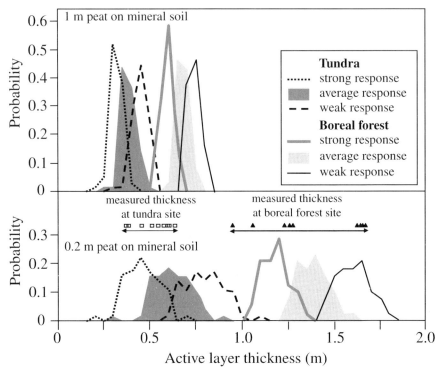

Fig. 4. Estimated probability of annual maximum active layer thaw depths for saturated soils with a thin (0.2 m) and a thick (1.0 m) organic surface layer, for tundra and boreal locations. Each site provides three probability distributions, for the cases of strong, average and weak ground surface heating responses to air temperature, corresponding to the upper, middle and lower curves shown in Fig. 3. Also shown are the ranges of active layer thickness measured at GSC's active layer monitoring sites with thin peat on mineral soil

4.2 Effect of Organic Layer Thickness

Ice-rich organic soil consumes much ground heat for ice melt (i.e., it has a high latent heat requirement) and when thawed and saturated, it has a lower thermal conductivity than saturated mineral soil (Woo and Xia 2005). Both factors retard the penetration of the thaw front in organic soils. Figure 4b demonstrates that the thaw front is unable to pass through the 1 m thick organic layer in both the tundra and the boreal site. Even when the extreme Tg/Ta ratio curves (upper and lower envelops in Fig. 3) are used in the simulation, the maximum annual thaw falls within a narrow range, being 0.29–0.42 m for the Delta site and to 0.58–0.72 m for the boreal site.

Without a thick organic layer to insulate the mineral substrate, ground thaw can descend deeper into the frozen soil. The maximum thaw depth is more variable than under a thick organic layer, becoming more responsive to the year-to-year fluctuations in air temperature (see Fig. 4a, which shows a range of 0.47–0.77 m for the Delta site and 1.16–1.53 m for the boreal site). These simulated results demonstrate the significant role of organic materials in controlling the probability distribution of maximum annual thaw in permafrost areas.

5 Conclusion

A method is presented to calculate annual maximum thaw of the active layer, using a combination of Stefan's algorithm for thaw penetration and empirical functions to evaluate the starting day of thaw and to obtain near-surface ground temperature from daily air temperature. Since only air temperature is used as the forcing, the method can be applied to most remote regions where data are scarce.

The method was used to derive probability estimate of the depths to which maximum ground thaw can attain. Long term weather station data provide the requisite input to run the model, to simulate maximum thaw for multiple years to allow the estimation of their probabilities. This approach was applied to two sites, one in the boreal and the other in a tundra environment, both under saturated conditions and with typical soil profiles of organic overlying mineral soils.

Simulation results confirm the effect of location on the probability distribution of maximum annual thaw, with shallower thaw at the tundra site. The effects of surface organic layer are also demonstrated. A thick organic soil cover retards thaw penetration so that ground thaw is shallower and the range of maximum thaw depth is smaller than for soils with a thin surface organic layer.

The method discussed can be used to generate information on the range of thaw depth that may occur under current climate conditions for representative locations and terrain types in the Mackenzie Valley. This provides baseline permafrost conditions within development areas which is required for both infrastructure design and environmental impact assessment. In addition, the model can be applied to predict thaw depth under various climate scenarios such as those projected in response to climate warming (Woo et al. 2007). The empirical equations can be modified to reflect changes in the relationship between air and ground surface temperatures, for prediction of change in thaw depth due to surface disturbance (e.g., vegetation clearing) associated with development. Given an increase in hydrocarbon exploration and development activities in the North, there is a strong demand for information on active layer thaw to facilitate sound engineering design and to ensure sustainable development. For such areas of data scarcity, the robust model presented will be highly suitable.

Acknowledgements

Support for the establishment and maintenance of the permafrost monitoring network in the Mackenzie region and associated data analysis is provided by the Department of Indian and Northern Affairs, GSC/Natural Resources Canada, the Panel for Energy Research and Development, and the Canadian government's Climate Change Action Plan 2000.

References

Aylsworth JM, Burgess MM, Desrochers DT, Duk-Rodkin A, Robertson T, Traynor, JA (2000) Surficial geology, subsurface materials, and thaw sensitivity of sediments. In: Dyke LD, Brooks GR (eds) The physical environment of the Mackenzie Valley, Northwest Territories: a base line for the assessment of environmental change. Geol Surv Can B 547, pp 41–47

Aylsworth JM, Kettles IM (2000) Distribution of peatlands. In: Dyke LD, Brooks GR (eds) The physical environment of the Mackenzie Valley, Northwest Territories: a base line for the assessment of environmental change. Geol Surv Can B 547, pp 49–55

Carey SK, Woo MK (1998) A case study of active layer thaw and its controlling factors. Proc 7th International Permafrost Conference, Yellowknife, NWT, pp 127–132

Farouki OT (1981) Thermal properties of soils. U.S. Army CRREL Monograph 81–1

Goodrich LE (1982) Efficient numerical technique for one-dimensional thermal problems with phase change. Int J Heat Mass Tran 21:615–621

Henry K, Smith M (2001) A model-based map of ground temperatures for the permafrost regions of Canada. Permafrost Periglac Process 12:389–398

Jumikis AR (1977) Thermal geotechnics. Rutgers University Press, New Brunswick, NJ.
Mackay JR (1973) A frost tube for determination of freezing in the active layer above permafrost. Can Geotech J 10:392–396
Nelson FE (1986) Permafrost distribution in central Canada: applications of a climate-based predictive model. Ann Assoc Am Geogr 76:550–569
Nixon FM, Tarnocai C, Kutny L (2003) Long-term active layer monitoring: Macknezie Valley, northwest Canada. In: Phillips M, Springman SM, Arenson LU (eds) Proc 8th International Conference on Permafrost, July 2003, Zurich, Switzerland. A.A. Balkema, Lisse, the Netherlands, pp 821–826
Oelke C, Zhang T, Serreze MC, Armstrong RL (2003) Regional-scale modeling of soil freeze/thaw over the Arctic drainage basin. J Geophys Res 108:D100, doi:10.1029/2002JD002722
Romanovsky VE, Osterkamp TE (1995) Interannual variations of the thermal regime of the active layer and near-surface permafrost in northern Alaska. Permafrost Periglac Process 6:313–315
Smith SL, Burgess MM, Heginbottom JA (2001) Permafrost in Canada, a challenge to northern development. In: Brooks GR (ed) A synthesis of geological hazards in Canada. Geol Surv Can B 548, pp 241–264
Tarnocai C, Nixon FM, Kutny L (2004) Circumpolar-Active-Layer-Monitoring (CALM) sites in the Mackenzie Valley, Northwestern Canada. Permafrost Periglac Process 15:141–153
Taylor AE (2000) Relationship of ground temperatures to air temperatures in forests. In: Dyke LD, Brooks GR (eds) The physical environment of the Mackenzie Valley, Northwest Territories: a base line for the assessment of environmental change. Geol Surv Can B 547, pp 111–117
Woo MK, Arain A, Mollinga M, Yi S (2004) A two-directional freeze and thaw algorithm for hydrologic and land surface modelling. Geophys Res Lett 31: L12501, doi:10.1029/2004GL019475
Woo MK, Mollinga M, Smith SL (2007) Climate warming and active layer thaw in the boreal and tundra environments of the Mackenzie Valley. Can J Earth Sci (in press)
Woo MK, Xia ZJ (1996) Effects of hydrology on the thermal conditions of the active layer. Nord Hydrol 27:129–142
Young KL, Woo MK (2003) Thermo-hydrological responses to an exceptionally warm, dry summer in a High Arctic environment. Nord Hydrol 34:51–70
Zoltai SC, Taylor S, Jeglum JK, Mills GF, Johnson JD (1988) Wetlands of boreal Canada. In: Wetlands of Canada. Ecological Land Classification Series No. 24. Environment Canada and Polysciences Publications, Montreal, pp 98–154

i

Chapter 8
Climate–Lake Interactions

Wayne R. Rouse, Peter D. Blanken, Claude R. Duguay,
Claire J. Oswald and William M. Schertzer

Abstract Lakes cover an estimated 11% of the Mackenzie River Basin (MRB) and they are an integral part of the Basin's energy and water cycling regime. For convenience they are classified into small, medium and large in terms of surface area. MRB lakes are subjected to a wide range in air temperature from south to north, and to substantial differences in precipitation. The ice-covered period is a function of lake size and location. Small lakes have a longer ice-covered period (6-9 months) than large lakes (4-7 months) and northern lakes have a longer ice-covered period than their southern counterparts. Lake ice duration and thickness is strongly influenced by both air temperature and overlying snow depths. During the open water season, for small shallow lakes, the seasonality of convective fluxes is similar to the surrounding land surfaces, but for medium and large lakes there is a large time lag. Their total seasonal evaporation is significantly greater than for terrestrial surfaces and their considerable heat storage capacity accounts for the temporal lag in energy and water cycling. A large lake such as the Great Slave Lake is highly sensitive to interannual climate variability and achieves substantially greater heat storage, higher temperatures and greater evaporative and sensible heat fluxes during a warmer year than during an average year. Instead of exhibiting strong diurnal evaporation cycles, its thermal and evaporative behavior is dominated by synoptic systems that approach a three-day cycle. Mass transfer methods of calculating evaporation works well, but are specific to size of the lake and its exposure to atmospheric forcing. Slab models employed to describe the energy cycles of different size lakes have met with some success. Lakes of all sizes are strongly impacted by climate variability and change and this has large influences on the regional hydrologic regimes.

1 Introduction

Lakes are an integral part of the Mackenzie River Basin (MRB) energy and water cycling regime (Woo et al. 2007) due to their high frequency of occurrence and large areal coverage (Fig. 1). A total lake surface area of 198,000 km^2 (11% of the Basin) is estimated from the recently available Canada Centre for Remote Sensing water fraction data derived from satel-

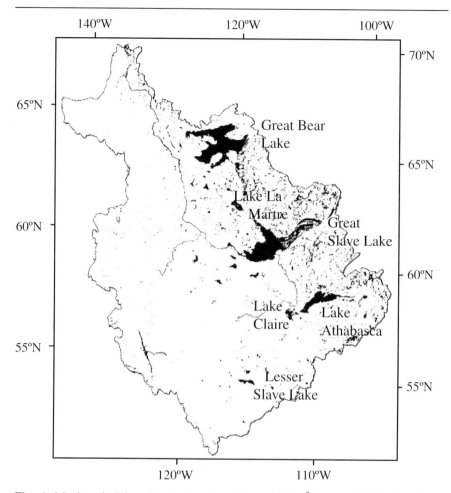

Fig. 1. Mackenzie River Basin showing lakes >10 km² in area. High density of lakes in the eastern portion of the basin coincides with the underlying Canadian Shield whose boundary with the Interior Plains to the west passes through central Great Slave Lake (after Bussières and Schertzer 2003, with permission of the authors)

lite analysis (Bussières, personal communication). Lake size ranges from small ponds to some of the largest fresh-water lakes in the world, including from north to south, Great Bear Lake (31,153 km^2), Great Slave Lake (28,450 km^2) and Lake Athabasca (7,850 km^2). Lake size and frequency are inversely related (Table 1). The largest concentration of lakes is in the Precambrian Canadian Shield and these are generally much deeper than their counterparts in the Interior Plains. Little is known about the average depths of small and medium lakes, but larger lakes are usually deeper than the smaller ones. The average depths of Great Bear Lake, Great Slave Lake and Lake Athabasca are about 76, 88 and 26 m respectively.

Table 1. Size categories of lakes in the Mackenzie River Basin and the depth ranges, approximate number of lakes, cumulative area and percentage of total cumulative basin lake area for each category

	Small	Medium	Large	Total
Size range [km^2]	<1	1–100	>100	
Depth range [m]	≤5	5–50+	>25–70+	
Number of lakes	25,651	9,590	261	35,503
Cumulative area [km^2]	6,549	63,827	127,624	198,000
Percent of total area	3	32	65	100

Lakes in the MRB experience a large range of climatic conditions. Annual precipitation in the MRB ranges from 170 mm at the Arctic Ocean coast to 620 mm in the mountainous southwest (Stewart et al. 1998; Woo et al. 2007). Most of the lakes are located in the drier, eastern parts of the Basin (Fig. 1). The MRB has a wide range of absorbed solar radiation, temperature and humidity. The annual course of absorbed solar radiation as calculated from AVHRR satellite images as described in Feng et al. (2002) indicate that only the most southerly portions of the MRB absorb any solar radiation during the winter months (DJF). During the summer (JJA), however, absorbed solar radiation exceeds 400 W m^{-2} throughout most of the Basin except for the cloudy areas along the eastern slopes of the Mackenzie Mountains and along the Arctic coast. Mean January air temperatures range from -12°C in the south to -28 °C in the north, but the mean July temperature difference between the north and the south is small, being about 15°C over the Basin except for the mountainous areas where temperature decreases with altitude. The largest mean annual temperature ranges are in the north-central MRB. Here, the largest lakes in the basin are subjected to annual ranges of up to 45°C.

The thermal and water storage and energy exchange capabilities of lakes differ from those of adjacent land surfaces. Prior to the Mackenzie GEWEX Study (MAGS), thermodynamic research was limited to temperature studies of Great Slave Lake (Melville 1997; Rawson 1950) and Great Bear Lake (Johnson 1975). Research conducted under MAGS has greatly advanced our understanding of energy and water cycling of MRB lakes. Many of the results from the MAGS lake studies can be applied to other circumpolar high latitude lakes. The Great Lakes of the Mackenzie have much in common with the Laurentian Great Lakes situated in the temperate region of eastern North America, but the Mackenzie lakes are exposed to larger climatic extremes. Lake size is critical to climate-lake interactions, so cross-region similarities will be closely linked to lake size, as well as to similarities in their climatic regions. This chapter provides a synopsis of MAGS findings pertaining to interactions of climate and lakes in this high latitude Basin.

2 Seasonal Thermal and Moisture Cycles in MRB Lakes

The size of a lake influences its seasonal cycles in many ways and dictates how much they will differ from those of the surrounding land. The lakes are arbitrarily distinguished into small (area ≤ 1 km^2 ; depth ≤ 5 m), medium (area 1–100 km^2; depth>5-50+ m) and large (area>100 km^2; depth>25–70+ m) categories (Rouse et al. 2005). The open water period and vigorous energy cycle of the small lakes starts in early spring with the onset of thaw (Table 2). This normally coincides with long daylight periods and above-freezing daytime temperatures. The melting of the snow and ponding of water on lake ice dramatically reduces the albedo of the lakes, typically from >0.6 to <0.25. Melting can proceed rapidly with the combination of atmospheric warmth and the strong absorption of solar radiation. Solar heating of the shallow lakes raises their temperatures rapidly

Table 2. Seasonal cycle of different size lakes in the central Mackenzie River Basin. Z, T, W, C, and F designate Frozen, Thaw, Warming, Cooling and Freeze-up respectively

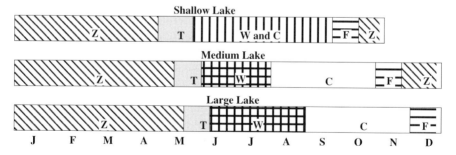

after the final ice melt, allowing evaporation to proceed vigorously. With the arrival of winter, a combination of subfreezing temperatures, short daylight periods and low lake volume encourages rapid freezing. Formation of lake ice creates an effective lid on outgoing turbulent heat fluxes and the shallow lake evaporation cycle abruptly ends. The period of substantial evaporative and sensible heat loss from shallow lakes in southern and central MRB lasts about four months (Table 2), and this duration decreases poleward.

With larger surface areas and greater depths, the volumes of medium lakes are also bigger than those of small lakes. These lakes display much greater time lags in their heat and moisture cycles than the smaller ones and shorter lags than the large lakes, though the cycles of many medium lakes appear to approximate more closely to those of the large lakes (Table 2). They are able to store substantial amounts of heat because solar radiation can penetrate to considerable depths depending on water clarity. Their surface heating generally lags behind that of small shallow lakes, leading to delays in the onset of latent and sensible heat fluxes in the spring. However, large heat storage allows these fluxes to continue vigorously into the fall and early winter.

Seasonal cycling of the large lakes in the MRB is in many respects similar to that of the Laurentian Great Lakes but in some respects their high latitude position puts them in a class of their own. The thermal and hydrologic regime of Great Slave Lake (Rouse et al. 2003; Schertzer et al. 2003, 2007) is probably representative of the large deep high latitude lakes, at least to a first approximation. The thermal cycle shows a prolonged spring heating period, a summer period of maximum heat storage, a fall-winter cooling phase and a winter minimum in temperature and heat storage (Table 2). Ice break-up ranges from late May to late June. Freeze-back occurs from late November to the end of December. Great Slave Lake and Lake Athabasca undergo biannual density-driven vertical turnover, exchanging surface and deep waters and are dimictic (i.e., they undergo vertical overturning at the temperature of maximum density ($4°C$) during spring warming and again during autumn cooling). Great Bear Lake has been described as cold monomictic with no vertical convective mixing (Johnson 1975; Melville 1977) but recent evidence indicates that this is not the case, at least in some parts of the lake (Rouse et al. 2005, 2007). During spring, the total heat flux is dominated by high net radiation that contributes primarily to lake heating (Schertzer et al. 2000). Heating proceeds slowly because of the large vertical mass and deep thermal mixing. During this heating phase the turbulent heat fluxes are often negative (Blanken et al. 2000) and positive fluxes usually do not commence until mid-summer. The evaporative and sensible heat fluxes reach their maxima during fall and early winter (Rouse et al. 2003) and remain large as long as there is open water. Significant evaporation from Great Slave Lake occurs over a period ranging from 6 to 7 months (Blanken et al. 2007).

3 Lake Ice Duration and Thickness

Freeze-up and breakup dates that bracket the duration of ice cover have been shown to be robust indicators of climate variability and change at northern latitudes (e.g., Duguay et al. 2006). Ice cover thickness, and in particular the duration of ice cover also play a significant role in the annual energy and water balance of large river basins such as the MRB. Ice cover on water bodies prohibits evaporative exchanges with the atmosphere for several months each year.

Ice cover duration varies as a function of lake size and location. Air temperature is the principal determinant of ice cover duration while ice thickness is also strongly influenced by the presence of snow on the ice (Ménard et al. 2003). To examine the effects of air temperature, snow cover, and lake depth on the ice regime of MRB lakes, maps were produced using output generated from the one-dimensional thermodynamic lake-ice model CLIMo (Duguay et al. 2002), forced with NCEP/NCAR Reanalysis and the gridded snow data (Brown et al. 2003) at a 2.5°x2.5° grid-resolution for the overlapping period of the two data sets (1980–96). Isolines of ice cover duration and maximum ice thickness given on these maps (Figs. 2 and 3) represent mean values computed over the 16-year period for typical small shallow lakes (3 m) and medium to large deep lakes (30 m).

Figure 2 shows the general spatial patterns as well as variations in ice cover duration with lake depth (3 and 30 m) across the MRB, assuming that the accumulated snow does not get redistributed by wind. Such conditions may be encountered at sheltered lake sites found in more densely forested areas of the Basin but are unlikely to be experienced on large lakes or in areas characterized by sparse a vegetation canopy (open forest and tundra). Figure 3 illustrates the results for completely snow-free lakes. In reality, conditions normally lie between the "full snow" and the "no snow" scenarios.

The ice cover period ranges from about 6 to 9 months for small lakes and 4 to 7 months for large lakes over the entire MRB. For lakes located at the same latitude, the shallower lakes remain ice covered for a longer period (by 40–50 days) than the deep lakes. The difference in ice cover duration between the two classes of lakes is due mainly to the larger heat storage capacity of the deep lakes which freeze at a later date than shallow lakes (Ménard et al. 2002). Maximum ice cover thickness on the other hand shows differences of only about 10 to 20 cm for lakes of different depths (being thinner for deeper lakes due to later ice formation) at the

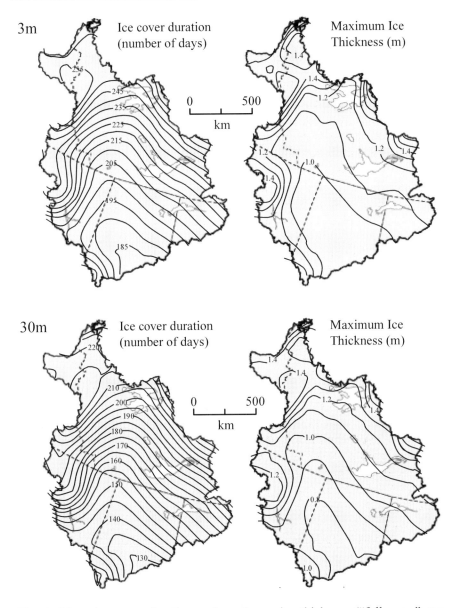

Fig. 2. Mean ice cover duration and maximum ice thickness ("full snow" scenario) for 1980–96 for hypothetical 3-m and 30-m deep lakes across the Mackenzie River Basin as modeled using the CLIMo model (after Duguay et al. 2003)

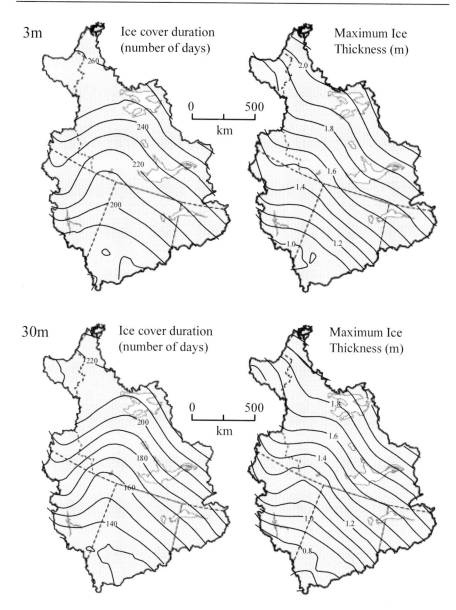

Fig. 3. Same as Fig. 2 but for "no snow" scenario (after Duguay et al. 2003)

same latitude. When both "full snow" and "no snow" scenarios are considered, the thickness varies from about 0.8 m in the south to 2 m in the north.

This experiment clearly demonstrates the significant impact of snow cover on lake ice thickness and duration.

4 Energy Fluxes and Lake Size

Lakes of all sizes have large net radiation and convective heat fluxes during their open water periods and their seasonal evaporation is significantly greater than for other high latitude surfaces (Rouse et al. 2007). The semi-transparent nature of lake water allows incident solar radiation to penetrate to considerable depths where it is largely absorbed. Hence lakes have low surface albedos. This penetration can reach the bottom of shallow lakes. In deeper lakes the penetration can be deep depending on transparency, and the absorbed heat energy is redistributed to greater depths by density- and wind-driven overturning. The total absorbed heat energy determines how long a lake will remain ice-free. It also determines how much heat is available to drive the evaporative and sensible heat fluxes. Thus large deep lakes remain ice-free the longest and evaporate the most water. Evaporation for medium and large lakes is almost double that of upland Canadian Shield surfaces and significantly larger than for wetlands and small lakes (Table 3).

Table 3. Seasonal energy balance components and evaporation for different size lakes compared with uplands and wetlands. (adapted from Rouse et al. 2005)

	Q^*	Q_S	Q_E	Q_H	E	max-Q_S	Days of open water
Upland	1094	0	560	534	227	57	
Wetland	1199	25	774	400	314	94	
Small lake	1386	28	855	503	346	268	154
Medium lake	1259	27	1002	230	406	562	170
Large lake	1290	10	1041	240	422	1048	228

Q^* net radiation,, Q_H sensible heat flux, Q_E latent heat flux (all in MJ m^{-2} d^{-1}), E evaporation (mm), Q_S and max-Q_S heat storage and maximum heat storage during the ice-free season (MJ m^{-2})

The net radiation of small lakes is large and comparable to medium and larger lakes (Table 3). Although they have shorter open water seasons than larger lakes, their earlier ice melt allows them to absorb more solar radiation during the high sun season around the summer solstice. Vertical gradients of air temperature and vapor pressure control their surface to air heating, cooling and evaporation rates. During the course of a day the lake surface warms less rapidly than the air above so that temperature and vapor pressure gradients are more moderate than the nearby terrestrial surfaces. This suppresses sensible and latent heat fluxes. During evening and night the lake surface cools less rapidly than the overlying atmosphere and this enhances the sensible and latent heat loss. From day to day and over periods of several days the overlying air masses exert a similar influence on fluxes of small lakes. With cold overlying air, vertical temperature and humidity gradients are large and the sensible and latent heat fluxes are large. Shallow lakes cool rapidly and vapor fluxes are large. With warm overlying air, vertical temperature and humidity gradients are suppressed, sensible and latent heat fluxes are reduced, and lakes can have a net gain in heat storage. Small lakes tend to reach their greatest heat storage in mid-summer, but they cool off quickly in the fall and their sensible and latent heat fluxes decline rapidly through to freeze-up.

The magnitudes of vertical heat and moisture fluxes to and from medium-size lakes are influenced temporally by several factors that can differ from small lakes. Their larger heat storage (Table 3) often results in little daily variation in their surface temperatures so that the diurnal variation of the convective fluxes is mainly controlled by air temperature and humidity of the overlying air. As well as responding to regional air mass activity, the vertical temperature and humidity gradients may be influenced by lake-land breezes (Oke 1987). By day, these enhance the vertical temperature and humidity gradients to promote evaporative and sensible heat losses, and by night suppress the vertical near-surface gradients to inhibit such moisture and heat losses. Seasonally, the greater frequency of cold air masses over relatively warm lake surfaces in late summer, fall and early winter strongly defines the vertical temperature and humidity gradients and drives especially strong evaporative heat fluxes through to final freeze-up. Warm dry air mass incursions during this period can lead to temperature inversion with downward directed sensible heat flux that helps accelerate evaporation.

Many factors influencing medium lakes are magnified when applied to large lakes. Diurnal and day to day variability in fluxes is small during fair weather days. Storm activity and seasonality exert strong influences. Major mid-summer storms can stir large lakes to great depth and bring cold water to the surface and this can dampen evaporation and sensible heat loss for days afterward. Seasonally, the long lag due to spring and early summer heating, enhanced by the vertical convective turnover, promotes inversion or weak lapse gradients at the lake-atmospheric interface. After final breakup, this can dampen evaporation and sensible heat loss for weeks. The large lakes heat gradually and tend to reach their highest temperatures in mid to late summer, coinciding with the period with the warmest overlying air masses so vertical gradients are moderate and fluxes are similarly modest. In fall, early winter and often into mid-winter cool to cold air masses dominate the overlying atmosphere but the ice-free lakes remain relatively warm. This drives very large fluxes. Figure 4 shows that the net radiation and evaporation for Great Slave Lake are out of phase. For 3.5 months after September 15, a period when average net radiation is negative, the evaporation magnitude is twice as large as the evaporation for the 3.5 months prior to that date, when average net radiation is large and positive.

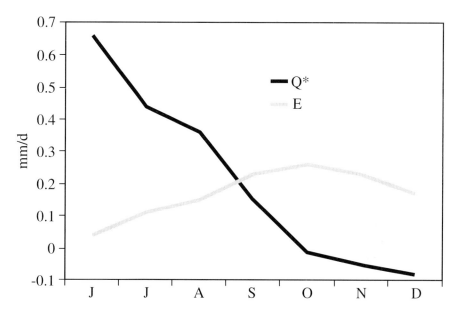

Fig. 4. Average monthly net radiation (Q^*) and evaporation rates (E) for Great Slave Lake (values converted in water equivalent unit)

Not only do large lakes control their own environment but because of their considerable volumes and heat storage (Table 3), they exert pronounced effects on their surrounding landscape. Strong lake breezes cool the peripheral terrestrial environment in spring and early summer. The large evaporation in fall and early winter can trigger downwind snow squalls as seen in satellite images over the Great Slave and Great Bear Lakes region.

5 Large Lake Thermodyamic and Hydrodynamic Processes

5.1 Seasonal Patterns

The annual temperature cycle of Great Slave Lake shows the following pattern and characteristics. This lake is dimictic and its minimum temperature (<1°C) usually occurs in mid-winter. Based on average condition between 1988 and 1999, the central basin of the lake is expected to be ice-free by June 18 and to be frozen by December 7 (Walker et al. 2000). After spring melt, the penetration of solar radiation and wind mixing largely determine the lake thermal structure. Following the spring turnover, a vertical thermal structure develops with the warmest water in the upper layer and the coldest water in the lower layer and the transition characterized by sharper temperature gradients representing the thermocline. The thermocline generally deepens to a maximum in mid-summer when the lakes heat content is at a maximum. Subsequent cooling proceeds through to freeze-up. During the stratified summer season the colder deeper portion of the lake undergoes much smaller temperature variation than the upper mixed layer. Because of its large size and high latitude location, Great Slave Lake is prone to high winds and storm waves that lower the thermocline and reduce the surface and subsurface temperatures (Schertzer et al. 2007). This reduces both the evaporative and sensible heat fluxes.

Interannual variability in the thermal regime of Great Slave Lake is substantial (Schertzer et al. 2007), as is evident in the sequential years of 1998 and 1999 (Table 4). Walker et al. (2000) showed that extensive open water areas were evident by May 27, 1998, about three weeks earlier than the average. This, combined with late freezing in the fall, provided an ice-free season of 213 days. 1999 was closer to normal with 175 ice-free days. The seasonal maximum heat storage was 18% larger, maximum daily air temperature 31% higher and maximum daily water surface temperature 24%

Table 4. Ice-free days, heat content, air temperature and surface temperature of Great Slave Lake (adapted from Schertzer et al. 2003; Walker et al. 2004)

	1998	1999
Ice-free days	218	172
Heat content [MJ x 10^{19}]	3.1	2.0
Air temperature [°C]	23.8	17.5
Surface temperature [°C]	21.2	16.7

higher in 1998 than in 1999. The timing of the maximum heat content between years is a clear signal of climate variability (Schertzer et al. 2007).

5.2 Time Scales of Evaporation

Recent analysis (Blanken et al. 2007) provides consistent insights that for Great Slave Lake, on average, 50% of total annual seasonal evaporation occurs during the first 60% of the seasonal cycle and 50% during the remaining 40% of the cycle that leads up to final freeze-up of the lake. These evaporation cycles are in distinct contrast to evapotranspiration cycles over land. For the large lake, there is a shift from a heat sink in the summer when radiation inputs are at a maximum, to a heat source in the fall/winter when the adjacent terrestrial surfaces are a heat sink. Short term behavior of the turbulent fluxes over the lake also differs from that of the land environment. Unlike the diurnal evaporation cycles over land where the peaks occur around the same time each day, neither the evaporative nor sensible heat fluxes from the Great Slave Lake display such a symmetrical, diurnal pattern. Rather, for this lake, latent and sensible heat fluxes remain quiescent, often for several days, then steadily increase over several hours to a maximum. Subsequently, the fluxes often return to their quiescent magnitudes until the process repeats itself. The dominant time period for the cycle from quiescent to maximum to quiescent is 2.7 days (Table 1 in Blanken et al. 2007). Synoptic scale processes exert important controls over the convective heat exchange cycles. For example, the passage of a synoptic-scale cold front will reduce transpiration from boreal vegetation over a three-day period, while the cold, dry air mass is in place. However, a distinct diurnal pattern normally is superimposed on the overall reduction in evapotranspiration. In contrast, the lake increases its evaporation on the same time-scale as the synoptic event without any imposed diurnal response. The response of Great Slave Lake to relatively long-term events is likely pronounced since incursions of cold, dry air are common at these latitudes (Serreze and Barry 2005). Blanken et al. (2003) report that these three-day-long evaporation events comprise an aggregation of sweeps of warm, dry air from above the lake surface that persist for a few seconds. Thus, there appears to be an integration or aggregate effect of the superimposed events happening at vastly difference time scales.

6 Modeling Lake Thermodynamic and Hydrodynamic Processes

Blanken et al. (2007) demonstrate that the product of the horizontal wind speed and the difference in vapor pressure between the lake surface and atmosphere explains most of the variance in evaporation. They also show that varying the sampling interval from one to six hours in one-hour increments has no significant affect on the relationship between evaporation and its controls. This implies that mass transfer techniques are applicable for calculating the evaporative and sensible heat fluxes. Oswald et al. (2007) used the Dalton-type equations to compute evaporation for lakes. This approach has the attractiveness that wind speed and vapor pressure are often available from measurement sites near lakes and surface temperatures can be derived from satellite data. Alternatively, the necessary data can be obtained from regional climate models. The drag coefficient is obtained experimentally using measured data of various types. Oswald et al. (2007) present a theoretically-based analysis of results of the application of mass transfer equations for four different-size lakes in the central MRB. The drag coefficients are shown to differ from lake to lake and from year to year. When averaged multi-year exchange coefficients are applied to ensemble measured data, their performance is variable (Table 5). As other investigators have found, drag coefficients appear to be lake specific.

A small-lake thermodynamic model that utilizes these drag coefficients was developed to investigate the thermal and evaporation regimes of MRB lakes (Binyamin et al. 2006). This is a whole-lake or slab model that allows input of the local meteorological forcing variables for computation of the required radiative and turbulent heat fluxes. The simulated temperatures represent lake-wide averages at each time step. As vertical mixing is easily achieved for small and medium-size lakes of shallow depths, the model can accurately approximate the temperature and surface flux regimes under such predominantly isothermal conditions.

To capture the spatial and temporal variations of temperature characteristics and currents of large lakes, the extensive database collected during MAGS was applied to develop and verify hydrodynamic and thermodynamic models for Great Slave Lake. A 1-D Dynamic Reservoir Model (DYRESM) was applied to derive the lake-wide averaged temperature

Table 5. Locations, dimensions and mass transfer coefficients of four research lakes. C_D is the drag coefficient and σ is its standard error. Absolute percent error, averaged for all years, is reported for cumulative latent (Q_E) and sensible heat loss (Q_H). Correlation (r^2) for linear regression of modeled and measured daily fluxes is averaged for all years (adapted from Oswald et al. 2007)

	Small lake 1	Small lake 2	Medium lake	Large lake
Latitude [°N]	63.6	62.5	62.9	61.9
Area [km^2]	0.1	0.3	5.5	1.9×10^4
Mean depth [m]	3.2	0.5	12	32.2
Mean volume [m^3]	1.8×10^5	1.4×10^5	4.2×10^7	6.0×10^{11}
Instrument height [m]	1.8	1.8	1.8	18.0
Years of measurement	3	3	3	6
$C_D \pm \sigma$ [x10^{-3}]	1.30 ± 0.07	1.26 ± 0.09	2.14 ± 0.20	1.10 ± 0.15
Absolute error for Q_E [%]	10	3	10	42
Absolute error for Q_H [%]	15	6	23	44
Average r^2 for Q_E	0.76	0.86	0.49	0.88
Average r^2 for Q_H	0.62	0.81	0.73	0.88

structure during the ice-free season (McCormick and Lam 1999). Both the simulation and observations show the beginning of thermal stratification in mid- to late June (Fig. 5). The simulated upper mixed layer and temperatures are similar to observations throughout the summer and show general deepening of the upper layer in the fall. Simulated temperature below the upper layer is more diffuse than in observations indicating that accurate approximation of the mean vertical light attenuation may be critical in affecting the vertical distribution of heating in this lake.

Hydrodynamic 3-D models involve principles of conservation of mass, momentum and energy, incorporating many meteorological and limnological processes. These models produce output for lake currents, water temperature, long-term and short-term water levels, and pollutant transport. Momentum and heat fluxes are the dominant factors on time scales from days to seasons, while the hydrologic balance is important on longer time scales. With respect to potential climate change, prediction of long-term fluctuations in lake circulation becomes important since variations in wind stress, heat flux, or hydrological balance will cause changes in the lake's thermal structure and circulation, and eventually influence the whole ecosystem.

The ELCOM 3-D hydrodynamic model (Hodges et al. 2000) was applied to Great Slave Lake (Leon et al. 2005). Results show that the dominant circulation characteristics for the upper layer of this lake are the presence of a counter-clockwise circulation pattern around the lake and a large

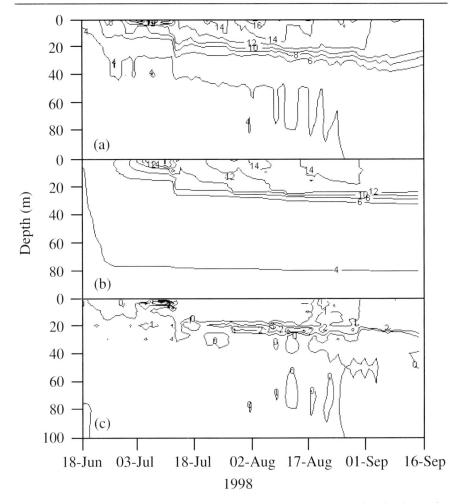

Fig. 5. Comparison of (a) observed, (b) simulated and (c) simulated minus observed lake wide mean vertical temperature for Great Slave Lake based on the DYRESM 1-D thermal model

central basin counter-clockwise gyre (Fig. 6). Similar patterns occur in the Laurentian Great Lakes (Schertzer 1999, 2003). A quasi-verification of the circulation characteristics in the eastern part of the central basin of Great Slave Lake is inferred from the spatial distribution of suspended sediments conveyed by the flow from the Slave River, following a rapid release of reservoir water during the repair of the Bennett Dam (see Woo and Thorne 2003, pp 334–335 for description of the event). While the particulate plume is affected largely by surface wind-driven circulation rather than the

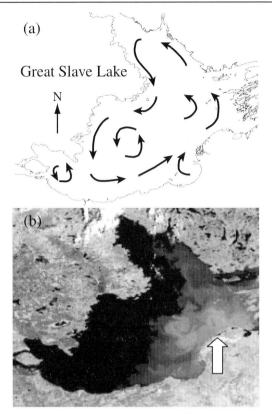

Fig. 6. (a) Generalized dominant circulation pattern for the upper layer of Great Slave Lake based on ELCOM 3-D hydrodynamic model simulations (Leon et al. 2005), and (b) example of the distribution of suspended sediments in Great Slave Lake after release from the Bennett Dam. The open arrow shows the point of input of particulates via the Slave River

surface integrated flow, it shows dominant flows along the coastal zone up into the north arm as well as the complex currents in the deeper eastern sections of the lake. Future hydrodynamic simulations are needed to incorporate the major inflow and outflow volumes through the Slave and Mackenzie Rivers as well as to account for the exchange between the central basin and the eastern arm of the lake. The 3-D hydrodynamic model is also useful for simulating the spatial (horizontal and vertical) temperature structure of a large deep lake in order to approximate the lake heat content (Schertzer et al. 2007).

7 Impacts of Climate Warming

Lakes in high latitudes are more susceptible to impacts from climate variability and climate change than are terrestrial surfaces. This arises from their strong absorption of solar radiation during the ice-free period. Any factor that increases this absorption will increase the net radiation, the heat storage, and the evaporative and sensible heat fluxes. Since final ice melt coincides with the annual period of longest days and most intense solar radiation, any factor that changes the date of ice melt has the potential to alter the total radiation absorbed and seasonal magnitude of net radiation. Higher temperatures will advance the date of lake ice melt and increase the net radiation in spring. Such increases in net radiation will increase the heat content of the lake, as was shown for Great Slave Lake in 1998. The increased heat storage ultimately must be dissipated, and this will increase evaporation and the sensible heat flux. For the large lakes this increase is especially notable in fall and early winter.

While the warm year of 1998 showed that even large lakes like the Great Slave Lake are sensitive to atmospheric warming in a single season, the same responsiveness applies to lakes of all sizes. A warmer year with earlier thaw gives rise to larger seasonal net radiation for all surfaces in a region (Table 6). Most of this increase goes into greater evaporation for the lakes, in contrast to the terrestrial surfaces where it is more equally divided with the sensible heat flux (Table 7). This is because uplands and wetlands can and do become drier under the impetus of higher temperature and greater net radiation, whereas lakes provide an unlimited water source. An exception is that shallow lakes can change to wetlands or even disappear if greater evaporation exceeds increases in precipitation and ground-ice meltwater input from permafrost degradation. This has happened in the post-glacial past and there is recent evidence of disappearance of thermokarst lakes (Smith et al. 2005). Large and medium size lakes are unlikely to be thus affected. Their level, however, may be drawn down considerably to impact the discharge regime of rivers issued from these lakes. Both the magnitude and timing of flow will be seriously altered. This is an important hydrologic consideration for a lake-rich system such as the MRB.

8 Summary and Conclusions

With areal coverage of about 11% of the MRB, lakes of various sizes are

Table 6. Surface energy balance comparisons for the warmest year to the averages of all measurement years for the 5 regional surfaces.

		Q^* [W m^{-2}]	Q_E [W m^{-2}]	Q_H [W m^{-2}]
Upland	Warmest	115	58	57
	Average	88	45	43
Wetland	Warmest	113	86	45
	Average	119	77	42
Small lake	Warmest	168	106	62
	Average	136	86	50
Medium lake	Warmest	144	117	27
	Average	124	100	24
Large lake	Warmest	151	123	28
	Average	130	106	24

Q^* net radiation, Q_E latent heat flux, Q_H sensible heat flux.

Table 7. Increase in the surface energy balance (all values in W m^{-2}) between the annual highest and average measured values (High minus Average) for Lake (small, medium and large averaged) and Terrestrial (upland and wetland averaged) surfaces

	Lake surfaces	Terrestrial surfaces
Net radiation	24	20
Latent heat flux	18	11

subjected to a wide range in temperature and to substantial differences in precipitation. Small lakes have a longer (by 1 to 2 months) ice-covered period than larger lakes and northern lakes have a longer (by 2.5 to 3 months) ice-covered period than their southern counterparts. For the open water period, the seasonality of convective fluxes for small shallow lakes is similar to the surrounding land surfaces, but for medium and large lakes there is a substantial temporal lag due to heat storage. The larger the lake the greater is its evaporation.

Intensive study of Great Slave Lake indicates that it is highly sensitive to interannual climate variability and achieves substantially greater heat storage, higher temperatures, and greater evaporative and sensible heat fluxes during a warmer year than during an average year. Unlike the surrounding terrestrial environment, this lake does not have a pronounced diurnal evaporation cycle. Rather its thermal and evaporative behavior is dominated by synoptic systems that approach a 3 day cycle. During these synoptic events, turbulence episodes lasting from seconds to minutes exchange cold dry air with the water surface and drive large evaporative fluxes.

Mass transfer methods of calculating evaporation and sensible heat fluxes yield drag coefficients that are specific to lake size and exposure. Smaller lakes can be adequately modeled with relatively simple whole-lake models but for large deep lakes, atmospheric interactions and in-lake processes are best modeled with 1-D or 3-D lake models linked to regional climate models.

Climate warming will cause early melting of lake ice. This will greatly increase the seasonal net radiation to the lakes, ultimately increasing the latent and sensible heat fluxes. The effect is most strongly felt in fall and early winter when enhanced evaporation from medium and large lakes can augment lake-effect snowfall by depositing more snow on downwind terrestrial locations. Increased evaporation losses from lakes will affect hydrological cycling and reduce lake outflows. Many of the findings in this study are applicable to other high latitude lake-rich regions in the northern hemisphere.

Acknowledgements

Many agencies and individuals have helped in various aspects of this project. Special acknowledgment goes to Chris Spence who was involved in aspects of planning and field work.

References

Binyamin J, Rouse WR, Davies JA., Oswald C, Schertzer WM (2006) Verification of a small lake thermodynamic model for energy fluxes and temperature. Int J Climatol 26:2261–2273

Blanken PD, Rouse WR, Culf AD, Spence C, Boudreau LD, Jasper JN, Kotchtubajda B, Schertzer WM, Marsh P, Verseghy D (2000) Eddy covariance measurements of evaporation from Great Slave Lake, Northwest Territories, Canada. Water Resour Res 36:1069–1078

Blanken PD, Rouse WR, Schertzer WM (2003) The enhancement of evaporation from a large northern lake by the entrainment of warm, dry air. J Hydrometeorol 4:680–693

Blanken PD, Rouse WR, Schertzer WM (2007) The time scales of evaporation from Great Slave Lake. (Vol. II, this book)

Brown RD, Brasnett B, Robinson D (2003) Gridded North American monthly snow depth and snow water equivalent for GCM evaluation. Atmos Ocean 41:1–14

Bussières N, Schertzer WM (2003) The evolution of AVHRR-derived water temperatures over lakes in the Mackenzie Basin and hydrometeorological applications. J Hydrometeorol 4:660–672

Duguay CR, Flato GM, Jeffries MO, Ménard P, Morris K, Rouse WR (2002) Ice cover variability on shallow lakes at high latitudes: model simulations and observations. Hydrol Process 17:3465–3483

Duguay CR, Prowse TD, Bonsal BR, Brown RD, Lacroix MP, Ménard P (2006) Recent trends in Canadian lake ice cover. Hydrol Process 20:781–801

Feng J, Leighton HG, Mackay MD, Bussières N, Hollman R, Stuhlman R (2002) A comparison of solar radiation budgets in the Mackenzie River Basin from satellite measurements and a regional climate model. Atmos Ocean 40:221–232

Hodges BR, Imberger J, Saggio A, Winters KB (2000) Modelling basin-scale internal waves in a stratified lake. Limnol Oceanogr 45:1603–1620

Johnson L (1975) Physical and chemical characteristics of Great Bear Lake, Northwest Territories. J Fish Res Board Can 32:1971–1987

León LF, Lam DCL, Schertzer WM, Swayne D (2005) Lake and climate models linkage: a 3-D hydrodynamic contribution. Adv Geosci 3:1–6

McCormick MJ, Lam DCL (1999) Potential climate change effects on great lakes hydrodynamics and water quality. In: Lam DCL, Schertzer WM (eds) Lake thermodynamics. Am Soc Civil Eng (ASCE), Reston, Virginia, Chap 3, pp 1-20

Melville GT (1997) Climate change and yield considerations for cold-water fish based on measures of thermal habitat: lake trout in the Mackenzie Great Lakes. Aqu Ecosys Sect, Environ Br, Sask Res Council, Mackenzie Basin impact study final report, pp189–200

Ménard P, Duguay CR, Flato GM, Rouse WR (2002) Simulation of ice phenology on Great Slave Lake, Northwest Territories, Canada. Hydrol Process 16:3691–3706

Oke TR (1987) Boundary layer climates. Methuen, London and New York

Oswald CJ, Rouse WR (2004) Thermal characteristics and energy balance of various-size Canadian Shield lakes in the Mackenzie River Basin. J Hydrometeorol 5:129–144

Oswald CJ, Rouse WR, Binyamin J (2007) Modeling lake energy fluxes in the Mackenzie River Basin using bulk aerodynamic mass transfer theory. (Vol. II, this book)

Rawson DS (1950) The physical limnology of Great Slave Lake. J Fish Res Board Can 8:3–66

Rouse WR, Binyamin J, Blanken PD, Bussières N, Duguay CR, Oswald CJ, Schertzer WM, Spence C (2007) The influence of lakes on the regional energy and water balance of the central Mackenzie River Basin. (Vol. I, this book)

Rouse WR, Oswald CJ, Binyamin J, Blanken PD, Schertzer WM, Spence C (2003) Interannual and seasonal variability of the surface energy balance and temperature of central Great Slave Lake. J Hydrometeorol 4:720–730

Rouse WR, Oswald CJ, Binyamin J, Spence C, Schertzer WM, Blanken PD, Bussières N, Duguay C (2005) The role of northern lakes in a regional energy balance. J Hydrometeorol 6:291–305

Schertzer WM (1999) Physical limnological characteristics of Lake Erie and implications of climate changes. In: Munawar M, Edsall T, Munawar IF (eds) Ecovision world mono series. Backhuys Publ, Lieden, The Netherlands, pp 31–56

Schertzer WM (2003) Physical limnology and hydrometeorological characteristics of Lake Ontario with consideration of climate impacts. In: Munawar M (ed) State of Lake Ontario: past, present and future. Ecovision world monograph series. Goodword Books Pvt Ltd, New Delhi

Schertzer WM, Rouse WR, Blanken PD (2000) Cross-lake variation of physical limnological and climatological processes of Great Slave Lake. Phys Geogr 21:385–406

Schertzer WM, Rouse WR, Blanken PD, Walker AE (2003) Over-lake meteorology, thermal response, heat content and estimate of the bulk heat exchange of Great Slave Lake during CAGES (1998–99). J Hydrometeorol 4:649–659

Schertzer WM, Rouse WR, Blanken PD, Walker AE, Lam D, León L (2007) Interannual variability in heat and mass exchange and thermal components of Great Slave Lake. (Vol. II, this book)

Serreze MC, Barry RG (2005) The arctic climate system. Cambridge University Press, Cambridge

Smith LC, Sheng Y, MacDonald GM, Hinzman LD (2005) Disappearing arctic lakes. Science 308, p 1429

Stewart RE, Leighton HG, Marsh P, Moore GWK, Ritchie H, Rouse WR, Soulis ED, Strong GS, Crawford RW, Kochtubajda B (1998) The Mackenzie GEWEX Study: the water and energy cycles of a major North American river basin. B Am Meteorol Soc 79:2665–2683

Walker A, Silis A, Metcalf JR, Davey MR, Brown RD, Goodison BE (2000) Snow cover and lake ice determination in the MAGS region using passive microwave satellite and conventional data. In: Strong GS, Wilkinson YML (eds) Proc 5th Sci Workshop Mackenzie GEWEX Study, Edmonton, AB, Canada, pp 39–42

Woo MK, Rouse WR, Stewart RE, Stone J (2007) The Mackenzie GEWEX Study: a contribution to cold region atmospheric and hydrologic sciences. (Vol. I, this book)

Woo MK, Thorne R (2003) Streamflow in the Mackenzie Basin, Canada. Arctic 56:328–340

Chapter 9

Modeling Lake Energy Fluxes in the Mackenzie River Basin using Bulk Aerodynamic Mass Transfer Theory

Claire J. Oswald, Wayne R. Rouse and Jacqueline Binyamin

Abstract Multiple years of micrometeorological and energy flux measurements for four Canadian Shield lakes were used to develop bulk aerodynamic mass transfer coefficients (C_D) for each lake and for groups of lakes. Transfer coefficients determined from multiple years of data for the two smallest lakes were similar (1.26 x 10^{-3} and 1.30 x 10^{-3}) while that for the largest lake was slightly smaller (1.10 x 10^{-3}). The coefficient for the medium-size lake was erroneously high (2.14 x 10^{-3}) likely due to generalizations in the calculation of heat storage. No strong relationships were found between the coefficient values and morphometric parameters. The linear regression comparison of measured and modeled daily fluxes using multi-year coefficients gave an average r^2 of 0.78. The same coefficients performed the best at estimating cumulative latent and sensible heat loss. Absolute percent errors suggest that the multi-year coefficients give acceptable results for small and medium-size lakes only, and that these coefficients cannot be transferred from one lake to another, unless the lakes are similar in size.

1 Introduction

This study uses multiple years of micrometeorological and energy balance data for four northern lakes to evaluate the potential for using the bulk aerodynamic mass transfer method to model surface turbulent fluxes. The ultimate goal is to provide lake and climate modelers with bulk aerodynamic mass transfer coefficients (C_x) that allow modeling of latent (Q_E) and sensible (Q_H) heat fluxes from lakes using meteorological data, such as horizontal wind speed, air, and water temperatures.

On the Canadian Shield portion of the Mackenzie River Basin (MRB), where this study was carried out, there are extensive systems of interconnected lakes and rivers. Lakes are high-energy exchange systems that respond readily to climate variability and display a distinctive seasonality in their energy budgets (Rouse et al. 2007). Q_E and Q_H are integral components of the surface energy balance. These fluxes can be measured directly using the eddy covariance method, but the instrumental requirement limits it from being widely used. Due to the expense of equipment, field logistics, access to lakes, complexity of water–atmosphere interaction and the effect of lake size on the wind profile, this approach is not always feasible. Instead of direct measurements, micrometeorological methods are often used to estimate the turbulent fluxes. Profile methods that estimate daily flux values include the Bowen Ratio Energy Balance (BREB) and aerodynamic methods. The bulk aerodynamic mass transfer method outlined in this chapter is the simplest way to estimate Q_E and Q_H. Horizontal wind speed, relative humidity, and air temperature are measured at one height above the surface, and surface water temperature is used to complete the profile. No radiation, heat storage, or complex wind measurements are necessary. A previously determined dimensionless transfer coefficient is then used to express these turbulent fluxes.

This study provided quality data on the latent and sensible heat fluxes over multiple lake surfaces. The dataset is unique in that measurements were obtained concurrently for multiple lakes and years in the same subarctic climatic region. When considering heat fluxes, the size of lakes is of concern because of their differences in heat storage capability and surface roughness which affects the wind profile. The design of this study addresses the question of whether bulk transfer coefficients can be generalized based on lake size.

2 Study Area

The study area is located north of Yellowknife, Northwest Territories (see Fig. 1 in Rouse et al. 2007). The region has subarctic continental climate, with relatively short, cool summers and long, cold winters (Environment Canada 1993). The Shield is characterized by exposed bedrock, subarctic boreal forest, discontinuous permafrost, and many lakes and wetlands formed by glacial scour and deposition (Spence and Woo 2007). Lakes, not including Great Slave Lake, cover approximately 23% of the total area. Of this 23%, small (<1 km^2), medium-size (1 to 50 km^2), and large (>50 km^2) lakes make up an estimated 39%, 52%, and 9% of the total lake area, respectively. The experimental lakes are typical of the range of lake sizes and shapes found in the region (Table 1; Fig. 1). Gar Lake (GR) is a small lake located within the city limit of Yellowknife and is the only lake that freezes to the bottom during the winter. Skeeter Lake (SK) is about 130 km north of Yellowknife and the shape of the basin is oval and regular.

Sleepy Dragon Lake (SD), located 95 km northeast of Yellowknife, has a maximum depth of ~30 m in its eastern basin and a maximum depth of ~12 m in its western basin. For the large Great Slave Lake (GS), the study site is at the rocky Inner Whaleback Islands (IWI) located in the main body of the lake. The depth of water surrounding the site is about 75 m (Schertzer 2000). The island is approximately 10 m above the mean water surface, and its width and length are 100 m and 180 m, respectively. Fetch to the nearest shoreline exceeds 12 km in all directions (Blanken et al. 2000). The main body of the lake typically becomes ice-free between June 1 and June 25 and refreezes between December 1 and January 1 (Rawson 1950; Walker et al. 2000).

Table 1. Locations and dimensions of experimental study lakes. GS values are for the main-lake, excluding the East Arm (GS values from Schertzer, 2000). Asterisk (*) indicates maximum seasonal values

	GS	SD	SK	GR
Latitude	61°55′ N	62°55′ N	63°35′ N	62°31′ N
Longitude	113°44′ W	112°55′ W	113°53′ W	114°22′ W
A_0 [km^2]	1.85×10^4	5.461	0.047	0.296
z_{max} [m]	168.7	35	*6.6	*0.9
z_{mean} [m]	32.2	12	*3.2	*0.5
V [m^3]	5.96×10^{11}	4.20×10^7	*1.80×10^5	*1.40×10^5

A_0 surface area, z_{max} maximum depth, z_{mean} mean depth, and V volume.

3 Methods

3.1 Instrumentation and Measurements

Climate towers were set up on each of the study lakes in specific locations to optimize fetch requirements necessary to achieve surface boundary layer adjustment (Pasquill 1972). The height of the instruments above the homogeneous lake surface did not exceed 1% of the fetch in most directions. Meteorological instrumentation had been installed on the IWI since 1997 (Schertzer et al. 2000) and the measurements are considered representative of a fully adjusted boundary layer at GS; but the small and medium-size lakes have developing boundary layers when the wind blows from directions of limited fetch. At GR and SK the towers were mounted on floating platforms securely anchored to the lake bottom to prevent drift

(b)

Fig. 1. Photographs of four study lakes: (a) Gar, (b) Skeeter, (c) Sleepy Dragon, and (d) Great Slave with Whaleback Island study site shown. (Photos: C. Oswald)

(c)

(d)

Fig. 1. (cont.)

or sway. The SD and GS towers were anchored to rock islands that were small enough not to influence the micrometeorological measurements.

Air temperature, relative humidity, and horizontal wind speed required by the bulk transfer method were recorded at 1.8, 1.8, 3.3, and 18.0 m above the water surface for GR, SK, SD, and GS, respectively. Surface water temperature measurements were taken at each site in the mid-lake area using Tidbit sealed optical communication self-logging temperature sensors. The saturation vapor pressure at height z was calculated according to Bolton (1980) and vapor pressure was calculated as proportional to the relative humidity. Vapor pressure at the water surface was calculated as the saturation vapor pressure at the temperature of the water surface. All sensors were calibrated prior to each field season.

A bathymetric map of each lake provided the morphometric data necessary for inter-lake comparisons (Oswald and Rouse 2004). Bathymetric maps were already available for the main body of GS (Schertzer 2000) and for SK (Spence et al. 2002). Bathymetric surveys were carried out at GR and SD in August 2000.

3.2 Estimation of Turbulent Heat Fluxes

The BREB method was used to estimate the convective energy fluxes for GR, SK, and SD (Oswald and Rouse 2004). This method invokes the gradient–flux relationships for heat and water vapor and requires measurements of radiation, heat storage within the water body, and meteorological parameters at multiple heights above the lake surface. The convective fluxes were estimated by apportioning the available energy between the latent and sensible terms by using their ratio. The eddy covariance method was used for GS (Blanken et al. 2000). This method involved measuring components of the wind vector with a three-dimensional sonic anemometer, air temperature with a fine wire thermocouple, and water vapor concentration with an open path sensor, all at the same height. Water vapor fluxes were calculated as the covariance between the vertical wind speed and the water vapor concentration.

3.3 Method for Determining Transfer Coefficients

The aerodynamic method uses the concept of eddy motion transfer of a property such as water vapor or sensible heat. The aerodynamic equations for latent and sensible heat fluxes are written as:

$$Q_E = -L_v k^2 z^2 \left(\frac{\overline{\Delta u}}{\Delta z} \cdot \frac{\overline{\Delta \rho_v}}{\Delta z} \right) (\phi_M \phi_V)^{-1} \qquad (1)$$

and

$$Q_H = -C_a k^2 z^2 \left(\frac{\overline{\Delta u}}{\Delta z} \cdot \frac{\overline{\Delta T}}{\Delta z} \right) (\phi_M \phi_H)^{-1} \qquad (2)$$

where L_V is the latent heat of vaporization (J kg^{-1}), k is the von Karman's constant ($\cong 0.40$), z is height (m), ρ_v is vapor density (kg m^{-3}), C_a is the heat capacity of air (J m^{-3} K^{-1}), u is the horizontal wind speed (m s^{-1}), T is air temperature (°C or K), ϕ_V and ϕ_H are dimensionless stability functions for water vapor and heat, and ϕ_M is the dimensionless stability function to account for curvature of the logarithmic wind profile due to buoyancy effects (Oke 1999, p. 381). Since $C_a = \rho C_P$ and $L_v = C_a/\gamma = \rho C_p/\gamma$ where γ is the psychrometric constant (g m^{-3} K^{-1}), the aerodynamic Eqs. (1) and (2) become

$$Q_E = \left(\frac{\rho C_p}{\gamma} \right) C_D u_z (e_0 - e_z) \qquad (3)$$

and

$$Q_H = (\rho C_p) C_D u_z (T_0 - T_z) \qquad (4)$$

where T and e are temperature and vapor pressure, and subscripts 0 and z denote water surface and a height z above water. Then,

$$C_D = \frac{k^2}{[\ln(z/z_0)]^2 \phi_M \phi_V, \phi_H} = \left(\frac{\gamma}{\rho C_P} \right) \frac{Q_E}{u_z(e_0 - e_z)} = \left(\frac{1}{\rho C_p} \right) \frac{Q_H}{u_z(T_0 - T_z)} \qquad (5)$$

C_D is the dimensionless aerodynamic drag coefficient for momentum and is dependent on the reference height z. The unmodified aerodynamic method described by the above equations only applies under neutral atmospheric stability where the stability functions are set to 1. In reality there are also specific transfer coefficients for water vapor (C_E) and sensible heat (C_H). These coefficients represent the efficiency of turbulent transfer between the surface and the reference height. Studies over large lakes indicate that $C_D \geq C_E \approx C_H$ (Heikinheimo et al. 1999). Table 2 gives several examples of C_{xz} values where x represents the subscripts for latent heat (E), sensible heat (H), or momentum (D). The values in Table 2 are suitable for the sites and time periods for which they were developed, however, errors are to be expected when the equations are applied to other studies without recalibrating the coefficients. Daily C_E and C_H can be calculated from both of the ratios shown in Eq. (5) using measurements of temperature and vapor pressure, and independent estimates of Q_E and Q_H. Daily transfer coefficients vary with surface roughness and atmospheric stability so that the frequency distribution of the data is skewed. Because the data are not normally distributed, median C_E and C_H values were calculated for each lake for each year and were combined as follows:

- Lake-specific *yearly* coefficients (C_D, for simplicity) were derived from weighted averages of C_E and C_H values.
- Lake-specific *multi-year* coefficients were derived from weighted averages of yearly C_D values.
- A weighted average was applied to the lake-specific multi-year coefficients, which were grouped in two ways: (1) small: GR and SK; medium-size: SD; large: GS; and (2) small/medium-size: GR, SK, and SD; large: GS.
- Standard error of the median and weighted standard error of the mean (S_e) were calculated to quantify the precision of C_D values (Freund and Simon 1997, p. 306).

Averages were weighted according to the number of days for which there were reliable data.

Table 2. Selected mass transfer coefficients for momentum, latent heat and sensible heat (C_{Dz}, C_{Ez}, and C_{Hz}, respectively; z is the height of the measurements above the water surface in meters). These coefficients and others are summarized in Heikinheimo et al. (1999)

Reference	$C_{xz} \times 10^{-3}$
Brutsaert (1982)	C_{E10} = 1.18 to 1.34
Heikinheimo et al. (1999)	C_{D3} = 1.81
	C_{E3} = 1.07
	C_{H3} = 1.40
Smith et al. (1996)	$C_E \approx C_H \approx 1.10$
Stauffer (1991)	C_{E10} = 1.35
Strub and Powell (1987)	C_{E10} = 1.90

4 Results and Discussion

4.1 Coefficient Development

The C_E, C_H, and C_D values are calculated for each lake for each year (Table 3) and the main results are listed below.

- Based on S_e, C_E and C_H are similar for each lake, except for four datasets (GR 2000, 2001 and GS 1998, 1999).
- As expected, there is no apparent temporal trend in the yearly coefficients (Fig. 2).
- For all lakes there are at least two years with similar yearly coefficients (based on S_e) (Table 3 and Fig. 2).
- The multi-year C_D for SD is 65 to 95% larger than the multi-year C_D values for the other three lakes (Fig. 3). There is no statistical difference between the multi-year C_D values for GR, SK and GS.
- No strong patterns in multi-year coefficients exist that can be attributed to morphometric factors (Fig. 4).
- The multi-lake C_D value for small lakes is 67% lower than the coefficient for SD, but is statistically similar to the coefficient for GS. The multi-lake C_D for the small/medium-size grouping is 39% higher than the C_D for GS.

The values of the transfer coefficients are consistent with typical reference values for water surfaces (Table 2). However, the transfer coefficient for SD seems too high given the dependence of the coefficient on lake surface area as proposed by Harbeck (1962). Smaller lakes are more likely to be influenced by advection of air from the surrounding land than are larger lakes. Consequently, the small lake boundary layer is often in a state of development, as opposed to being fully equilibrated with the surface as is often the case for large lakes. It follows that more heat will be lost from a lake with a developing boundary layer, and hence the exchange coefficient should be smaller for larger lakes (Heikinheimo et al. 1999).

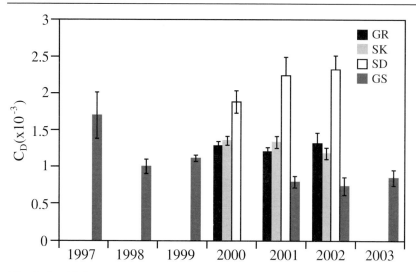

Fig. 2. Multi-flux weighted average bulk mass transfer coefficients (C_D) for Gar Lake (GR), Skeeter Lake (SK), Sleepy Dragon Lake (SD), and Great Slave Lake (GS). Error bars represent weighted standard error of the mean

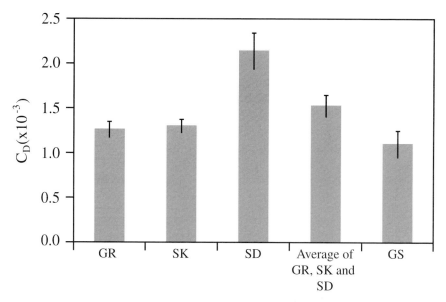

Fig. 3. Multi-year weighted average bulk mass transfer coefficients (C_D) for all lakes. AVG (GR, SK, SD) and GS are multi-lake coefficients. Error bars represent the weighted standard error of the mean

Table 3. Experimentally-determined transfer coefficients (CE or CH) measured at a height of 1.8 m above the water surface, for latent and sensible heat fluxes. Columns (a), (b) and (c) refer respectively to the yearly, multi-year and multi-lake weighted averages of the coefficients and their associated standard errors (in brackets)

Lake	Year	Coeff.	No. days	Median and stand. error ($\times 10^{-3}$)	Mean and standard error ($\times 10^{-3}$)			
					(a)	(b)		(c)
GR	2000	C_E	101	1.26 (0.03)	1.26 (0.06)	1.26 (0.09)	1.28 (0.07)	1.53 (0.12)
		C_H	25	1.39 (0.08)				
	2001	C_E	48	1.24 (0.06)	1.21 (0.06)			
		C_H	45	1.17 (0.06)				
	2002	C_E	16	1.33 (0.13)	1.32 (0.15)			
		C_H	15	1.31 (0.16)				
SK	2000	C_E	53	1.34 (0.06)	1.35 (0.06)	1.30 (0.07)		
		C_H	52	1.37 (0.06)				
	2001	C_E	22	1.34 (0.08)	1.33 (0.08)			
		C_H	24	1.33 (0.08)				
	2002	C_E	28	1.14 (0.09)	1.18 (0.08)			
		C_H	34	1.22 (0.07)				
SD	2000	C_E	66	1.94 (0.15)	1.88 (0.15)	2.14 (0.20)	2.14 (0.20)	
		C_H	60	1.82 (0.16)				
	2001	C_E	31	2.21 (0.20)	2.24 (0.25)			
		C_H	28	2.28 (0.30)				
	2002	C_E	73	2.29 (0.19)	2.33 (0.19)			
		C_H	66	2.36 (0.19)				
GS	1997	C_E	44	1.76 (0.17)	1.70 (0.31)	1.10 (0.15)	1.10 (0.15)	1.10 (0.15)
		C_H	17	1.54 (0.48)				
	1998	C_E	13	1.31 (0.12)	1.00 (0.10)			
		C_H	16	0.76 (0.05)				
	1999	C_E	27	1.24 (0.05)	1.12 (0.04)			
		C_H	33	1.02 (0.03)				
	2001	C_E	24	0.78 (0.07)	0.79 (0.08)			
		C_H	28	0.80 (0.08)				
	2002	C_E	21	0.77 (0.09)	0.74 (0.12)			
		C_H	11	0.67 (0.16)				
	2003	C_E	15	0.84 (0.10)	0.85 (0.11)			
		C_H	12	0.86 (0.11)				

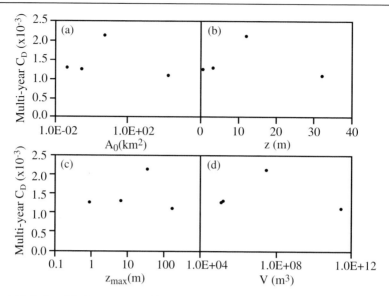

Fig. 4. Relationship between multi-year C_D values and the morphometric parameters (a) surface area (A_0), (b) mean depth ($\bar{z}$), (c) maximum depth (z_{max}), and (d) volume (V). Note the logarithmic scale used on the horizontal-axis

4.2 Coefficient Verification

Splitting the datasets into independent test periods to minimize bias allowed for rigorous verification of the derived coefficients. Lake-specific yearly coefficients were averaged for all years except one and the result was used to model data from the year not included. For example, a weighted average of the C_D values of years 2000 and 2001 for GR (from Table 3) resulted in a multi-year C_D of 1.25×10^{-3}, which was used to model daily fluxes in 2002. The performance of the mass transfer approach and the yearly coefficients was evaluated by comparing the measured and modeled daily fluxes. Figure 5 shows reasonable agreement between measured and modeled fluxes for all lakes except SD. Linear regression analysis was used to quantify the relationship (Table 4). If an r^2 value of 0.60 is used as an arbitrary level of acceptance, all coefficients give satisfactory results except for four datasets (SK Q_H 2000, SD Q_E 2000, and SD Q_E and Q_H 2002). The average r^2 value of 0.78 indicates that 78% of the variability in the modeled data can be explained by the variability in the

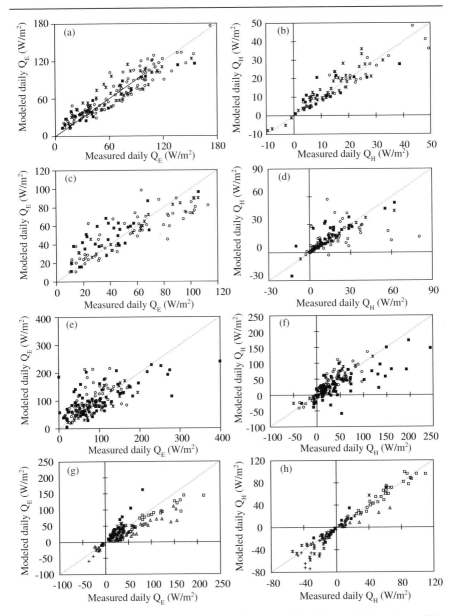

Fig. 5. Comparison between daily measured and modeled latent (Q_E) and sensible (Q_H) heat fluxes for (a) GR Q_E, (b) GR Q_H, (c) SK Q_E, (d) SK Q_H, (e) SD Q_E, (f) SD Q_H, (g) GS Q_E, and (h) GS Q_H for all years. (Δ 1997; + 1998; □ 1999; ○ 2000; * 2001; ■ 2002; - 2003)

Table 4. Intercept and slope values of linear regression equations and r^2 values for comparison of modeled (dependent variable) and measured (independent variable) daily sensible (Q_H) or latent (Q_E) heat fluxes. Modeled data use transfer coefficient values averaged from multiple years

Lake	Flux	Year	Intercept	Slope	r^2
GR	Q_E	2000	9.77	0.87	0.88
		2001	15.5	0.80	0.79
		2002	13.0	0.72	0.91
	Q_H	2000	2.49	0.83	0.87
		2001	1.09	1.03	0.86
		2002	5.08	0.62	0.69
SK	Q_E	2000	15.6	0.66	0.72
		2001	13.6	0.72	0.92
		2002	19.7	0.73	0.64
	Q_H	2000	9.48	0.31	0.17
		2001	1.63	0.80	0.97
		2002	5.18	0.84	0.73
SD	Q_E	2000	45.2	0.66	0.29
		2001	19.0	0.75	0.70
		2002	54.7	0.50	0.49
	Q_H	2000	8.65	1.03	0.74
		2001	0.84	0.98	0.87
		2002	2.10	0.60	0.59
GS	Q_E	1997	-1.98	0.60	0.85
		1998	-8.53	1.06	0.94
		1999	8.09	0.70	0.89
		2001	1.27	1.25	0.88
		2002	-5.49	1.85	0.90
		2003	5.26	1.16	0.83
	Q_H	1997	-4.48	0.65	0.94
		1998	-0.51	1.43	0.77
		1999	1.21	0.77	0.94
		2001	-3.38	1.11	0.79
		2002	1.23	1.13	0.92
		2003	-1.39	1.14	0.92

measured. It is important to note that the variability in the modeled data is largely a function of errors in the meteorological input.

As well as verifying the use of mass transfer theory for determining daily turbulent fluxes, the ability of the method to estimate cumulative latent and sensible heat loss over the measurement period was tested. Relative error was used to indicate the accuracy of the model where

$$\text{Relative error (\%)} = \left[\frac{\text{modeled value} - \text{measured value}}{\text{measured value}}\right] \times 100 \qquad (6)$$

Table 5 shows the results of this testing using yearly coefficients (column a), multi-year small vs. medium-size lakes (column b), and multi-year small/medium-size lake (column c). The coefficients used for column a are the same as those for the regression analysis. For column b, the weighted average coefficient for small lakes (Table 3) was tested on SD and the multi-year coefficient for SD was tested on GR and SK. For column c, a weighted average of multi-year coefficients from two of the three lakes in the small/medium-size grouping was used to model data from the lake not included.

Column a in Table 5 yields absolute percent errors of at best 1%, on average 11%, and at worst 36% for the small/medium-size lakes and at best 1%, on average 43%, and at worst 101% for the large lake, GS. Average absolute percent errors increase with increasing lake depth and volume (GR: 5%, SK: 13%, SD: 17%, GS: 43%) indicating that this method may perform better for smaller lakes. There are factors operating on GS that render the use of a multi-year coefficient unreliable. This may be related to large lake boundary layer processes that contradict the fully adjusted boundary layer assumptions of aerodynamic theory (Blanken et al. 2003, 2007). As well, the multi-year transfer coefficient for GS may not be applicable for all time periods covered by the datasets. Strubb and Powell (1987) found different C_E values depending on the time of year and associated stability conditions. Column b in Table 5 shows absolute percentage errors of at best 25%, on average 60%, and at worst 94%, which are all higher than column c that inidcates absolute percentage errors of at best 5%, on average 37%, and at worst 60%. If an average relative error of ±20% is used as an arbitrary measure of acceptance, the lake-specific multi-year coefficients yield satisfactory results for all lakes except GS. Both groupings of multi-year coefficients give unsatisfactory results except for SK 2000 for Q_H. These findings indicate that an acceptable level of confidence cannot be achieved when transferring coefficients from one lake to another, unless the lakes are similar in size, such as GR and SK.

Table 5. Transfer coefficient ($\overline{C}_D$) verification quantified by error (in %) in cumulative latent (Q_E) and sensible heat (Q_H) loss estimates. Column (a) Multi-year values where coefficients for a particular lake from all years except one were averaged and then tested on data from the year not included. Column (b) Multi-lake values derived for small group are used for the medium-size lake and vice versa. Column (c) Multi-lake values derived for the small/medium-size group are applied by averaging two lake coefficients and using the result to model fluxes from the lake not included. A negative value indicates that the modeled data underestimates the measured data; N/A = not applicable

Lake	Year	Flux	(a) Multi-year		(b) Multi-lake (S grouped)		(c) Multi-lake (S/M grouped)	
			$\overline{C}_D$	Error	$\overline{C}_D$	Error	$\overline{C}_D$	Error
GR	2000	Q_E	1.26	1	2.14	71	1.72	37
		Q_H		-5	(SD)	62	(SK	30
	2001	Q_E	1.30	6		74	/SD)	40
		Q_H		12		84		48
	2002	Q_E	1.25	-4		65		33
		Q_H		2		75		40
SK	2000	Q_E	1.26	-5	2.14	61	1.70	28
		Q_H		-15	(SD)	44	(GR	15
	2001	Q_E	1.27	-4		62	/SD)	29
		Q_H		-9		54		22
	2002	Q_E	1.34	22		94		54
		Q_H		21		93		53
SD	2000	Q_E	2.28	22	1.28	-32	Same as (2)	
		Q_H		33	(GR	-25		
	2001	Q_E	2.10	-4	/SK)	-42		
		Q_H		1		-39		
	2002	Q_E	2.06	5		-35		
		Q_H		-36		-60		
GS	1997	Q_E	0.90	-44	N/A		N/A	
		Q_H		-101				
	1998	Q_E	1.04	-46				
		Q_H		45				
	1999	Q_E	1.02	-22				
		Q_H		-1				
	2001	Q_E	1.08	33				
		Q_H		33				
	2002	Q_E	1.09	62				
		Q_H		50				
	2003	Q_E	1.07	43				
		Q_H		32				

4.3 Modeling Considerations

4.3.1 Representativeness of Lakes

One major question to consider is whether the lakes in this study are representative of all lakes in the MRB. Although each lake size category is represented, the number of samples is obviously not optimal. There are also characteristics of each lake that may be site-specific. For example, lily pads grow over the surface of GR throughout the summer, and this may affect the penetration of solar radiation and the amount of heat stored. Presumably, not all lakes of similar area and depth have the same vegetation growth as GR. SK is surrounded by steep bedrock uplands; hence it is subject to the advection of warm air from the surrounding land. This could influence measured results although efforts were made to minimize the effects by eliminating data when wind direction and air temperature suggested advective conditions. SD has two distinct basins that differ in maximum depth by 23 m and hence have different heat storage capacities. Turbulent energy fluxes may also vary between basins, which may not be the case for single basin medium-size lakes. The only other large lake near the size of GS in the MRB is Great Bear Lake, which is located north of GS and experiences very different climatic conditions.

4.3.2 Potential Sources of Error

The approach used in this study assumes neutral stability conditions and similarity of the coefficients for Q_E and Q_H (Oke 1999). Differences between C_E and C_H in Table 3 may be due to errors in the flux measurements or the use of different days to calculate the coefficients. Annual variability in the coefficients may be due to the representation of lake storage by a point measurement in the center of the lake and the propagation of this generalization through to the calculation of the turbulent fluxes using the BREB approach (only at small and medium-size lakes). Any errors associated with the calculation of Q_E and Q_H that arise from measurement inaccuracies will ultimately affect the derived coefficients.

The BREB method of calculating Q_E and Q_H invokes the measurement of the change in lake heat storage (ΔQ_S). In this study, ΔQ_S was estimated using temperature measured over the entire depth profile from one location in each lake. There are possibilities of over or underestimating ΔQ_S depending on the depth of the lake at the point of measurement and the representativeness of that depth for the rest of the lake. An underestimation of ΔQ_S would cause an overestimate of the energy available for exchange with the atmosphere and an erroneously high exchange coefficient would result. The eddy covariance method of determining the turbulent heat fluxes, which was used at GS, is preferable since it does not require the measurement of heat storage. However, there remains the problem of using a point measurement of surface temperature for an entire lake as surface temperature is a key element that influences the determination of Q_E and Q_H (LeDrew and Reed 1982). This is especially problematic on large lakes, where surface temperatures can differ greatly (Schertzer et al. 2000, 2007) and the point of measurement is not always representative of the upwind region. The measurement errors engendered by these factors are difficult to estimate accurately.

5 Conclusions

Aerodynamic mass transfer coefficients were developed and evaluated using climatological data from four different-sized lakes in the MRB. Yearly, multi-year, and multi-lake coefficients were calculated using bulk aerodynamic mass transfer theory. The coefficients are consistent with other values reported in literature; but generalizations made in the calculation of heat storage at the medium-size lake may be responsible for the inflation of its transfer coefficient. A comparison of measured and modeled daily fluxes indicates that lake-specific multi-year coefficients provide acceptable results. Further verification of the coefficients suggests that once the coefficients are determined using historical data they can be used with sufficient confidence to estimate cumulative latent and sensible heat loss over the ice-free period for future years for the same lake. However, the transferability of multi-year coefficients to another lake or the use of a multi-lake coefficient cannot attain an acceptable level of confidence unless the lakes are of similar size.

Unique to this study is the simultaneous measurement of meteorological variables for the computation of turbulent fluxes for four nearby lakes in the subarctic Shield. Although these lakes are generally representative of their size categories found in the MRB, further research on medium-size lakes is warranted. It is expected that the findings for small lakes in this study will be applicable to other northern regions. The lake-specific coefficients for the small lakes should prove useful to those attempting to model daily fluxes or seasonal heat loss. The transfer coefficient for Great Slave Lake should also provide a starting point for lake modelers, though comparison with other large lake values is recommended.

Acknowledgements

The authors wish to thank all individuals who helped with aspects of this work, in particular John Davies, Chris Spence, and Peter Brown.

References

Blanken PD, Rouse WR, Culf AD, Spence C, Boudreau LD, Jasper JN, Kochtubajda B, Schertzer WM, Marsh P, Verseghy D (2000) Eddy covariance measurements of evaporation from Great Slave Lake, Northwest Territories, Canada. Water Resour Res 36:1069–1077

Blanken PD, Rouse WR, Schertzer WM (2003) Enhancement of evaporation from a large northern lake by the entrainment of warm, dry air. J Hydrometeorol 4:680–693

Blanken PD, Rouse WR, Schertzer WM (2007) The time scales of evaporation from Great Slave Lake. (Vol. II, this book)

Bolton D (1980) The computation of equivalent potential temperature. Mon Weather Rev 108:1046–1053

Environment Canada (1993) Canadian climate normals 1961–90. Ministry of Supply and Services Canada, Ottawa

Freund JE, Simon GA (1997) Modern elementary statistics. Prentice Hall, New Jersey

Harbeck GE (1962) A practical field technique for measuring reservoir evaporation utilizing mass-transfer theory. US Geological Survey Professional Paper 272-E, pp 101–105

Heikinheimo M, Kangas M, Tourula T, Venalainen A, Tattari S (1999) Momentum and heat fluxes over lakes Tamnaren and Raksjo determined by the bulk-aerodynamic and eddy-correlation methods. Agr Forest Meteorol 98:521–534

LeDrew EF, Reid PD (1982) The significance of surface temperature patterns on the energy balance of a small lake in the Canadian Shield. Atmos Ocean 20:101–115

Oke TR (1999) Boundary layer climates. Routledge, Cambridge

Oswald CJ, Rouse WR (2004) Thermal characteristics and energy balance of various-size Canadian Shield lakes in the Mackenzie River Basin. J Hydrometeorol 5:129–144

Pasquill F (1972) Some aspects of boundary layer description. Q J Roy Meteor Soc 98(417):469–494

Rawson DS (1950) The physical limnology of Great Slave Lake. J Fish Res Board Can 8:3–66

Rouse WR, Binyamin J, Blanken PD, Bussières N, Duguay CR, Oswald CJ, Schertzer WM, Spence C (2007) The influence of lakes on the regional energy and water balance of the central Mackenzie River Basin. (Vol. I, this book)

Rouse WR, Oswald CJ, Binyamin J, Spence C, Schertzer WM, Blanken PD, Bussières N, Duguay C (2004) The role of northern lakes in a regional energy balance. J Hydrometeorol 6:291–305

Schertzer WM (2000) Digital bathymetry of Great Slave Lake. National Water Research Institute Report, Contrib no 00–257

Schertzer WM, Rouse WR, Blanken PD (2000) Cross-lake variation of physical limnological and climatological processes of Great Slave Lake. Phys Geogr 21:385–406

Schertzer WM, Rouse WR, Blanken PD, Walker AE, Lam D, León L (2007) Interannual variability in heat and mass exchange and thermal components of Great Slave Lake. (Vol. II, this book)

Smith SD, Fairwall CW, Geernaert GL, Hasse L (1996) Air-sea fluxes, 25 years of progress. Bound-Lay Meteorol 78:247–290

Spence C, Rouse WR, Worth D, Oswald CJ, Reid R (2002) Energy budget processes of a small northern lake. J Hydrometeorol 4:694–701

Spence C, Woo MK (2007) Hydrology of the northwestern subarctic Canadian Shield. (Vol. II, this book)

Stauffer RE (1991) Testing lake energy budget models under varying atmospheric stability conditions. J Hydrol 128:115–135

Strubb PT, Powell TM (1987) Exchange coefficients for latent heat and sensible heat flux over lakes; dependence upon atmospheric stability. Bound-Lay Meteorol 40:349–361

Walker A, Silis A, Metcalf JR, Davey MR, Brown RD, Goodison BE (2000) Snow cover and lake ice determination in the MAGS region using passive microwave satellite and conventional data. Proc. 5th Scientific Workshop for the Mackenzie GEWEX Study, Edmonton, Alberta, pp 39–41

Chapter 10

The Time Scales of Evaporation from Great Slave Lake

Peter Blanken, Wayne Rouse and William M. Schertzer

Abstract Direct measurements of evaporation from Great Slave Lake, Northwest Territories, Canada, over a six-year period are compared in terms of the influence of time scales on the magnitude and control of the process. Based on measurements using the eddy covariance method, both the latent and sensible heat fluxes were consistently small following ice break up, but increased dramatically as ice formation commenced. Half of the total evaporative water loss occurred over only 20% of the observation periods. Large evaporation events occurred over periods typically lasting nearly three days, and were controlled by the product of the horizontal wind speed and the difference between vapor pressure at the water surface and atmosphere. Both a stepwise linear regression and principal component analysis were used to determine the sensitivity of the relationship between evaporation and its controls to various sampling intervals. Varying the sampling interval from one to six hours in 1-hr steps had no significant effect on the coefficients used to predict the 24-hr evaporation totals. The dichotomy between lake and land surfaces in terms of temporal changes in heat source and sink, and the longer duration for processes occurring over the lake than land are discussed.

1 Introduction

The large number and areal coverage of lakes in the Mackenzie River Basin (MRB) indicate that likely play an important role in the Basin's annual water and energy cycles (Rouse et al. 2007). For large, deep lakes such as Great Slave and Great Bear, the linkages and feedbacks between the lakes and the surrounding terrestrial areas result in a dynamic coupled system especially sensitive to climate change. The objective of this study is to better understand the role lakes play in this coupling and the climate system of northern ecosystems.

Pioneering studies were made over the past several years on Great Slave Lake, building upon a foundation study by Rawson (1950). In this chapter, evaporation data from Great Slave Lake are analyzed. The first direct measurements described by Blanken et al. (2000) showed that evaporation occurred over events typically lasting 45 hours, and that 4 °C above-average air temperatures associated with the 1997-98 El Niño caused a delay in ice formation and a substantial increase in evaporative water loss. Schertzer et al. (2000) detailed the cross-lake spatial variation in the bathymetry, water temperature, and meteorology. These studies were followed by investigations of the specific processes behind turbulent exchange (Blanken et al. 2003), the interannual and seasonal variability of the surface energy balance (Rouse et al. 2003a), and the lake's heat content and thermal mixing properties (Schertzer et al. 2003). Studies examining the role of lakes in the regional context of the MRB are provided by Rouse et al. (2003b), Rouse et al. (2005), and Rouse et al. (2007).

These studies conducted over Great Slave Lake have revealed special and unique issues pertaining to the temporal scale of observations related to evaporation and its controls. Rather than the diurnal patterns typically exhibited over terrestrial surfaces, Great Slave Lake shows large evaporative water losses over a typical time scale of nearly three days, yet embedded within these long-duration evaporation events are fast "pulses" of evaporation generated by the entrainment of warm, dry air above the water. This raises the question of whether the relationship between evaporation and its controls varies with the time scale of observation. Such a question has relevance to the popular use of the mass transfer approach for spatial and temporal extrapolation of evaporation measurements. Since the mass transfer coefficient is dependent upon the sampling interval and averaging period of the observations, the sensitivity of the mass transfer coefficient to sampling intervals needs to be addressed. This chapter investigates the influence of time scales on the magnitude and controls of evaporation from Great Slave Lake.

2 Study Area

Physical characteristics of Great Slave Lake can be found in Rawson (1950) and Schertzer et al. (2007). Thermal stratification occurred between mid-July and late August, with a thermocline depth of roughly 15 m. Thermal mixing occurred around July 1 and again around October 1 (dimictic). Lake ice conditions vary significantly from year to year (Schertzer et al. 2007, Walker et al. 2000), but the lake is generally ice-free from early June through mid- to late December.

Field measurements were made at a location central to the main body of Great Slave Lake. Instruments were placed on a small island (Inner

Whalebacks; 61.92°N; 113.73°W) surrounded by waters with a depth that matches the average depth of 50 m (main body of the lake excluding the deep Eastern Arm; Rawson 1950) located roughly 12 km from the nearest shore. A turbulent flux footprint model (Schuepp et al. 1990) was used to quantify the upwind sampling distance contributing to the turbulent flux measurements (Blanken et al. 2000). These calculations indicated that 80% of the turbulent fluxes originated from within 4.9, 5.9, and 8.4 km of the site (daytime, neutral, and nighttime atmospheric stability conditions, respectively). Turbulent flux measurements were most sensitive to upwind distances of 557, 657, and 933 m (daytime, neutral, and nighttime atmospheric stability conditions, respectively), well beyond the maximum island dimensions of roughly 100 by 180 m. Measurements were made for as long as possible during the years 1997, 1998, 1999, 2001, 2002 and 2003.

3 Methods and Instrumentation

The eddy covariance (or correlation) method was used to directly measure the latent (λE) and sensible (H) turbulent heat fluxes from the lake over the upwind distances described above. This method calculates λE and H using the covariance between fluctuations (primes) relative to the temporal means (overbars) between high-frequency measurements of the vertical wind speed (w) and either vapor density (ρ_v) for λE; $\lambda E = \lambda \overline{w' \rho_v'}$ or air temperature (T) for H: $H = \rho_a c_p \overline{w'T'}$ where ρ_a and c_p is the air density and specific heat, respectively, and λ is the latent heat of vaporization.

During 1997, 1998, and 1999, the Mk2 Hydra eddy covariance system (Shuttleworth et al. 1988) was used. The Hydra's one-dimensional sonic anemometer, fine-wire thermocouple, and infrared hygrometer were mounted on a telescopic mast 6.9 m above the ground surface (roughly 17 m above the water surface), well away from any spray or splash from the lake. Signals were sampled at a frequency of 20 Hz with fluxes calculated over a 1-hr averaging period. Additional details of the Hydra system use at Great Slave Lake are provided in Blanken et al. (2000) and Blanken et al. (2003).

In 2001, the Hydra was replaced with another eddy covariance system consisting of a three-dimensional sonic anemometer (model CSAT 3, Campbell Scientific Inc. (CSI), Logan, UT), and an open-path gas analyzer (model LI-7500, Li-Cor Inc., Lincoln, NB). Signals were sampled at 10 Hz with fluxes calculated over a 0.5-hr period using a data logger (model 23X,

CSI, Logan, UT). The instruments were mounted 7 m above the ground surface from a triangular-constructed tower. Standard corrections were applied to the eddy covariance measurements, including coordinate rotation (Kaimal and Finnigan 1994), virtual air temperature correction (Schotanus et al. 1983), and the Webb adjustment (Webb et al. 1980). For both systems, turbulent fluxes away from the surface are considered positive; toward the surface, negative.

Supporting meteorological measurements were made from a separate 9-m tall tower, roughly 18 m above the water surface. Horizontal wind speed (U_{18}) was measured with a prop-vane anemometer (model 5310, R.M. Young, Traverse City, MI), and relative humidity and air temperature were measured with a resistance-based hydrometer (model HMP-35D, Vaisala, Helsinki, Finland). Water surface temperature was measured with an infrared thermometer (model 4000.GL, Everest Interscience, Tucson, AZ) with a 15° field of view. The infrared thermometer was mounted at the end of a horizontal boom which extended 14 m from the shore at a height of 2 m above the water surface over water with a mean depth of 9 m. Signals were sampled every 5 seconds and 15-minute means were stored in a data logger (model CR10X, CSI, Logan, UT). The vapor pressure at a height of 18 m above the water surface (e_{18}) was calculated from the relative humidity and air temperature measurements. The vapor pressure at the water surface (e_0) was calculated as the equilibrium vapor pressure evaluated at the water surface temperature (Buck, 1981).

The plots showing the flux "fingerprints" (Fig. 1) were created by calculating the mean hourly λE and H for each day for each year. Isolines were interpolated and the areas between isolines were filled with different colors. Power spectral densities (Fig. 2) were estimated using Welch's averaged periodogram method (Math Works, Inc., Lowell, MA). A fast Fourier transform length of 2^9 was used with non-overlapping Hanning windows of the same length. Since the length of observations varied for each year, the power was divided by the maximum power to enable comparisons between years. To allow comparison between years of different observation periods, the cumulative evaporation (Fig. 3) was calculated as the cumulative daily total evaporation (E in mm), in descending order, divided by the total E for each year plotted against the number of days (percent) in each year's observations.

The 1-hr time series provided the base data for the calculation of 24-hr means evaluated at various time intervals. For example, the 24-hr mean based on a 1-hr sampling interval was determined simply as the average of 24 numbers (for E, the 24-hr totals were then calculated). The 24-hr mean based on a 2-hr sampling interval was determined by averaging the values

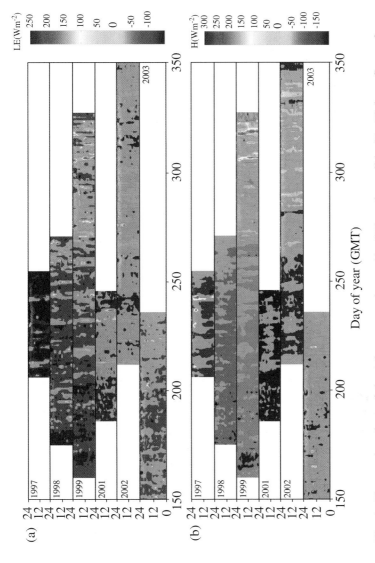

Fig. 1. Fingerprint plots of the 1-hr average latent (A: *LE*) and sensible (B: *H*) heat fluxes from Great Slave Lake over a six-year period

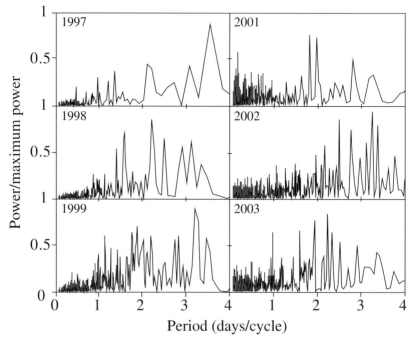

Fig. 2. Power spectrums for the latent heat flux from Great Slave Lake over a six-year period

selected from every other number in the 1-hr time series; the 24-hr mean based on 3-hr sampling interval was derived by sampling every third number, and so on. Means aggregated from means obtained over shorter periods are not affected by the calculations.

Stepwise linear regressions were performed to determine the coefficients and the correlation between the 24-hr total E for each sampling interval and the independent variables (i.e., U_{18}, e_0-e_{18}, and the product of the two). Principal component analysis was used to determine the "loadings" or importance of U_{18}, e_0-e_{18}, and their product in determining the 24-hr total E as a function of different sampling intervals. Since the variables have different units, they were standardized by dividing by their respective standard deviations.

4 Results

Flux "fingerprints" for λE (Fig. 1a) and H (Fig. 1b) for each of the measurement years illustrate how the turbulent fluxes did not follow a consistent, regular diurnal pattern. These fingerprint plots show the hourly averaged fluxes (vertical axis) for each day of each year (horizontal axis), with the magnitude of the flux indicated by the scale on the right. In general, λE was small and positive, and H was small and often negative during roughly half of the ice-free period (ca. before day-of-year 250 or September 7). As the ice-free period progressed (ca. after day-of-year 250), both λE and H increased dramatically. This was shown notably in 1999 and 2002 when the instrumentation continued to operate during the harsh late autumn and early winter (other years possibly show a similar increase in λE and H had the instrumentation remained operational).

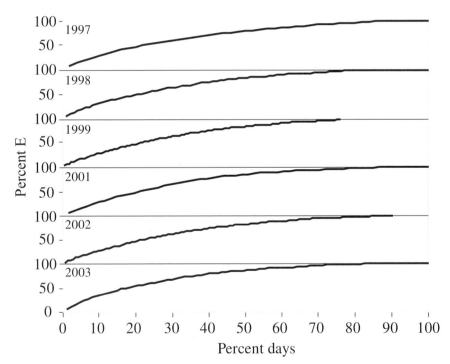

Fig. 3. Percent of cumulative evaporation plotted against percent of the observation days for each year

In addition to shifting from a heat sink in the summer when radiation inputs are at a maximum, to a heat source in the fall/winter when the adjacent terrestrial surfaces are a heat sink, the lake surface also differs from its terrestrial counterparts in terms of the temporal behavior of its turbulent fluxes. Figure 1 shows that neither λE nor H displayed a symmetrical, diurnal pattern which is common to the terrestrial surfaces. Rather, λE and H were quiescent often for several days, then rose steadily over a period of several hours to reach a maximum. After that, the fluxes often subsided and the process repeated itself.

Figure 2 shows the power spectral densities of λE for each year. This type of analysis converts a time series to frequencies using a Fast Fourier transformation to reveal the periodicity of the signal (λE in this case). Figure 2 and Table 1 show that the dominant time period for the λE events for all six years of observations was 2.7 days with a standard deviation of 0.74 days (64.8 hrs ± 17.8 hrs). This is nearly three times the dominant time period for terrestrial turbulent events. These late-season episodic evaporation events were concentrated over short durations. Figure 3 indicates that about 50% of the evaporation in each year occurred over only 20% of the days. This episodic behavior is similar to Lake Ontario, where 50% of the evaporation took place over 18% of the days (Quinn and den Hartog 1981).

Table 1. The dominant period for the latent heat flux from Great Slave Lake for each of the six observation years as indicated by the power spectrums shown in Figure 2

Year	Dominant period [days]
1997	3.6
1998	2.2
1999	3.2
2001	1.8
2002	3.3
2003	2.2
Mean (SD)	2.7 (0.74)

The meteorological forcing that regulates evaporative water loss from this and other large lakes includes the extent of open water (ice-free or covered), horizontal wind speed, and the difference in vapor pressure between the water surface (equilibrium vapor pressure) and the atmosphere. Since periods of high winds are often associated with the intrusion of cold, dry air, the product of U_{18} and e_0-e_{18} is highly correlated with λE (Blanken et al. 2000). Figure 4 shows the linear relationship between $U_{18}(e_0-e_{18})$ and E where E is the total mm of evaporation over 24 hours using data from the first three years when simultaneous measurements of all the variables were available (for clarity, individual data points are not shown). There are six overlapping lines in Fig. 4, each representing a different sampling internal from which the 24-hr means were determined. The sampling intervals of 1, 2, 3, 4, 5 and 6 hours from the 1-hr time series of each variable represent, for example, a measurement that was available at a coarser resolution than perhaps is required to resolve higher frequency processes. Since the dominant period of E events was nearly three days, sampling below this frequency should not dilute any details in the variables necessary to reconstruct the behavior of E.

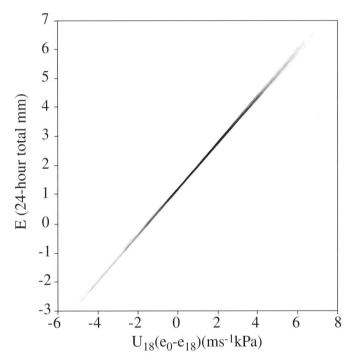

Fig. 4. Linear regression lines showing the relationship between total evaporative water loss over a 24-hr period and the product of horizontal wind speed at 18 m above the water surface and the difference between the vapor pressures at the water surface and at 18 m above the water surface. Six overlapping lines are shown, each representing sampling intervals of 1, 2, 3, 4, 5 and 6 hours. For clarity, individual data points are not shown

Figure 4 and Table 2 show that increasing the sampling interval from one to six hours had no significant effect on the linear regression coefficients for the 24-hr mean U_{18}, e_0-e_{18}, or $U_{18}(e_0-e_{18})$. Therefore, estimates of the 24-hr total E as a function of $U_{18}(e_0-e_{18})$ were not significantly dependent on the sampling interval chosen. Table 3 shows, however, that for each variable, the R^2 decreased and the RMSE increased as the sampling period increased because the sample size decreased. Overall, the highest R^2 and lowest RMSE were obtained when $U_{18}(e_0-e_{18})$ was used to predict the 24-hr total E.

Table 2. Coefficients based on a stepwise linear regression analysis using sampling intervals of various lengths

Sampling interval [hours]	Coefficients		
	U_{18} [m s^{-1}]	$e_0 - e_{18}$ [kPa]	$U_{18}(e_0 - e_{18})$ [m s^{-1} kPa]
1	0.414	4.74	0.795
2	0.408	4.86	0.804
3	0.406	4.64	0.792
4	0.394	4.60	0.766
5	0.395	4.74	0.782
6	0.378	4.70	0.792

Table 3. Statistics associated with the stepwise linear regression analysis using sampling intervals of various lengths

Sampling interval [hours]	U_{18} [m s^{-1}]		$e_0 - e_{18}$ [kPa]		$U_{18}(e_0 - e_{18})$ [m s^{-1} kPa]	
	R^2	RMSE	R^2	RMSE	R^2	RMSE
1	0.350	1.32	0.458	1.20	0.629	0.994
2	0.337	1.34	0.474	1.19	0.638	0.989
3	0.316	1.38	0.436	1.26	0.601	1.06
4	0.394	1.37	0.435	1.24	0.586	1.07

The Principal Component Analyses (PCA) showed that 71% of the variance in the 24-hr total E was explained by the first principal component regardless of the sampling period (Table 4). The second component explained 27% of the variance, again regardless of the sampling interval. For the first principal component, $U_{18}(e_0-e_{18})$ had the largest coefficient, indicating its large contribution in predicting E (Table 5), followed by e_0-e_{18} and U_{18}. The order changed for the second principal component, with U_{18} having the largest coefficient followed by $U_{18}(e_0-e_{18})$ then e_0-e_{18}. For both components, the sampling interval had virtually no effect on the component coefficients (Fig. 5).

Table 4. Percent of the variance explained by the first and second principal components in relation to the sampling interval

Sampling interval [hours]	Variance explained by each principle component [%]		
	first	second	first + second
1	71	27	98
2	71	27	98
3	71	27	98
4	71	27	98
5	71	27	98
6	71	27	98

Table 5. First and second principal component coefficients in relation to the sampling interval

	Principal component coefficients					
Sampling interval [hours]	U_{18} [m s^{-1}]		$e_0 - e_{18}$ [kPa]		$U_{18}(e_0 - e_{18})$ [m s^{-1} kPa]	
	first	second	first	second	first	second
1	0.391	0.909	0.635	-0.380	0.667	0.171
2	0.392	0.908	0.634	-0.383	0.667	0.171
3	0.396	0.901	0.633	-0.386	0.666	0.171
4	0.401	0.903	0.630	-0.394	0.665	0.171
5	0.386	0.912	0.636	-0.375	0.668	0.169
6	0.390	0.908	0.634	-0.385	0.668	0.166

6 Discussion and Conclusions

There is a large seasonal disparity between lakes acting as a heat sink during the spring/summer when the terrestrial surface is a heat source, and between the lakes acting as a heat source during the fall/winter when the terrestrial surface is a heat sink. The behavior of escalating increases in λE and H due to increased gradients in vapor pressure and temperature has been well documented and discussed for Great Slave Lake (Blanken et al. 2000; Rouse et al. 2003a) and the Laurentian Great Lakes (Schertzer 1997). The incursion of cold, dry air over the relatively warm, humid open water late in the ice-free season results in massive evaporation and sensible heat losses to the atmosphere. This process is especially pronounced over large cold region lakes that remain ice-free late into the winter due to their high volumetric heat capacity.

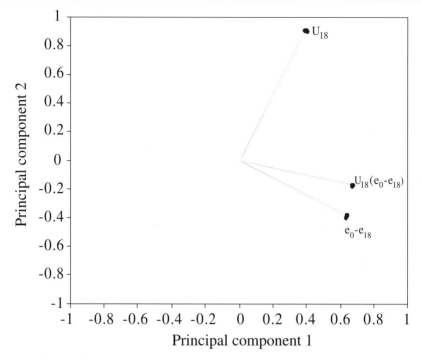

Fig. 5. First and second principal components for each the horizontal wind speed at 18 m above the water surface (U_{18}), and the difference between the vapor pressures at the water surface and 18 m above the water surface (e_0-e_{18}), and their product. The components for the six sampling intervals (1 to 6 hours) are plotted, but they overlap and cannot be distinguished individually in the diagram

Large lakes act as a heat sink during the summer, with the subsequent heat release late in the fall and winter likely having a large impact on their adjacent land surface (Rouse et al. 2007). The terrestrial area influenced by the lake is likely a function of its surface area to volume ratio, thus the climatological and ecological terrestrial footprint of Great Slave and the other large lakes in the MRB is probably large. For example, lake-effect snow would alter the hydrology and ecology of the region; increased summer cloud cover would alter the radiation balance with a shift from direct to diffuse short-wave radiation, which would enhance photosynthesis and thus affect the carbon cycle in boreal regions (Gu et al. 1999); lake-breezes would alter the air temperature and local wind patterns; and cyclogenesis may be modified by the presence of the large lakes (Gyakum 1997). All these potential effects are influenced by the ice-free duration, especially late in the ice-free season when the turbulent fluxes are largest.

Whereas the terrestrial surface often displays a distinct diurnal pattern in energy and radiation balance components, the lake's λE and H late in the ice-free season typically had a dominant frequency of about three days, suggesting that the lake and the land respond to forcings operating at different time scales. For example, the passage of a cold front over the region would likely reduce transpiration from the boreal vegetation over the typical three-day period, but probably with a diurnal pattern superimposed on the overall reduction in evapotranspiration. In contrast, the lake response to this synoptic-scale event would be an increase in λE over the same time scale as the synoptic-scale event without any imposed diurnal response. Thus, we hypothesize that transpiration responds to both the radiation cycle (diurnal time scales) and air temperature and humidity (diurnal and longer time scales) whereas the lake responds primarily to air temperature and humidity (longer time scale of three days).

A periodicity of 2.7 days for λE is typical of the passage of synoptic-scale systems (e.g., cold, dry air associated with the passage of cold fronts). This response to such relatively long-term events is likely pronounced in the MRB since such incursions of cold, dry air are common at these latitudes (Gyakum, 1997). Blanken et al. (2003) reported that these evaporation events were an aggregation of brief (i.e., a few seconds) sweeps of warm, dry air from above the lake surface. Thus, there appears to be an integration or aggregate effect of the superimposed events happening at vastly difference time scales. If the processes controlling λE are acting in a fractal manner (i.e., the physics behind the response of λE is the same regardless of the time scale; processes are independent of time), then the chosen observational time scale (such as seconds, minutes or days) should not affect the relationships between λE and its controls. Lake evaporation, whether driven by intermittent sweeps of warm dry air from above, or driven by synoptic-scale cold fronts lasting three days, is controlled by the same variables.

Over Great Slave Lake, as over other large lakes, evaporation can be expressed adequately in terms of Fick's diffusion law: $E = K_E U z (e_0 - e_a)$ where K_E is the eddy diffusivity for water vapour (Blanken et al. 2000; Oswald et al. 2007; Winter et al. 1995). The sensitivity of E to the product of U_Z and $e_0 - e_a$, expressed by the slope of the linear regression between the two variables (K_E) should not vary significantly if these controls on E behave in a fractal manner. The "mass transfer coefficient", analogous to K_E, is often derived from measurements of E, U and e, with K_E obtained at a time scale different from that for the measurements of U and e (temperature and humidity) and from different locations or at different times. For example, after the transfer coefficient is determined from 1-hr measurements, one may wish to estimate E from the same location using buoy or meteorological data that are only available at 6 or 12-hr intervals. This analysis has shown that the sampling interval has little effect on the mass transfer coefficient when used to estimate 24-hr E totals. This result implies that the fast, short-term events reported by Blanken et al. (2003) are reflected in the longer-term averages, thus the sensitivity of E to U and e does not change. Also, because the dominant period for the large E events that are responsible for the majority of E is about three days, sampling at or below the Nyquist frequency of roughly 1.5 days is sufficient to describe these events. Therefore, as these analyses have shown, sampling at intervals of between one and six hours had no significant affect on the sensitivity of E to U_{18}, e_0-e_{18}, or $U_{18}(e_0-e_{18})$.

Principal component analysis was used to discern which combination of variables explained most of the variance in E, and to ascertain whether the order of importance of these variables changed as the sampling interval changed. In agreement with the stepwise linear regression analysis, 71% of the variance in the 24-hr total E was explained by the first principal component with $U_{18}(e_0-e_{18})$ having the largest coefficient. Importantly, the coefficients did not vary significantly when the sampling intervals were less than the dominant averaging period of roughly three days.

The role of large northern lakes in the water and energy cycles of the MRB has only recently been appreciated. Two dichotomies of lakes versus land are the seasonal temporal shifts between heat sinks and sources, and the longer time scales involved in lakes compared to land. A landscape such as the MRB with a large coverage of lakes should exhibit dynamic coupling and feedback between the two surfaces. The magnitude, controls and seasonal dynamics of several processes important to understanding this lake-land coupling are now known. What remain to be discovered are the spatial extent of the coupling, its ecological and environmental importance, and how both will be affected by climate change.

Acknowledgements

This large study undertaken under difficult logistical conditions would not be possible if it were not for the dedicated and appreciate efforts of the following groups: Great Slave Lake branch of the Canadian Coast Guard; Environment Canada, Yellowknife, NT; Institute of Hydrology, Wallingford, U.K.; MSC, Toronto, ON; National Water Research Institute, Burlington, ON; Royal Canadian Mounted Police Marine Unit, Yellowknife, NT; Water Survey of Canada, Yellowknife, NT. In addition to the network grant for MAGS, financial support was provided by Natural Science and Engineering Research Council of Canada (NSERC) grants and a research grant from NASA, USA.

References

Blanken PD, Rouse WR, Culf AD, Spence C, Boudreau LD, Jasper JN, Kotchtubajda B, Schertzer WM, Marsh P, Verseghy D (2000) Eddy covariance measurements of evaporation from Great Slave Lake, Northwest Territories, Canada. Water Resour Res 36:1069–1078

Blanken PD, Rouse WR, Schertzer WM (2003) The enhancement of evaporation from a large northern lake by the entrainment of warm, dry air. J Hydrometeorol 4:680–693

Buck AL (1981) New equations for computing vapor pressure and enhancement factor. J Appl Meteor 20:1527–1532

Gu L, Fuentes JD, Shugart HH, Staebler RM, Black TA (1999) Responses of net ecosystem exchange of carbon dioxide to changes in cloudiness: results from two North American deciduous forests. J Geo Res 104:31,421–31,434

Gyakum J (1997) Cyclones and their role in high latitude water vapour transport. In: Proc 3rd Scientific Workshop for the Mackenzie GEWEX Study (MAGS), University of Saskatchewan, Saskatoon, Saskatchewan, Canada, pp 28–30

Kaimal JC, Finnigan JJ (1994) Atmospheric boundary layer flows: their structure and measurement. Oxford University Press

Oswald CW, Rouse WR, Binyamin J (2007) Modeling lake energy fluxes in the Mackenzie River Basin using bulk aerodynamic mass transfer theory. (Vol. II, this book)

Quinn FH, den Hartog G (1981) Evaporation synthesis. In: Aubert EJ, Richards TL (eds) IFYGL – The international field year for the Great Lakes. National Oceanic and Atmospheric Administration, Ann Arbor, Michigan, pp 221–245

Rawson DS (1950) The physical limnology of Great Slave Lake. J Fish Res Board Can 8:3–66

Rouse WR, Blyth EM, Crawford RW, Gyakum JR, Janowicz JR, Kochtubajda B, Leighton HG, Marsh P, Martz L, Pietroniro A, Ritchie H, Schertzer WM, Soulis ED, Stewart RE, Strong GS, Woo MK (2003) Energy and water cycles in a high-latitude north-flowing river system. B Am Meteorol Soc 84:73–87

Rouse WR, Binyamin J, Blanken PD, Bussières N, Duguay CR, Oswald CJ, Schertzer WM, Spence C (2007) The influence of lakes on the regional energy and water balance of the central Mackenzie River Basin. (Vol. I, this book)

Rouse WR, Oswald CJ, Binyamin J, Blanken PD, Schertzer WM, Spence C (2003a) Interannual and seasonal variability of the surface energy balance and temperature of central Great Slave Lake. J Hydrometeorol 4:720–730

Rouse WR, Oswald CJ, Binyamin J, Spence C, Schertzer WM, Blanken PD, Bussières N, Duguay C (2005) The role of northern lakes in a regional energy balance. J Hydrometeorol 6:291–305

Schertzer WM (1997) Freshwater lakes. In: Bailey WG, Oke TR, Rouse WR (eds) The surface climates of Canada. McGill-Queens Press, Montreal, pp 124–148

Schertzer WM, Rouse WR, Blanken PD (2000) Cross-lake variation of physical limnological and climatological processes of Great Slave Lake. Phys Geog 21:385–406

Schertzer WM, Rouse WR, Blanken PD, Walker AE (2003) Over-lake meteorology, thermal response, heat content and estimate of the bulk heat exchange of Great Slave Lake during CAGES (1998–99). J Hydrometeorol 4:649–659

Schertzer WM, Rouse WR, Blanken PD, Walker AE, Lam DCL, León L (2007) Interannual variability of the thermal components and bulk heat exchange of Great Slave Lake. (Vol. II, this book)

Schotanus P, Niewstadt FTM, DeBruin HAR (1983) Temperature measurements with a sonic anemometer and its application to heat and moisture fluxes. Bound-Layer Meteorol 26:81–93

Schuepp PH, Leclearc MY, MacPherson JI, Desjardins, RL (1990) Footprint prediction of scalar fluxes from analytical solutions of the diffusion equations. Bound-Layer Meteorol 50:355–373

Shuttleworth WJ, Gash JHC, Lloyd D, McNeil DD, Moore CJ, Wallace JS (1988) An integrated micrometeorological system for evaporation measurements. Agr Forest Meteorol 43:295–317

Walker A, Silis A, Metcalf JR, Davey MR, Brown RD, Goodison BE (2000) Snow cover and lake ice determination in the MAGS region using passive microwave satellite and conventional data. In: Strong GS, Wilkinson YML (eds) Proc 5th Sci Workshop Mackenzie GEWEX Study, Edmonton, AB, Canada, pp 39–42

Webb EK, Pearman GI, Leuning R (1980) Correction of flux measurements for density effects due to heat and water vapor transfer. Q J Roy Meteor Soc 106:85–100

Winter TC, Rosenberry DO, Sturrock AM (1995) Evaluation of 11 equations for determining evaporation for a small lake in north central United States. Water Resour Res 31:983–993

Chapter 11

Interannual Variability of the Thermal Components and Bulk Heat Exchange of Great Slave Lake

William M. Schertzer, Wayne R. Rouse, Peter D. Blanken, Anne E. Walker, David C.L. Lam and Luis León

Abstract Intensive year-round observations (1998–2003) revealed that Great Slave Lake is highly responsive to variations in both short-term meteorological and longer-term climatic conditions. The lake is dimictic and exhibits spatial and temporal variability in air temperature and wind speed that impacts water surface temperature and lake heat flux. Storm events with high winds were responsible for short-term highly dynamical responses of the lake, with deep vertical mixing of temperature affecting thermal stratification characteristics. Significant interannual variability was observed. The longest ice-free period over a 16-year period occurred in 1998 (213 days) compared to a cooler year in 2002 (174 days). Maximum surface temperature was in 1998 (21.2°C) compared to 2002 (14°C). Simulation of surface temperature using a 3-D hydrodynamic model demonstrated marked nearshore and offshore temperature differences. Variation of the 8°C isotherm effectively demonstrated interannual variability in the lake heating. Annual heat content ranged from 2.61×10^{19} J in 1998 to 2.13×10^{19} J in 2002. Five-year mean heat content maximum was 2.24×10^{19} J. The bulk heat exchange, ranging between 267 W m^{-2} and -338 W m^{-2}, compared well with conventional heat flux calculations at Inner Whaleback Island located in the central basin of the lake.

1 Introduction

There is a growing concern regarding potential impacts of climatic change on freshwater resources (Schertzer et al. 2004). In Canada, observations and model projections suggest that some of the regions most vulnerable to change are in the north (Rouse et al. 2002; Stewart et al. 1998). Knowledge regarding the thermal and energy responses of deep northern lakes such as Great Slave Lake to climate is lacking and requires quantification.

Pioneering research on Great Slave Lake conducted by Rawson (1950) was based largely on ship surveys in 1947 and 1948, sampling at a grid of stations to provide information on lake depth and its temperature structure.

These enabled the first lake-wide description of the temperature structure and heat content, but the limited number of depth soundings permitted only a general description of the lake bathymetry for use in applications to thermal or hydrodynamic problems. Also, lake surveys were conducted only during the ice-free period which precluded derivation of an annual temperature cycle. Temperature observations were conducted at ~30 lake sites at approximately weekly intervals, necessitating use of mean values for heat content analysis. Dynamics of air-lake interactions, important to understanding the role of lakes in a northern environment (Rouse et al. 2007a), were not examined.

During the Mackenzie GEWEX Study (MAGS), intense field observation and modeling effort were made on Great Slave Lake. The objective was to quantify the interannual variability of key over-lake meteorological variables, the thermal structure, heat content, and heat exchange of the lake. The multi-year observations established baseline thermal characteristics and their interannual variability, thus enabling insight to be gained on the responsiveness of this large, deep northern lake to climatic variability.

2 Study Area

Great Slave Lake is one of three large lakes in the Mackenzie River Basin (Fig. 1 in Rouse et al. 2007b). The lake has a surface area of 27,200 km^2 and a volume of 1,070 km^3 (van der Leeden et al. 1990). It consists of a main basin, a north arm, and an east arm. A 2 x 2 km^2 grid bathymetry was derived for Great Slave Lake (Schertzer 2000), and Fig. 1 shows the depth contours of the central basin which is the main area of this study. A large part of the central basin was less than 60 m deep though the maximum depth sounding of the central basin attained 187 m. The maximum depth of the 2 x 2 km^2 grid was 167.8 m and the mean depth was 32.2 m. Based on the grid bathymetry, the central basin has a surface area of 18,500 km^2 and an integrated volume of 596 km^3.

The measurement program for a large lake should ideally include a combination of in situ moorings and lake-wide surveys over a grid of stations to sample spatial variability. Such a program was precluded by logistical constraints to cover a lake of such large size. Instead, a cross-lake transect of in situ moorings was deployed along the lake main-axis (Fig. 1), including land-based, island-based, and lake-based moorings. This array configuration was used to provide the first intensive measurements of meteorological, radiation, lake temperature, and hydrodynamic variables on this lake.

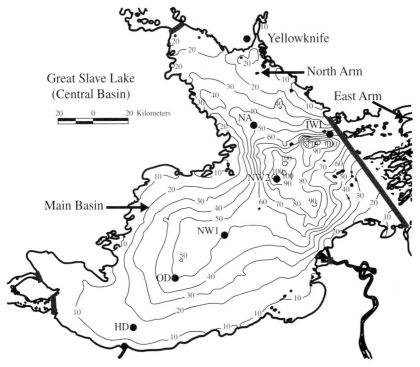

Fig. 1. Bathymetry of Great Slave Lake and location of cross-lake meteorological and temperature mooring observation sites. Depths are in meters

3 Methods

Cross-lake measurements using meteorological buoys and thermistor moorings were made (e.g., Blanken et al. 2000; Schertzer et al. 2000). Figure 1 provides the station names and instrumentation used at each location during the summer and winter periods (described in detail in Schertzer et al. 2003). The basic measurement configuration was maintained throughout 1998 to 2003, allowing examination of interannual variability. In 2000, a temperature mooring was added to sample the temperature structure in the north arm and in 2003, currents were sampled with acoustic doppler current meters (ADCP) at two sites in support of hydrodynamic model development and verification.

3.1 Meteorology

Meteorological observations included air temperature, water surface temperature, relative humidity, wind speed, wind direction, and solar and long-wave radiation. Measurements were recorded at 10-minute intervals at sites HR, NW1, NW2, IWI, and hourly at OD (Fig. 1). All data were processed to hourly and daily averages. Wind speed along the lake transect, measured at different heights, was converted to a common height of 8 m based on Shore Protection Manual (1984) engineering procedures. The characteristics of selected variables such as over-lake air temperature, wind speed and water surface temperature were depicted as means and ranges of daily values from individual sites averaged over the study period.

3.2 Water Temperatures

Water temperature moorings in 1998 included a combination of temperature loggers (StowAway Optic, StowAway Tidbit, and Brancker loggers). In 1999–2003, moorings were equipped with StowAway Tidbit sensors exclusively. Sensor depths were standardized to allow for comparability of thermal profiles (Schertzer et al. 2003) and observations were recorded at 15-minute intervals. A bulk surface temperature was measured at each of the meteorological buoys (Schertzer et al. 2000, 2003). This is not the "skin" temperature which is often derived from infrared thermometer; rather, the observations represent a bulk water surface temperature measurement common in physical limnology and oceanography.

Winter temperature moorings were installed at OD in 1998–99 and at OD and NW2 in 1999–2003. To avoid ice rafting, thermistors were deployed from a depth of 12 m below the surface to 1 m above the lake bottom. Isothermal conditions are assumed for the water layer above the 12 m depth for the winter experiment in order to facilitate computation of the winter heat content over the whole lake volume. Vertical temperature profiles and heat content over the lake volume were computed using procedures applied to the 2 x 2 km^2 grid bathymetry (Neilson et al. 1984).

3.3 Ice Freeze-up and Breakup Dates

Passive Microwave Imagery (SSM/I) at the 85 GHz range obtained over Great Slave Lake enabled discrimination between areas of ice cover and open water (Walker and Davey 1993). The SSM/I technique is applicable at any time of the day, in all weather conditions and provides a resolution of 12.5 km^2 from the 85 GHz channel. One major application of the technique was in monitoring the dates of complete freeze-up or lake-ice breakup for Great Slave Lake (Walker et al. 1999) to permit determination of the ice-free duration.

3.4 Heat Content and Bulk Heat Exchange

The heat content of Great Slave Lake was approximated using lake bathymetry data, lake volume, and with temperature profile observations:

$$H = \iiint_V \rho_W c_P T_W \, dxdydz \tag{1}$$

where H is the lake-wide mean heat content, ρ_w is the density of water, C_p is the specific heat of water, and T_w is the water temperature at depth. On the Laurentian Great Lakes, the mean lake-wide heat content is usually derived based on many years of short duration lake temperature surveys over a grid of stations (Schertzer et al. 1987), only in ice-free conditions (Schertzer 1997). The present investigation included continuous vertical temperature observations along the cross-lake transect as well as in the north arm. Since observations were conducted through a summer and winter program, an annual heat content cycle can be derived, allowing interannual comparison.

With heat content information, the daily bulk heat exchange (Q_{BH}) of Great Slave Lake can be obtained:

$$Q_{BH} = (dH/dt)/A_s \tag{2}$$

where H is the heat content over the time interval of 1 day, and A_s is the surface area of the lake. Q_{BH} represents the total heat gain or loss from the lake, assuming that all possible exchanges at the air-water interface and through hydrologic flows are accounted in the thermal observations used to derive the daily heat content curves.

3.5 Surface Heat Flux

The heat budget for a lake is computed as:

$$Q_T = Q^* - Q_E - Q_H - Q_A \qquad (3)$$

where Q_T is the total surface heat flux. The net radiative flux, Q^*, is the net balance of incoming and reflected solar radiation, and incoming and emitted long-wave radiation. Turbulent exchanges are represented by the latent (Q_E) and sensible (Q_H) heat fluxes. At the buoy locations, incoming solar and longwave radiation are measured and the turbulent exchange is determined from temperature, wind, and humidity observations (Schertzer et al. 2003). At IWI, the turbulent heat fluxes are determined directly through the eddy correlation technique (Blanken et al. 2000, 2003, 2007; Rouse et al. 2003). Evaluation of the advective component (Q_A) is complicated because information on hydrologic flow and temperature is limited. However, on the Laurentian Great Lakes such as Lake Erie, the advective components are small compared to the radiative and turbulent exchanges (e.g., Derecki 1975).

4 Variability of Lake Thermal and Heat Flux Components

4.1 Over-lake Meteorology, Lake Surface Temperature, and Radiation

General characteristics of the over-lake meteorology in Great Slave Lake for the Enhanced Canadian GEWEX Study period were presented in Schertzer et al. (2003). Table 1 provides a summary of the means and ranges for key over-lake meteorological components for the June to October ice-free seasons in 1998 and 1999. These were two contrasting years with warmer conditions in 1998, possibly influenced by an intense El Niño event that resulted in significantly higher air and water temperatures but little change in average winds. The warm conditions prolonged the ice-free duration (Walker et al. 1999) that affected heat exchanges and lake heating (Schertzer et al. 2003).

This chapter enlarges upon that study using aggregated observations for 1998–2003. For the ice-free period, averages include over-lake observations from the meteorological buoys while the winter program included meteorology from Hay River, Inner Whaleback Island (IWI), and Yellowknife. Mean air temperature (Fig. 2a) ranged from -10 to -20°C in December–January and peaked at ~15°C in July–August. Larger air temperature

Table 1. Statistical summary of measured over-lake meteorological components during 1998 and 1999. Averages are computed for the Great Slave Lake stations over the June–October period. For comparison, June–October averages are provided for Yellowknife Airport

Variable	Year	Mean	Maximum	Minimum	YZF
Air temperature [°C]	1998	12.9	23.5	0.3	9.8
	1999	9.7	18.4	-2.7	
	Δ^a	3.2	5.1	3.0	
Surface temperature [°C]	1998	12.9	20.5	3.5	—
	1999	9.3	16.9	0.2	
	Δ^a	3.6	3.6	3.3	
Wind speed [m s^{-1}]	1998	6.8	15.6	1.5	4.3
	1999	6.7	13.7	2.7	
	Δ^a	0.1	1.9	-1.2	

YZF represents Yellowknife airport, NWT (June–Oct. climate normal 1942–90).
[a] Δ represents value 1998 minus 1999.

ranges over winter and in the heating phase leading up to August, compared to the cooling phase from August through November, indicate a moderating influence by the lake. Figure 2b indicates that lower wind speeds occurred during winter (recorded largely from land-based sites) with higher values recorded in the ice-free period. Mean values of 5 m s^{-1} were expected between August and October. Variability in the daily wind speeds was very high in the ice-free season, reaching maximum daily values of 15–20 m s^{-1} during intense storms. High wind events influenced both the surface temperature and vertical temperature structure as well as over-lake heat fluxes. Water surface temperatures (Fig. 2c) increased rapidly after ice-melt in June and attained a lake-wide mean of ~15°C in late July to early August. Certain characteristics between the lake response and atmospheric variables are noted. For example, mean water temperature follows increased air temperature to the peak in June–July. Strong winds in September–October are associated with the progressive decrease in lake heat as the surface water layer cools to the temperature of maximum density (fall overturn). The wide range of water surface temperature is an indication that Great Slave Lake is highly sensitive to weather conditions.

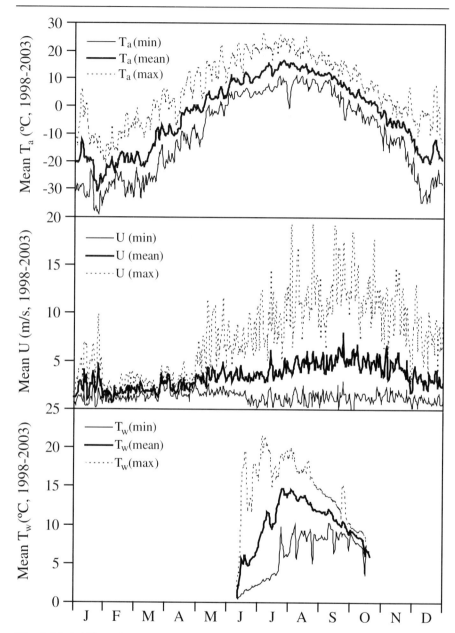

Fig. 2. Variability of over-lake air temperature (T_a), wind speed (U) and water surface temperature (T_w)

4.2 Lake Temperatures

4.2.1 Measured Characteristics

Monthly mean air temperatures for October to April at Yellowknife range from -1 to -25°C. These low temperatures cause the formation of a lake ice cover that effectively de-couples the lake–atmospheric exchange. The length of the ice-free season reflects climatic conditions. Based on a composite of ice breakup and freeze-up dates from 1988 to 2003 (Fig. 3), the central basin is expected to be ice-free by June 17 and ice freeze-over is expected on December 8 with an average ice-free duration of 174 days. The longest ice-free period of 213 days occurred in 1998 compared to only 174 days in 2002. Spring air temperature in 1998 was 3°C higher than normal, resulting in extensive open water areas about 3 weeks earlier than average (Walker et al. 1999), in contrast to the cooler 2002 in which ice breakup occurred nearly 21 days later than in 1998. In cooler years (e.g., 2002), ice freeze-up can occur significantly earlier than in very warm years such as 1998 (Fig. 3).

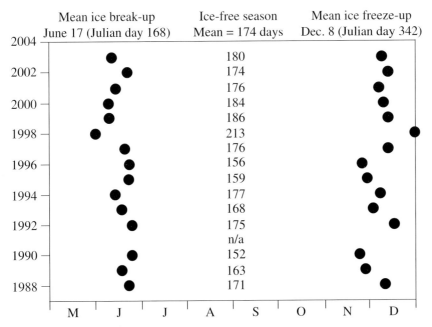

Fig. 3. Dates of ice breakup and freeze-up and corresponding ice-free days for the period 1988–2004 based on SSM/I techniques. Data from Walker et al. 1999 (reproduced from Schertzer et al. 2003)

The thermally stratified period for dimictic lakes such as Great Slave Lake is defined as the number of days between the spring and fall overturn (when the lake temperature passes the temperature of maximum density at 4°C, once in the spring and once in the fall). Table 2 shows large variability in the onset date of thermal stratification. The average onset occurred on day 178 (June 27); the earliest in 1998 and 2000 and the latest in the cooler 2002. Table 2 also shows the interannual variability in the maximum lake-wide mean surface temperature. The highest maximum surface temperature of 21.2°C occurred in 1998 and the lowest, 14.0°C, in 2002.

Table 2. Interannual variability of thermal components for 1998–2002 showing date of disappearance of the 4°C isotherm (onset of thermal stratification), the maximum lake-wide surface temperature and date, and the maximum heat content and date. Day of year is shown in parentheses

Year	Onset of thermal stratification	Max. surface temperature [°C]	Date	Max. heat content [x 10^{19} J]	Date
1998	Jun.19 (170)	21.2	Jul.6 (187)	2.61	Aug.5 (217)
1999	Jun.26 (177)	16.5	Aug.21 (233)	2.16	Aug.23 (235)
2000	Jun.18 (170)	18.4	Aug.5 (218)	2.41	Aug.12 (224)
2001	Jun.25 (176)	18.7	Jul.30 (211)	2.27	Sep.11 (254)
2002	Jul.15 (196)	14.0	Jul.26 (207)	2.13	Sep.9 (252)
Average	Jun.27 (178)	17.8	Jul.30 (211)	2.32	Aug.24 (236)

4.2.2 Modeled Characteristics

The spatial variability in surface temperature was simulated by León et al. (2004), applying the hydrodynamic model ELCOM (Hodges et al. 2000). Simulations assumed a no flux condition between the central basin and the east arm of the lake, and inflow from the Slave River and outflow through the Mackenzie River were set to zero. Figure 4 illustrates surface temperatures simulated for four time periods through the summer ice-free season. On July 10 2003, Great Slave Lake was still in the heating phase. Lake bathymetry played an important role in lake heating and in the surface temperature distribution. Nearshore areas with shallower water showed more rapid increases in temperature (e.g., >6–12°C) compared to the cooler mid-lake which was at the temperature of maximum density (4°C). On July 20 2003, heating has increased with 4°C water remaining only in the deepest part of the central basin. The disappearance of the 4°C isotherm at the end of July signaled the beginning of full summer thermal stratification for the central basin of the lake. On August 20, 2003, the central basin was near the period of maximum heat content of the lake. While the mid-lake temperature was 6–8°C, much of the nearshore coastal area had temperatures of 12–18°C. Cooling phase continued and on September 20, 2003 (Fig. 4d), cooling progressed more rapidly in the shallow nearshore than in the mid-lake.

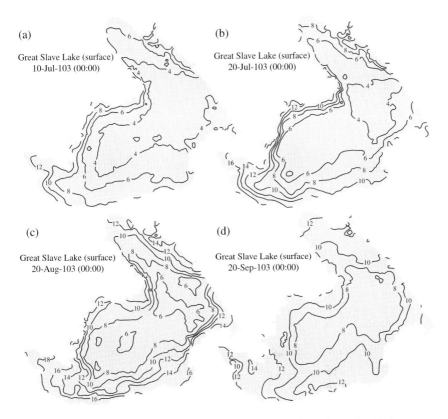

Fig. 4. Spatial variation of the surface temperature based on simulation results from the 3-D hydrodynamic model ELCOM

Figure 5 shows a time series of isotherms for temperature moorings across the lake for the summer of 2003. Evident in the time series are the differences in the timing of thermal stratification (Table 2), the depth of the upper warmer layers, and the relative differences in temperature. Such differences are largely attributed to inter-annual variations in climatic conditions and to depth effects: shallower parts of the lakes (e.g., site HR) warm faster and have generally higher temperatures than the deeper, cooler midlake (e.g., Site NW2). Weather events strongly influence the short-term dynamic responses of lakes and their water temperature. For example,

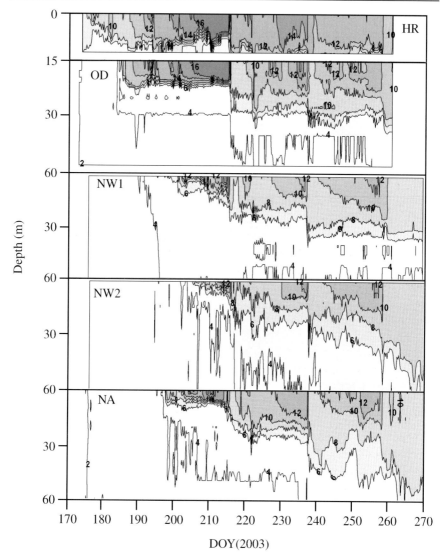

Fig. 5. Isotherms for Great Slave Lake temperature moorings for 2003. Intense mixing occurs in response to storms on August 3 (DOY 215) and August 24 (DOY 237)

large storms with high winds occur relatively frequently over large, deep lakes such as Great Slave Lake or the Laurentian Great Lakes (Schertzer and Croley 1999). Due to fetch characteristics, the nearshore winds are generally lower than in the mid-lake. Schertzer et al. (2003) found that storms with high winds occurring at the start of thermal stratification significantly mixed the water column in the vertical. This effectively lowered the thermocline (a transition zone separating the warm upper layer and colder lower layers) and reduced both the surface and subsurface temperatures. Two large storm events in 2003 offer illustrative examples (Fig. 5). Atmospheric pressure was relatively uniform (~980 hPa) across the lake on August 3, but on August 24 it was ~975 hPa on the west side of the lake and 971 hPa on the east side. Peak winds varied considerably across the lake in both events but were generally higher at either end of the cross-lake transect compared to the mid-lake. Particularly evident in these time-series is the deep vertical mixing in response to the two storms. On August 3, at the shallow nearshore site OD, temperatures approaching 20°C prior to the storm were mixed to 12°C. As the water at OD began to warm over the next three weeks, a second intense storm on August 24 caused further mixing and cooling of the water column, deepening the upper mixed layer to ~18–20 m depth, where it remained throughout the summer. Mixing also affected the offshore sections of the lake and the north arm, resulting in lowering their upper layer temperatures, which impacted on the seasonal thermal structure of the lake and affected the air–water heat exchange.

The colder, deeper portion of the lake was less responsive to diurnal changes than the upper mixed layer during the stratified summer period. Timing of the beginning and end of the thermally stratified period varies between years (Fig. 6). For most of the year, Great Slave Lake was below the temperature of maximum density. Thermal stratification began when the lake warmed to the temperature of maximum density at 4°C. In contrast to the temperature structure at specific sites where dynamic responses are discernible, the lake-wide temperature structure shows the averaged response to the meteorological forcing. Consequently, the thermocline region deepens almost monotonically from the start of stratification in June–July to the period of complete mixing achieved in late September–October as the temperature progressively decreases due to increased surface heat losses to the atmosphere (Rouse et al. 2007a).

The influence of the climate on lake heating can be seen by tracking the progression of an isotherm associated with, for example, the depth of the thermocline region (e.g., Simons and Schertzer 1987). Figure 7 shows the interannual variation of the depth of the 8°C isotherm during the stratified period between years in the central basin. Compared with other years in this study, in warmer 1998, the 8°C isotherm began earlier in the year, extended deeper in the water column during summer, and disappeared later in the fall. In the cooler 2002, the lake took longer to heat up. The 8°C iso-

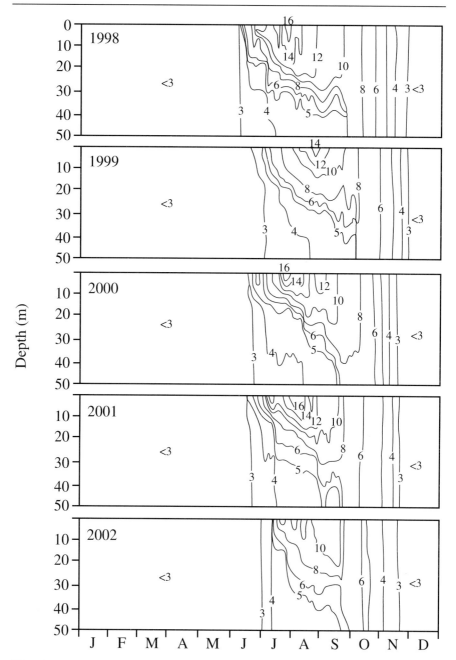

Fig. 6. Interannual variability in the lake-wide average thermal structure in the upper 50 m of Great Slave Lake

therm formed later, did not attain the vertical extent seen in the other years, and disappeared earlier, especially when compared to 1998.

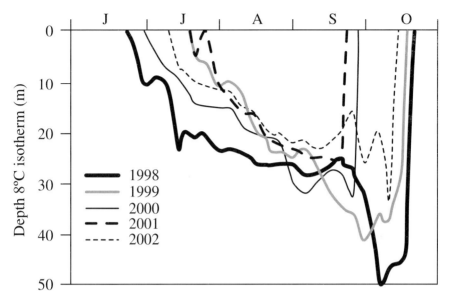

Fig. 7. Variation of the depth at the 8°C isotherm from 1998 to 2002.

4.3 Lake-wide Heat Content

The lake-wide heat content for 1998 was higher than the other years (Table 2). Its maximum value of 2.61×10^{19} J was significantly larger than those reached in 1999 and 2002. Interannual difference in the timing of the maximum is an important indication of the climate variability. With earlier ice cover loss in warm years, the lake has increased time to absorb solar radiation which results in higher heat content. In 1998, for example, maximum heat content was attained on August 5 (DOY 217), 37 days earlier than it did in cooler 2001 and 2002.

The spring heat income is defined as the amount of heat gained by the lake from its minimum temperature to the temperature of maximum density (4°C). The summer heat income represents heating from the time of temperature of maximum density to the date of maximum heat content. Table 3 shows that 1998 had both larger spring and summer heat incomes compared to the other years. The heat content range (minimum to maximum) was 2.54×10^{19} J in 1998 compared to 2.00×10^{19} J in 2002. The derived daily heat contents show a pronounced asymmetry between the heating and cooling rates especially for 1998 (Schertzer et al. 2003). During the ice-free period in fall, the high heat content contributes to increased heat and moisture transfer to the cooler overlying air (Rouse et al. 2007b; Schertzer et al. 1987).

Table 3. Spring and summer heat incomes, and annual heat content range for the period 1998–2002

Year	Spring heat income [x 10^{19} J]	Summer heat income [x 10^{19} J]	Annual heat content range [x 10^{19} J]
1998	0.78	1.75	2.54
1999	0.68	1.39	2.07
2000	0.59	1.66	2.29
2001	0.67	1.49	2.16
2002	0.71	1.16	2.00
Average	0.69	1.49	2.21

4.4 Heat Flux and Bulk Heat Exchange

Heat fluxes were computed at Inner Whaleback Island using detailed radiation and the eddy correlation technique (Blanken et al. 2000, 2007; Rouse et al. 2003). Under a stable atmosphere, generally from ice-melt to the time of maximum heat content, the sensible and latent heat fluxes are small and most of the net solar radiation is consumed in heating the lake. After mid-August until the ice freeze-up, the atmosphere becomes increasingly unstable and the sensible and latent heat fluxes begin to dominate the total surface heat flux as the lake loses heat to the atmosphere.

Heat fluxes computed from lake buoys or island platforms are essentially site specific and may not necessarily represent the average condition for the entire lake, especially for large lakes with considerable cross-lake variability in meteorology and water temperatures. One approach to approximate the lake-wide total heat flux is to derive a bulk heat exchange (Eq. 2) based on the daily lake-wide heat content. Figure 8 compares the 5-day mean bulk heat exchange determined from heat content computations with the total heat flux computed near Inner Whaleback Island in 1999 (Rouse et al. 2003). Both methods yield similar magnitudes of lake heat gains during the spring and heat losses during the fall. There was considerable spatial variability in hydrometeorologic variables across the lake (e.g., surface temperature as shown in Fig.4), and this accounts for the short-term (e.g., daily) differences between the lake-wide mean bulk heat exchange and the surface flux computed at single points. On a seasonal basis, however, their good agreement gives confidence in applying Eq. (2) to calculate the lake heat storage and provides verification of the computed heat fluxes from Inner Whaleback Island.

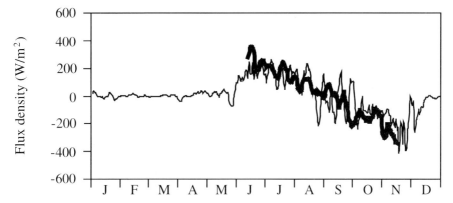

Fig. 8. Comparison of the computed heat flux from Inner Whaleback Island (thick line) and the bulk heat exchange (thin line) derived from lake temperature and heat content (reproduced from Schertzer et al. 2003)

4.5 Interannual Variability of the Heat Content and Bulk Heat Exchange

Figure 9 shows the between-year variations in the monthly bulk heat exchange of Great Slave Lake. Changes in lake heat content are very small during the ice covered period (January–June). There is virtually no exchange with the atmosphere, but only exchanges through such processes as hydrologic discharges, release of heat from the sediments, and possibly penetration of light through the ice cover. Ice breakup from end-May to early-June results in rapid heat gain to the lake, continuing through June to mid-August. From mid-August to mid-September, the monthly mean (and daily-averaged) bulk heat exchange alternates between periods of heat gain and loss in response to storms that induce progressive deepening of the upper mixed layer. Large negative bulk heat exchanges are more prevalent from mid-September to the beginning of December and these are related to deep mixing from strong winds and enhanced turbulent heat losses (Rouse et al. 2003). The interannual differences (Fig. 8) demonstrate the sensitivity of the lake heating and heat exchange process that can occur in the northern region.

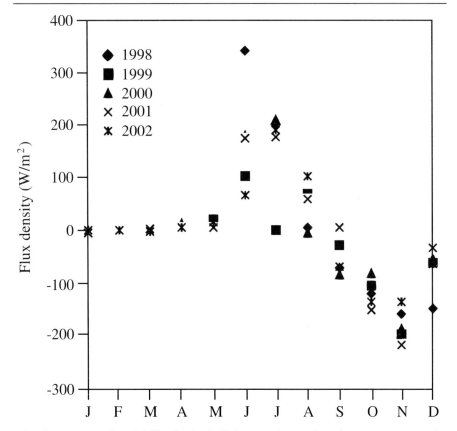

Fig. 9. Interannual variability in the bulk heat exchange based on temperature observations

A composite of the heat contents and derived bulk heat exchange over the 1998–2002 period is shown in Figs. 10a and 10b, respectively. Superimposed on the interannual distribution is a 5-year mean. The average heat content of Great Slave Lake is 2.08×10^{19} J, with a range between 2.24×10^{19} J and 0.16×10^{19} J. The mean bulk heat exchange ranges between approximately 267 W m^{-2} and -338 W m^{-2}. Errors in the computed heat content can be attributed to errors in lake bathymetry and temperature measurements. Schertzer (2000) indicates that ~1.4×10^9 m^2 of the deep midlake lacks depth soundings, requiring interpolation. Assuming that the estimated depth in the mid-lake area is in error by 1 m and that the temperature of this deep volume of water is in error by 1°C, then the expected error is < 0.1% of the maximum heat content.

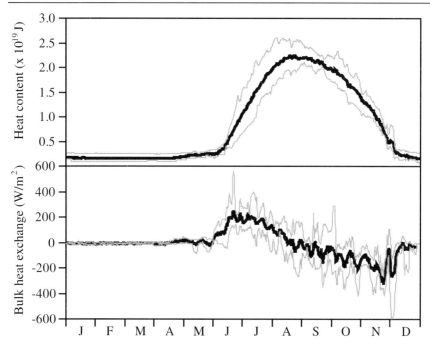

Fig. 10. Five-year mean and range of the heat content and bulk heat exchange. Thick lines represent means and the grey lines represent the range of values over the period 1998–2002

5 Discussion and Conclusions

This chapter presents the baseline characteristics of a large, deep northern lake, with a focus on the magnitude, variability, and interannual variability of its thermal responses to climate. The measurement program adopted is unique in physical limnological research for large, deep lakes. A cross-lake transect consisting of land-based and island-based stations, meteorological buoys and temperature moorings allowed year-round water temperature measurements from which daily lake-wide heat content and bulk heat exchange were derived. The computed bulk heat exchange is compared with traditional methods of evaluating the surface heat flux from a fixed location, and the cross-lake observations demonstrated considerable spatial variability in over-lake air temperature and winds. Shallow nearshore areas warm faster than the deeper mid-lake during spring and cool faster in the fall. Future studies should assess the winter temperature distribution especially in the lower water column using thermistors with higher sensitivity.

Great Slave Lake is dimictic in all years studied with considerable interannual variability in the dates of spring and fall overturn. It acquires a complete ice cover in winter. The length of the ice-free season dictates the available time that the solar radiation can heat the lake and consequently, its temperature and heat content characteristics. The lake-wide average temperatures of 1998–2002 show that the vertical temperature structure varies greatly from year to year. Large between-year differences are found in the magnitude of heat content, and in the timing of the temperature and heat content maxima.

Initial application of 3-D hydrodynamic models forced with meteorology and heat fluxes in this investigation provided surface temperature simulations that compared well with observed temperatures. The simulations required assumptions on water exchange and river flows. Based on Rawson (1950), the volume input of the Slave River is about 0.6% of the total central basin volume. With mean river water temperatures of ~10–14°C above the temperature of maximum density, the heat delivered by the Slave River would be sufficient to raise the temperature of the central basin by 0.6°C, and may be more important near the river mouth. Heat input from other, smaller tributaries may be important locally but insignificant in relation to the expansive central basin. Nevertheless, more work is required to quantify the heat contribution of rivers for the purpose of hydrodynamic modeling of the lake.

The lake-wide bulk heat exchange and the heat flux computed from Inner Whaleback Island exhibit comparable seasonal trends. However, there are differences in absolute values of the total heat flux and the bulk heat exchange. Differences are expected since meteorological conditions at the deeper, eastern part of the lake at may not be representative of conditions in the nearshore or the western part of such a large lake. The bulk heat exchange is based on temperature observations that included a sampling of conditions in the nearshore and deep-lake regions in the central basin. For evaluation of the lake-wide heat flux, future investigations should consider an expanded network of high-accuracy systems such as were employed at Inner Whaleback Island, and increased spatial representation of temperature observations. An expanded network of measurements including the east arm of the lake is required to provide a comprehensive assessment of the baseline and climate impact on Great Slave Lake as a whole.

Acknowledgements

Financial support was received from the Global Energy and Water Cycle Experiment on the Mackenzie Basin (GEWEX-MAGS), the National Water Research Institute (NWRI), Environment Canada, and from the Canadian Foundation for Climate and Atmospheric Sciences. The Research Support Branch (RSB) of NWRI, Engineering Services prepared the instrumentation and pre-processing of data. Deployment, refurbishing and retrieval of the moorings were conducted by the Technical Operations Services of RSB, NWRI. Special thanks to Barry Moore, Bob Rowsell and Irwin Smith of NWRI for Technical and Engineering support and Craig McCrimmon for writing some of the computer programs. Logistical support was provided by the Great Slave Lake Branch of the Canadian Coast Guard (Hay River). Thanks to Ron McLaren (MSC, P&Y Region) who kindly gave permission for placement of a relative humidity sensor on the OD buoy and provided access to data. Claire Oswald and Devon Worth participated in the measurement programs on Inner Whaleback Island.

References

Blanken PD, Rouse WR, Culf AD, Spence C, Boudreau LD, Jasper JN, Kotchtubajda B, Schertzer WM, Marsh P, Verseghy D (2000) Eddy covariance measurements of evaporation from Great Slave Lake, Northwest Territories, Canada. Water Resour Res 36:1069–1078

Blanken PD, Rouse WR, Schertzer WM (2003) On the enhancement of evaporation from a large deep northern lake by the entrainment of warm, dry air. J Hydrometeorol 4:680–693

Blanken PD, Rouse WR, Schertzer WM (2007) The time scales of evaporation from Great Slave Lake. (Vol. II, this book)

Derecki JA (1975) Evaporation from Lake Erie, NOAA Tech Report ERL 342-GLERL 3, Boulder, Colorado

Hodges BR, Imberger J, Saggio A, Winters KB (2000) Modeling basin-scale internal waves in a stratified lake. Limnol Oceanogr 45:1603–1620

León LF, Lam DCL, Schertzer WM, Swayne D (2004) Lake and climate models linkage: a 3D hydrodynamic contribution. J Adv Geosci 4:57–62

Neilson M, Stevens R, Hodson J (1984) Documentation of the averaging lake data by regions (ALDAR) program, Tech B No 130, IWD-ON, Burlington, Ontario

Rawson DS (1950) The physical limnology of Great Slave Lake. J Fish Res Board Can 8:3–66

Rouse WR, Binyamin J, Blanken PD, Bussieres N, Duguay CR, Oswald CJ, Schertzer WM, Spence C (2007a) The influence of lakes on the regional energy and water balance of the central Mackenzie River Basin. (Vol. I, this book)

Rouse WR, Blyth EM, Crawford RW, Gyakum JR, Janowicz JR, Kochtubajda B, Leighton HG, Marsh P, Martz L, Pietroniro A, Richie H, Schertzer WM, Soulis ED, Stewart RE, Strong, GS, Woo, MK (2002) Energy and water cycles in a high latitude, north-flowing river system: summary of results from the Mackenzie GEWEX Study – Phase I. B Am Meteorol Soc 84:73–87

Rouse WR, Oswald CM, Binyamin J, Blanken PD, Schertzer WM, Spence C (2003) Interannual and seasonal variability of the surface energy balance and temperature of central Great Slave Lake. J Hydrometeorol 4:720–730

Rouse WR, Blanken PD, Duguay CR, Schertzer, WM (2007b) Climate-lake interactions. (Vol. II, this book)

Schertzer WM (1997) Freshwater lakes. In: Bailey WG, Oke TE, Rouse WR (eds) The surface climates of Canada, McGill-Queens University Press, Montreal, Canada, Chap. 6, pp 124–148

Schertzer WM (2000) Digital bathymetry of Great Slave Lake, NWRI Contrib No 00–257, National Water Research Institute, Burlington, ON, Canada

Schertzer WM, Croley II TE (1999) Climate and lake responses. In: Lam DCL, Schertzer WM (eds) Potential climate change effects on Great Lakes hydrodynamics and water quality, American Society of Civil Engineers (ASCE) Press, Reston, Virginia, pp 2-1–2-74

Schertzer WM, Rouse WR, Blanken PD (2000) Cross-lake variation of physical limnological and climatological processes of Great Slave Lake. Phys Geog 21:385–406

Schertzer WM, Rouse WR, Blanken PD, Walker AE (2003) Over-lake meteorology and estimated bulk heat exchange of Great Slave Lake in 1998 and 1999. J Hydrometeorol 4:649–659

Schertzer WM, Rouse WR, Lam DCL, Bonin D, Mortsch LD (2004) Climate variability and change – lakes and reservoirs. In: Threats to water availability in Canada. National Water Research Institute, Burlington, ON, NWRI scientific assessment report series no 3 and ACSD science assessment series no 1, pp 91–100

Schertzer WM, Saylor JH, Boyce FM, Robertson DG, Rosa F (1987) Seasonal thermal cycle of Lake Erie. J Great Lakes Res 13:468-486

Shore Protection Manual (1984) Shore protection manual, vol 1. U.S. Army Corps of Engineers, Coastal Engineering Research, 4th ed, Vicksburg, Mississippi

Simons TJ, Schertzer WM (1987) Stratification, currents and upwelling in Lake Ontario, summer 1982. Can J Fish Aquat Sci 44:2047–2058

Stewart RE, Leighton HG, Marsh P, Moore GWK, Richie H, Rouse WR, Soulis ED, Strong GS, Crawford RW, Kochtubajda B (1998) The Mackenzie GEWEX Study: the water and energy cycles of a major North American river system. B Am Meteorol Soc 79:2665–2683 van der Leeden F, Troise FL, Todd DK (1990) The water encyclopedia. Lewis Publ Inc, Chelsea, MI, USA

Walker AE, Davey MR (1993) Observation of Great Slave Lake ice freeze-up and breakup processes using passive microwave satellite data. Proc 16th symposium on remote sensing, L'AQT/CRSS, pp 233–238

Walker AE, Silis A, Metcalf J, Davey M, Brown R, Goodison B (1999) Snow cover and lake ice determination in the MAGS region using passive microwave satellite and conventional data. Proc 4th Scientific Workshop, Mackenzie GEWEX Study, Nov 16–18, 1998, Montreal, Quebec, pp 89–91

Chapter 12

Flow Connectivity of a Lake–Stream System in a Semi-arid Precambrian Shield Environment

Ming-ko Woo and Corrinne Mielko

Abstract Many small lakes occupy Precambrian Shield and lowland areas in the boreal region. To investigate the processes causing seasonal severance of flow connection in the lake-stream system, a chain of lakes in northern Canada was studied in 2004. Water balance shows that rapid and substantial runoff from the local basin slopes during the snowmelt period led to a rise of lake levels above their outlet elevations to generate outflow. Continued summer evaporation caused drawdown of lake storage below the outflow thresholds, represented by the lake outlet elevations. Outflow ceased and the lakes became disconnected. Summer rainfall in a semi-arid environment was insufficient to overcome storage deficit to re-establish flow connectivity among all lakes. For the drainage system as a whole, streamflow interruption or continuity depends on linkage of its lake-stream sub-units. The principle of fill and spill of lakes should be considered in modeling Shield hydrology under semi-arid conditions, to take account of (1) antecedent storage in individual lakes, (2) their storage change calculated through water balance, and (3) the thresholds to be exceeded for outflows to occur.

1 Introduction

Many areas in the Canadian and Eurasian subarctic are occupied by Precambrian rocks, with landscapes comprising bedrock uplands and flat-bottomed valleys containing lakes and wetlands. Most investigations of lake hydrology were in humid environments where there is usually adequate precipitation and inflow to maintain continuous discharge from the lakes. During dry subarctic summers, however, lakes have been observed to experience a draw down that sometimes cuts off their outflow, isolating them into disjointed hydrologic entities. Spence (2000), for example, observed flow cessation in a small headwater lake that produced flow for only 13 days in an entire year. The processes relating surface flow connections with lake storage remain inadequately understood.

Attention has been paid to flow connectivity in relation to runoff generation in a variety of landscapes, including northern wetlands (Bowling et al. 2003; Quinton et al. 2003), bedrock upland with soil-filled valleys in temperate and boreal latitudes (Branfireun and Roulet 1998; Buttle et al. 2004; Spence and Woo 2003), and lake basins in the subarctic and the Arctic (FitzGibbon and Dunne 1981; Woo et al. 1981). Mielko and Woo (2005) investigated the runoff processes of a subarctic Shield lake and its catchment during the snowmelt season and confirmed the importance of lake storage status in streamflow generation. The study was restricted to the snowmelt season and referred to only a single lake. The findings need to be enlarged to encompass lakes and streams in a drainage system throughout the spring and summer periods. It is the purpose of this study to investigate the flow connectivity in a subarctic Shield setting, to understand the processes operating in a lake-stream drainage network. Results of this study will increase knowledge on the behavior of the subarctic hydrologic system that can lead to improved algorithm development to model Shield hydrology in a semi-arid environment abundant with lakes.

2 Study Area

This study was carried out from mid-April to the end of August 2004, in a headwater catchment (62°33'N; 114°21'W) located 15 km north of Yellowknife, Northwest Territories, Canada. It is situated in the subarctic Canadian Shield that is scattered with numerous small lakes. The climate is subarctic and semi-arid, with annual precipitation recorded at Yellowknife Airport of about 280 mm, 116 mm of which is snowfall (Environment Canada 2004). Mean January temperature falls to -26.8°C and mean July temperature reaches 16.8°C. The area is dominated hydrologically by a large snowmelt freshet followed by a gradual drying of the landscape. The study basin covers 4.76 km^2, consisting of a chain of four lakes with surface areas between 0.1 km^2 and 0.6 km^2, and a contributing basin of 3.15 km^2 draining to the lowest lake. The chain of lakes is the focus of the present study. The uppermost Melvin Lake drains into Shadow Lake, which in turn drains to Hazel Lake and finally to Lois Lake (Fig. 1 in Spence and Woo 2007). The lowest, Lois Lake, receives additional surface flows from Rater Lake that lies outside of the chain. These lakes are linked by channels that have intermittent flows. In addition to lakes, the basin includes crystalline bedrock outcrops, soil-filled valleys, and wetlands.

3 Methods

Detailed description of the field techniques and the instruments used are presented in Mielko and Woo (2005). The deployment of instrumentation was enlarged for this study (Fig. 1) and following is a summary of the methods employed.

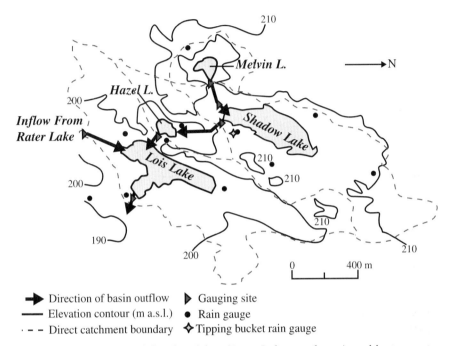

Fig. 1. Study area containing four lakes (Rater Lake not shown), and instrumentation sites

Lake levels for five lakes (Melvin, Shadow, Hazel, Lois, and Rater) were recorded at half-hour intervals, commencing after the onset of lake ice breakup, which enabled the installation of the water level gauges. Rainfall was measured using a tipping-bucket gauge and ten manual gauges located throughout the basin. Both the snow and ice on the lake are treated as antecedent storage in the lake, being solid inputs and attributed to phase change of the lake water in the winter. Lake stage reflects the hydrostatic pressure exerted by the snow and ice on the lake water, and therefore is an integration of both the solid and liquid water in the lake. Photographs were taken of the decaying lake ice covers during the spring to estimate the changing ice-free fraction. Evaporation was calculated half hourly by the

Priestley and Taylor (1972) method for the ice-free portion of each lake, using measured radiation, air, and water temperatures. Channel flow into and out of the lakes was gauged at high and moderate flows using a velocity-area method with velocity obtained by a pygmy current-meter, and at low flows by collecting a measured volume over a fixed time (several minutes). These periodically measured flows were related to lake levels to establish rating curves that allowed the conversion of lake level records into hourly discharges. As this study concentrates on the chain of four lakes from Melvin to Lois, the Rater Lake sub-basin was not instrumented and only the level and the outflow of Rater Lake were monitored.

4 Water Balance

Examination of lake water balance permits an understanding of how each of its components influences lake storage at different times of the thawed season. The water balance of a lake is (expressed in mm day^{-1}):

$$\Delta S = R - E + Q_i - Q_o + Q_b^* + \xi \qquad (1)$$

Here, rainfall on the lake (R), storage change based on lake level fluctuation (ΔS), flows along stream channels into (Q_i) and out of the lake (Q_o) were measured directly, and lake evaporation (E) was calculated by the Priestley and Taylor method using measured variables. Q_b^* is the net flux of water delivered to or leaked from the lake via overland or subsurface flows in its direct catchment area; and ξ is the error in the evaluation of the water balance components. Both Q_b^* and ξ cannot be assessed independently. The magnitude of ξ cannot be established but is assumed to be small relative to the magnitude of various components of the water balance. Q_b^* was then obtained as a residual term by rearranging Eq. (1). Water balance calculations were performed after much of the basin snow had melted, at the onset of lake ice breakup. Figure 2 is a plot of the daily values of water balance for the four lakes studied and Table 1 provides their total magnitudes for the study period.

Snowmelt on the basin slopes produced a large influx of Q_b^* that reached the lake through overland and subsurface flows (Mielko and Woo 2005). After the melt period (all the basin snow disappeared by 26 May), Q_b^* became negative, suggesting that there has been a continuous groundwater loss from the lakes. Despite the bedrock structure, seepage loss is highly plausible because the rock fractures can be effective conduits for water (Spence and Woo 2007; Thorne et al. 1999). Lake evaporation incr-

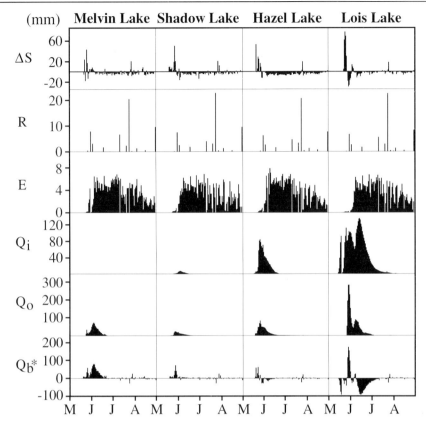

Fig. 2. Daily variation of water balance components of four study lakes from snowmelt to late summer, 2004

Table 1. Water balance (in mm) of four lakes for the period May 5 to August 31, 2004

	Melvin	Shadow	Hazel	Lois
Lake area [km^2]	0.009	0.087	0.007	0.108
Rainfall	54	53	50	53
Evaporation	349	310	343	326
Stream inflow to lake	0	91	980	3523
Stream outflow from lake	880	283	933	2413
Net exchange with basin	1052	343	-17.05	-821
Net storage change	-123	-107	-264	17

eased steadily in May when the lake ice cover diminished. Afterwards, lake evaporation averaged 3.3 mm d^{-1} and was the main process responsible for lake storage decline. A 20-22 mm rainstorm event in late July raised the water levels in all four lakes. Lake inflow and outflow were observed at all the lakes during the snowmelt period but summer rainfall events raised channeled outflow from only one lake (Lois).

Several generalizations can be drawn from the water balance study. At the beginning of the study period and prior to snowmelt, there was zero flow in the channels that enter or leave the lakes. This was a time when lake levels were below the elevation of their outlets. Slope runoff in the melt season was the dominant process that led to a sharp rise in storage. In this regard, both the intensity and the amount of flow are important, hence the rate and duration of snowmelt on the basin slopes are significant considerations (Mielko and Woo 2005). Evaporation, inhibited when a lake ice cover was present, continued throughout the summer to deplete lake storage. Although rainfall can produce a rapid rise in lake level, the magnitude can seldom match the rise due to snowmelt runoff to the lake because intense rain is seldom sustained in the semi-arid environment. Water balance calculations indicate that a lake can be recharged by slope runoff and can also lose water through subsurface conduits. Major hydrologic exchanges are between the lakes and their local catchment area, but except for the uppermost Melvin Lake, all the lakes receive periodic stream inflow from the lakes above them.

5 Lake Outflow Generation

The snowmelt season is the primary period for outflow generation. Of note is that the starting and ending dates differed among the lakes (Table 2). Hazel Lake, the third along the chain of lakes, was the first to generate outflow, followed by Shadow Lake above it. Thus, there was no systematic downstream sequencing in the starting or ending dates of flow, and outflow generation from a lake is not necessarily predicated upon the release of discharge from a lake upstream along the lake-stream system.

For the entire study period, Melvin Lake, the highest in elevation but with the smallest drainage area, had the lowest volume of outflow. On the other hand, Lois Lake, which receives inflows from both Hazel and Rater lakes, yielded the largest outflow. Shadow and Hazel lakes had intermediate and comparable volumes but different timings of outflow. For all the lakes, snowmelt was the dominant, if not the only period when lake outflow was produced. Rainfall was able to revive outflow only for Lois Lake.

Field observation of outflow occurrence shows that a lake has to rise above the lip of its outlet which marks the flow threshold. This threshold is normally the lowest point along the perimeter of the lake and can be a bedrock sill, a channel carved in soil, the bottom of culvert such as for Rater Lake, or a blockage by a beaver dam as in the case of Lois Lake. In the spring, snow drift and ice often raises the threshold elevations so that the lake water would be impounded to a greater height than in the summer (Fig. 3). Such situations have been reported for other lakes in the subarctic (FitzGibbon and Dunne 1981) and in the Arctic (Woo et al. 1981).

Table 2. Start dates, end dates, and total volumes of outflow from the five study lakes

	Melvin	Shadow	Hazel	Lois	Rater
Direct catchment area[a] [km^2]	0.116	0.565	0.168	0.760	3.150
Basin area[b] [km^2]	0.116	0.681	0.849	4.759	3.150
Lake area [km^2]	0.009	0.087	0.007	0.108	0.214
Lake/basin ratio [%]	7.8	12.8	0.8	2.3	6.8
Snowmelt period outflow					
Start date	May 24	May 22	May 20	May 24	May 30
End date	June 25	June 25	June 15	July 5	-
Total flow volume [m^3]	8,265	24,490	23,329	260,693	
Summer season outflow					
Start date				July 26	-
End date				July 31	Aug. 24
Total flow volume [m^3]				4	173,328[c]

[a]Direct catchment area refers to the basin area that feeds directly to the lake and exclusive of the area contributing to streamflow above the lake inlet.
[b]Total basin area that drains into a lake, including its direct catchment and the areas upstream.
[c]Outflow for Rater Lake was continuous between May 30 and August 25; number refers to total flow during this period.

6 Lake Storage and Outflow

Field information can be combined with water balance analysis to relate lake storage with outflow. The fill-and-spill concept, previously applied to soil-filled valleys in the Shield environment to explain flow generation

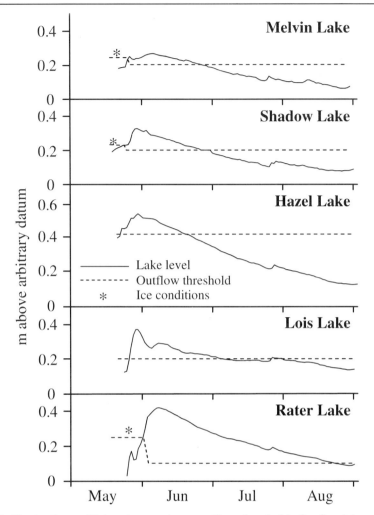

Fig. 3. Fluctuations of lake storage above outflow thresholds for five lakes. Note that threshold elevations can change when the lake outlets are blocked by snow and ice

(Spence and Woo 2003), can be extended to the lake-stream flow system. For a lake, outflow (Q_o) occurs when its storage level exceeds the flow threshold (S_T):

$$Q_o = 0, \text{ if } (S_{t-1} + \Delta S_t) < S_T \qquad (2)$$

where S_{t-1} is lake storage level at end of the past period t-1; ΔS_t is the change in storage for the current period t, obtained by water balance through Eq.

(1). Thus, whether outflow can be produced is predicated upon antecedent storage (S_{t-1}), the change in storage for time t, and the threshold that must be exceeded, and this threshold can be higher in the spring when the lake outlet is blocked by snow and ice.

It was previously thought that a lake with small area relative to its catchment size (<5%) will not significantly attenuate streamflow because lake storage then plays a relatively little role compared with the magnitude of basin runoff (FitzGibbon and Dunne 1981, p 282). Table 2 shows that Lois Lake with a low lake-to-basin ratio of 2% was able to arrest the flow from Rater Lake throughout most of the summer. This suggests that a low ratio is not a satisfactory indicator of flow attenuation; rather, it is the capacity of lake storage relative to water input that is of concern.

7 Flow Connectivity of Lake–Stream System

Outflow generation does not proceed systematically from the highest to the lowest lake, nor does outflow stoppage follow any ordered sequence along the chain of lakes. Hazel Lake in the middle of the chain was the first to generate and terminate outflow, whereas Shadow Lake above it continued to discharge for almost two more weeks. As flow connectivity does not progress systematically along the drainage network, each lake can form a subsystem that may or may not link with others after certain hydrologic events.

The physical setting of a lake plays an important role in terms of it hydrologic connectivity within the flow system. The Precambrian Shield rock is composed mainly of impervious granite and gneiss, with an occasional veneer of soil cover on the uplands, that allows efficient shedding of rain and meltwater, though infiltration is encouraged where fissures abound (Spence and Woo 2002). Fast delivery of water from the basin slopes causes rapid increase in lake storage, resulting in quick rise of lake level above the outlet threshold to produce outfow. Under cold subarctic conditions, a long duration of the lake ice cover shortens the evaporation season. Lake level drawdown due to evaporation is confined to the ice-free period but daily evaporation in the summer can be large because of long daylight hours. Finally, a semi-arid climate ensures that post-snowmelt evaporation loss exceeds rainfall input so that the summer water balance favors lake level decline that leads to periods of no outflow. Outflow cessation is less commonly reported for Shield lakes in humid temperate areas due to lake storage replenishment by high rainfall.

8 Implications on Streamflow

In a humid environment, discharge often increases downstream and this may be attributed to an enlarged area of flow contribution from the basin slopes that exceeds the retention capacity of channel storage. The presence of lakes represents a significant storage along the drainage channels so that lakes modify the shape of the hydrographs. In semi-arid and arid areas, the vertical water balance of the lake and its direct catchment area (here referred to as the basin area that feeds directly to the lake and exclusive of the area contributing to streamflow above the lake inlet) can overwhelm the magnitude of channel inflow-outflow to distort the downstream change in flow normally exhibited in rivers of the humid region. Flow interruption arises when connection between a lake and its adjacent streams is severed. The hydrograph of Lois Lake outlet offers such an example. Outflow from this basin was maintained for only 49 days in 2004.

The same fill-and-spill processes apply to other headwater catchments and to larger drainage systems in a lake-studded landscape under cold, semi-arid conditions. Baker Creek (62°30'48"N; 114°24'34"W) offers examples that a larger basin (area 121 km^2 or two orders of magnitude larger than the study basin) also experienced flow discontinuities in the summer (Water Survey of Canada 2003). This Shield basin has a number of lakes that form parts of a lake-stream network. Figure 4 shows that in 1998, flow ceased on August 19-24, but was revived by rainfall in early November. In 2001 and 2004, Baker Creek discharge fell below 0.01 m^3 s^{-1} after August 19 and remained so for half a month. Similarly, in 2004 its flow dropped below 0.01 m^3 s^{-1} in mid September. Such zero or negligible flows occurred late in the dry summers after much water was lost to evaporation. It is proposed that the severance of connection among the lakes effectively reduced the source areas that maintained flow for the Creek, as discussed by Spence (2006) who mapped the changing runoff contributing areas in the basin for two summers (Fig. 7 in Spence and Woo 2007).

Fill and spill of individual lakes represents the major mechanism for streamflow generation in the Shield lake-stream system. This concept can be applied to the modeling of flow connectivity along a chain of lakes. Hydrologic models usually use routing procedures to treat flow attenuation along channels, or reservoir schemes to represent flow retention and release. This study shows that for a natural lake-stream system, particularly in a dry environment, lake storage plays a deciding role in outflow production. Each lake and its direct catchment may be considered to operate as an independent unit along the chain, subject to variable channel flow linkages with the entities

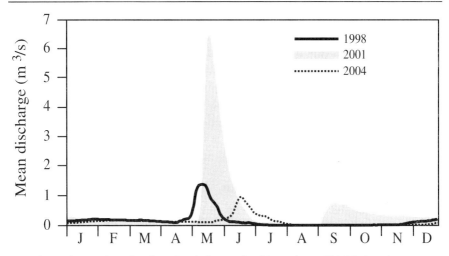

Fig. 4. Hydrographs of Baker Creek in semi-arid northern Shield showing occurrences of zero or negligible flows during open channel (ice-free) conditions

above or below. In modeling the lake-stream system, therefore, not only the channel flows but the storage capacity and the water balance of the direct catchment areas for each lake must be represented explicitly.

9 Discussion and Conclusions

Investigation of flow connectivity between lakes and streamflow in the semi-arid northern Canadian Shield demonstrates that outflow production is governed by (1) antecedent storage in the lake, which is a product of cumulative water balance in the past periods, (2) lake water balance of the current period, which includes vertical fluxes due to snowmelt, rainfall and evaporation, lateral exchanges with the slopes that drain directly to the lake, and inflow and outflow, and (3) lake storage capacity, which sets the threshold for lake outflow. Outflow occurs only when the threshold is exceeded, hence channel flow is variable in time and flow linkage need not be continuous along a chain of lakes. This confirms the significance of the fill-and-spill concept in outflow generation from small lakes. While the processes described are applicable to arid and semi-arid areas subject to alternating flooding and drying, the lake fill-and-spill principle is also relevant to the humid region, but with ample water inputs to the lakes, the severance of channel flow linkage between lakes in the humid Shield areas is uncommon. The concept can be applied to the development of algorithms for modeling Shield hydrology.

A process-based model of flow release and stoppage for lakes will permit improved prediction of flow reliability which can be an important consideration in water resource management in regions with many small lakes. This modeling approach is not limited only to lake-stream flow systems, but can be adapted to predict flow in other systems and environments where the fill-and-spill principle is at work, including those presented by Spence and Woo (2003) and Tromp-van Meerveld and McDonnell (2006).

Acknowledgements

The Yellowknife Dene First Nation generously granted permission to carry out this study in their territory. Logistical and instrumentation support was provided by the Prairie and Northern Region of Environment Canada, through the assistance of Jess Jasper and Chris Spence. We acknowledge the following for their assistance in the field and for offering supplemental data: Bob Reid, Paul Saso, May Guan, Paula Lucidi, Derek Faria, and Todd Pagett.

References

Bowling LC, Kane DL, Gieck RE, Hinzman LD, Lettenmaier DP (2003) The role of surface storage in a low-gradient Arctic watershed. Water Resour Res 39: 1087, doi:10.1029/2002WR001466

Branfireun BA, Roulet NT (1998) The baseflow and storm hydrology of a Precambrian shield peatland. Hydrol Process 12:57–72

Buttle JM, Dillon PJ, Eerkes GR (2004) Hydrologic coupling of slopes, riparian zones, and streams: an example from the Canadian Shield. J Hydrol 287:161–177

Environment Canada (2004) Canadian Climate Normals 1961–2000. Ministry of Supply and Services, Ottawa, Canada

FitzGibbon JE, Dunne T (1981) Land surface and lake storage during snowmelt runoff in a subarctic drainage system. Arctic Alpine Res 13:277–285

Mielko C, Woo MK (2005) Snowmelt runoff processes in a headwater lake and its catchment, subarctic Canadian Shield. Hydrol Process 20:987–1000

Priestley CHS, Taylor RJ (1972) On the assessment of surface heat flux and evaporation using large scale parameters. Mon Weather Rev 100:81–92

Quinton WL, Hayashi M, Pietroniro A (2003) Connectivity and storage functions of channel fens and flat bogs in northern basins. Hydrol Process 17:3665–3684

Spence C (2000) The effect of storage on runoff from a headwater subarctic Shield basin. Arctic 53:237–247

Spence C (2006) Hydrological processes and streamflow in a lake dominated water course. Hydrol Process 20:3665–3681

Spence C, Woo MK (2002) Hydrology of a subarctic Canadian shield: bedrock upland. J Hydrol 262:111–127

Spence C, Woo MK (2003) Hydrology of a subarctic Canadian shield: soil-filled valleys. J Hydrol 279:156–166

Spence C, Woo MK (2007) Hydrology of the northwestern subarctic Canadian Shield. (Vol. II, this book)

Thorne G, Laporte J, Clarke D (1999) Water budget of an upland outcrop recharge area in granitic rock terrane of southeastern Manitoba, Canada. Proc 12th International Northern Research Basins Symposium and Workshop, Reykjavik, Iceland, August 23–37, pp 317–330

Tromp-van Meerveld HJ, McDonnell JJ (2006) Threshold relationships in subsurface stormflow: 2. The fill and spill hypothesis. Water Resour Res 42: W02411, doi:10.1029/2004WR003800

Water Survey of Canada (2003) HYDAT: surface water and sediment data. CD-ROM, Environment Canada

Woo MK, Heron R, Steer P (1981) Catchment hydrology of a high Arctic lake. Cold Reg Sci Technol 5:29–41

Chapter 13

Hydrology of the Northwestern Subarctic Canadian Shield

Christopher Spence and Ming-ko Woo

Abstract The Canadian Shield occupies about one-third of Canada's land area, one-fifth of the Mackenzie Basin, and contains much of Canada's freshwater resource. The northwestern portion of the Shield has a relatively dry, subarctic climate and a heterogeneous landscape comprising exposed bedrock uplands, soil mantled slopes, wetlands and lakes. The water budget of bedrock uplands is influenced by rock fractures, slope aspect, precipitation or melt intensity, and storage capacity. The lateral transfer of runoff from exposed bedrock is crucial to the hydrology of downslope soil-filled areas where the water maintains storage to sustain evaporation and support streamflow generation. Headwater lakes are an important store of water that must be recharged to storage capacity prior to yielding outflow to higher order streams. Even in perennial streams, flow thresholds must be exceeded before runoff generated in headwaters can be transferred to basin outlets. The subarctic Shield environment has large spatial and temporal variations in available storage, where the hydrologic connectivity that dictates runoff response is controlled in part by the geometry and spatial distribution of landscape components. Streamflow regime may be dominated by snowmelt runoff from the land (nival regime) or modified by storage effects of lakes along the channel network (prolacustrine regime).

1 Landscape and Hydrologic Considerations

The Canadian Shield occupies about one-third of Canada's land area, extending from the humid temperate latitudes of southern Ontario and southern Quebec to the eastern Queen Elizabeth Islands of the High Arctic. The Shield constitutes the eastern fifth of the Mackenzie River Basin (MRB) where the Precambrian rock was glaciated by the Laurentide Ice Sheet to produce a rolling topography. The landscape includes undulating bedrock uplands, soil mantled slopes and soil-filled valleys that often contain wetlands and lakes, with open water accounting for about 25% of the Shield area (Fig. 1). Drainage networks are controlled by the bedrock structure to produce rectangular drainage patterns.

Fig. 1. Typical Canadian Shield landscapes with exposed bedrock uplands, soil-filled valleys and numerous beautiful lakes. Shadow Lake is shown at the center (Photo: C. Mielko)

The main bedrock types are granite, gneiss, migmatite, and metavolcanic and metasedimentary rocks of Archean age and less metamorphosed Proterozoic sedimentary rocks. Bedrock uplands tend to be moderately to highly fissured with joints and exfoliation fractures. Glaciofluvial deposits in the forms of drumlins, eskers and deltas appear throughout the physiographic province. Depressions left by glacial action are now filled predominantly by Dystric Brunisols. Turbic and Organic Cryosols are found in poorly drained frozen peat-filled depressions. The Canadian Shield has a wide range of soil climates with discontinuous permafrost present north of a broad arc extending from northern Saskatchewan, south of James Bay and east to southern Labrador. Large changes in topography, vegetation, winter snow accumulation, local hydrology and geology over short distances result in abrupt soil temperature transitions in the discontinuous permafrost zone. Permafrost is typically found in wetlands and peat plateaus but seldom occurs in exposed bedrock and well drained coarse thin overburden (Wolfe 1998). Permafrost generally becomes continuous north of the tree-line (Brown 1978).

Arboreal vegetation in the subarctic shield ecozone is predominantly open black spruce (*Picea mariana*) forest with periodic stands of white spruce (*Picea glauca*), jack pine (*Pinus banksiana*), birch (*Betula spp.*) and aspen *(Populus tremuloides)*. The open canopy allows for a prominent understory dominated by dwarf willow (*Salix* spp.), Labrador tea (*Ledum groenlandicum*), blueberry (*Vaccinium augustifolium*) and bog cranberry (*Vaccinium vitis-idaea*). Ground cover on bedrock outcrops includes moss and lichen (*Cladina* and *Cladonia* spp.). Peat wetlands are common. *Sphagnum* moss species cover expansive parts of wetlands, with lichen, grass (*Eriophorum* spp.) and sedges (*Carex* spp.) also present. Organic soils derived from *sphagnum* decomposition are commonly 0.5 m thick or more and are underlain by either bedrock or sandy till. Forest fires (Kochtubajda et al. 2007) create vegetation communities that vary widely in composition and age.

The climate of northwestern subarctic Shield may be characterized by the record at Yellowknife (1971–2000 average); see http://www.climate.weatheroffice.ec.gc.ca/climate_normals/index_e.html. It has short cool summers with a July daily average temperature of 17°C, and long cold winters with a January average temperature of -27°C. Annual unadjusted precipitation for the same period averages 280 mm, 55% of which falls as snow. Convective cells produce much of the summer precipitation, with large interannual variability (Szeto et al. 2007). The weather becomes cool and damp in autumn when periodic synoptic conditions allow for frequent travels of cyclones across the region (Spence and Rausch 2005). Annual snow cover begins in October and lasts until the end of April and beginning of May.

Hydrologic characteristics of the Shield region have been examined at different scales, from bedrock uplands to small headwater areas to lake-dominated basins. This chapter provides a synopsis of several studies conducted in the semi-arid Shield environment of the MRB.

2 Bedrock Upland Hydrology

2.1 Snow Processes

Snow accumulation patterns vary widely across bedrock uplands due to drifting events that often leave ridge tops free of snow or covered by shallow dense snow, and fill most depressions with deep snowdrifts. Local topography and slope orientation affect the timing and rate of snowmelt.

Normally, snow on south facing slopes and ridge tops melts first, followed by areas with average snow depth and finally the large drifts.

2.2 Storage

Landals and Gill (1972) found that near Yellowknife, the water storage capacity of bedrock terrain is about 13 mm, which is in agreement with the 8 mm (Peters et al. 1995) and 6mm (Allan and Roulet 1994) values noted for Shield areas in Ontario. Spence and Woo (2002) observed that only depressions deeper than 50 mm are likely to contain soil. The smallest depressions (<10 mm) are found exclusively on unfractured bedrock surfaces, while deeper depressions are often associated with fissures. The frequency distributions of mean depth and areal extent of depressions (Fig. 2) are both highly skewed, indicating that smaller depressions dominate.

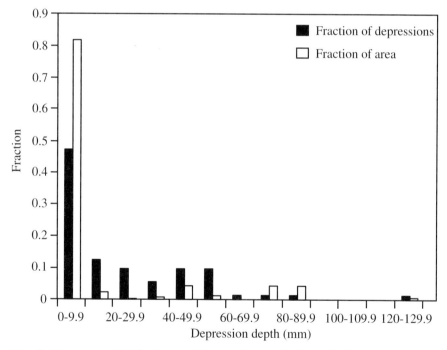

Fig. 2. Frequency distributions of depressions and size relative to depression depth in bedrock plots at Pocket Lake, NWT

Deep depressions reach their water storage capacity more slowly and less often than shallow ones, but the deeper depressions can retain water longer after rainfall (Fig. 3). Spence and Woo (2002) documented an example of a 0.025 m³ depression on a bedrock slope that, after drying out following spring snowmelt, did not reach capacity again for four months. The time resolution in Fig. 3b is too coarse to show that storage capacity in shallow depressions can be fully attained and the water can be lost completely to evaporation within a single day. Only those days in late summer and fall with low evaporation, or periods of intense rain in the midsummer allow storage in these depressions to remain longer than a day.

2.3 Evaporation

Evaporation occurs from both the exposed rock face and soil patches. Spence and Woo (2002) found that depending on water availability, summer rates of evaporation from exposed bedrock can be as high as 2.0 mm d^{-1}. Daily values of Priestley-Taylor α average 1.7 in the summer months, a value considerably higher than the 1.26 usually reported for northern locations (Rouse et al. 1977) but not unexpected given the inevitable advection of heat from the dry, warm granite surfaces surrounding the ponded water and wet soil.

As most summer rainfall in the western subarctic is of low intensity that lasts for several hours, much of the rain is evaporated from the water film on the bedrock surfaces. Water held in deeper depressions may sustain evaporation for several days after rainfall. However, deep depressions constitute only a small fraction of the surface area, and it is the larger areal coverage of the shallow depressions that dictates the overall plot evaporation (Fig. 3c). Moisture availability plays a greater role than energy in affecting evaporation. At a site near Yellowknife, a late summer (August and September of 2000) with lower temperature and higher humidity than the mid-summer accounted for 35% of the annual evaporation due to the frequently wet conditions of the rock surfaces (Spence and Woo 2002).

2.4 Infiltration

The groundwater flow system in Shield rock may be treated as impermeable blocks dissected by fractures, the latter being the only effective conduits of flow (Davison, 1984). A parallel plate analogue can be used to calculate fluid velocity (K_f in m s^{-1}) within individual fractures (Domenico and Schwartz 1998; Raven et al. 1985):

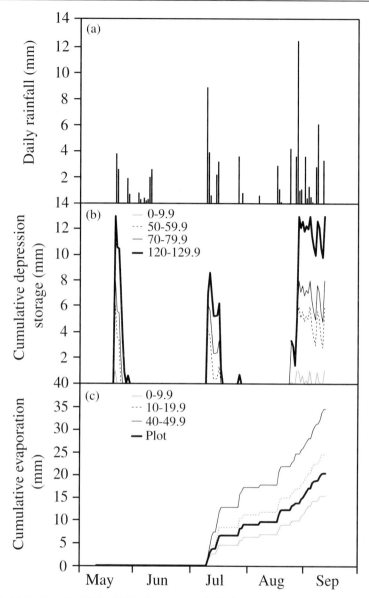

Fig. 3. (a) Daily rainfall in 1999, (b) storage in bedrock depressions of various depression-depth (mm) classes, and (c) cumulative evaporation from a bedrock plot and bedrock depressions; at a site near Pocket Lake, NWT

$$K_f = b^2 g (12\mu)^{-1} \tag{1}$$

where b is fracture width (m), g is acceleration due to gravity (m s^{-2}) and μ is the kinematic viscosity of water (m s^{-2}). Equation 1 shows that a few wide fractures will greatly enhance infiltration into bedrock. The influence of fracture width is pervasive, proving to be more important than fracture density (Spence and Woo 2002). The capacity of fractures to transfer water is crucial to infiltration. Both Spence and Woo (2002) and Thorne et al. (1999) documented cases when more intense storms and snowmelt events experienced less infiltration as the K_f of individual fractures was exceeded. Well connected fractures transfer water easily from the surface to depth, thus preventing the water from remaining in the surface cracks to freeze in the winter (Spence and Woo 2002). Frozen bedrock does not inhibit infiltration unless its fractures are filled with ice. Thorne et al. (1994) observed a water table rise in frozen Precambrian bedrock in southeastern Manitoba, Canada, corresponding with snowmelt each spring. Fracflow (1998) working in a mine below Yellowknife, found that meteoric water could infiltrate the mine workings, with infiltration controlled by snowmelt rate in the spring and by rainfall intensity in the summer.

2.5 Runoff

The magnitude of runoff during a given storm depends on the presence and distribution of soil on the bedrock and the meteorological and antecedent moisture conditions. The presence of a soil cover significantly reduces runoff because it greatly enhances water storage (Landals and Gill 1972). Spence and Woo (2002) documented a 0.44 difference in average runoff ratio (ratio of runoff to rainfall or snowmelt) between bare rock and slopes with soil patches (Fig. 4). Runoff ratios may double once soil patches on upper slopes become saturated (Spence and Woo 2002). Soil covered zones downslope of exposed bedrock can delay, reduce or even nullify runoff produced during some rain events (Allan and Roulet 1994; McDonnell and Taylor 1987; Spence and Woo 2002).

Evaporative demand affects the amount of water left for runoff generation (Landals and Gill 1972; Roulet 1990). Opportunity time is crucial. Long and steady rain events support more evaporation at the expense of runoff. However, high intensity rain allows storage requirements to be met quickly to permit runoff initiation. The magnitude of runoff is controlled by storage, as is supported by Spence and Woo's (2002) observation that the runoff ratio in bedrock slopes decreases with large storage capacity.

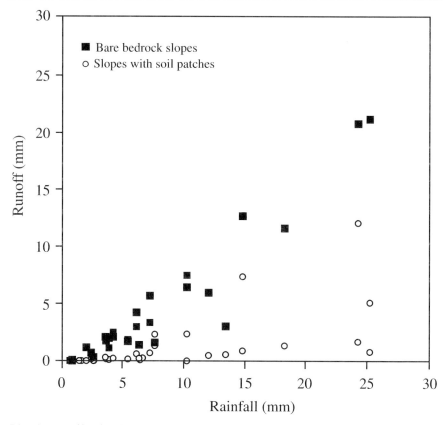

Fig. 4. Runoff ratios from bedrock plots near Pocket Lake, NWT

3 Soil-filled Valley Hydrology

Runoff produced on uplands is transferred downslope where it is often modified by the storage and flow delivery mechanisms in the soil-filled zones (Allan and Roulet 1994; Buttle and Sami 1992; Peters et al. 1995). There is a wide range of runoff ratios reported for such headwater areas (Branfireun and Roulet 1998; Landals and Gill 1972; Spence and Woo 2003). Variable soil depths across the Canadian Shield landscape result in a diverse range of storage capacity (Devito et al. 1996). The influence of deep soils on hydrologic connections depends on whether conditions are wet (Branfireun and Roulet 1998) or dry (Spence and Woo 2003). The presence of saturated frozen soils (Metcalfe and Buttle 2001) in some locations further complicates the hydrology.

3.1 Water Sources

Sources of water to soil-filled valleys include rainfall and snowmelt, as well as lateral transfer of runoff from upslope. Recharge occurs in the spring and fall when there are large amounts of water and small net evaporation losses, respectively (Thorne 1992; Thorne et al. 1994). Recharge during spring snowmelt depends on the condition of the frozen soil. Meltwater infiltration is likely in frozen soils with low ice content (Pomeroy et al. 2007), possible due to dessication or good drainage before the freeze-up. However, infiltration is limited where concrete frost is present, as in most valley wetlands.

Vertical recharge from summer rain depends on soil wetness. A ground cover of lichen and moss common in subarctic locations can intercept and withhold much rainwater, especially when dry (Bello and Arama 1980). Spence and Woo (2003) noted an interception fraction of 0.6 over a summer in a dry valley near Yellowknife. In dry regions of the Canadian Shield, the main source of recharge is from lateral inflow that enters the valley through seepage along the soil-bedrock interface. This process can be observed even under frozen-soil conditions because soils at the edges of the valleys tend to be well drained and do not have much pore ice to block the passage of water. Such a mode of inflow avoids interception loss from the vegetation and feeds directly to the valley soil storage.

3.2 Water Delivery and Storage

Storage tends to be highest soon after snowmelt recharge. Afterwards, water table in the valley declines and then flattens as dry summer conditions cause soil moisture deficits. When large rainfall occurs, lateral inputs are re-activated if the upland areas are large enough to produce sufficient runoff. As more water enters the valley along the bedrock surface, it gives rise to a saturated layer along the sides of the valley while the middle of the valley is still dry (Bottomley et al. 1986; Branfireun et al. 1996; Peters et al. 1995; Renzetti et al. 1992; Wells et al. 1991). Configuration of bedrock topography further facilitates a rapid rise and an earlier peaking of the water table along the valley sides where the soil tends to be thinner than in the center of the valley (Peters et al. 1995; Renzetti et al. 1992; Spence and Woo 2003). When the rain ceases and lateral inflow stops, the water table declines first at the sides and then in the middle, eventually flattening across the valley (Spence and Woo 2003).

Water budget of a valley north of Yellowknife illustrates the hydrologic importance of lateral inputs (Spence and Woo 2003). In the relatively dry summer of 2000, direct rainfall on the valley was 155 mm and inflow from the uplands was 106 mm. These amounts sustained the losses to evapotranspiration (164 mm) and outflow (34 mm), yielding a calculated soil moisture recharge of 64 mm (measured storage change was 73 mm).

3.3 Water Spillage

Flow along a valley is closely linked to where, when and how the soil storage capacities are satisfied. Surface flow occurs only where the water table rises above ground, otherwise only subsurface flow may be maintained through the soil matrix and the macropores. Even subsurface flow may be arrested if blocked by bedrock sills. Field observation showed that as surface flow moves down a valley, it may encounter a non-saturated lower valley segment. The water will infiltrate along its path and if all the surface flow is lost to seepage, the stream becomes intermittent. This runoff mechanism is captured by the "fill-and-spill" process proposed by Spence and Woo (2003). Its central tenant is that the spatially variable valley storage needs to be satisfied before water spills to generate either surface or subsurface flow. Storage capacity in the valley is variable because of surface and bedrock topography, soil heterogeneity and unevenness, and seasonal presence of ground frost. Valley storage status is dynamic, being enriched by rainfall, snowmelt and lateral inflow, but lost to evaporation and downstream drainage. Thus, along segments of a valley, soil water storage will continue to be filled until (1) the local storage spills over the threshold created by any bedrock sills to permit subsurface flow, and (2) the water table rises above the ground surface to generate overland flow.

Runoff from slopes varies considerably in a semi-arid Shield environment, producing highly uneven lateral input to different segments of a valley. A Shield valley may be considered to comprise a series of storage reservoirs with inflows from adjacent slopes and from upstream, filling individual reservoirs to satisfy their deficits until their thresholds are reached. Then spillage resumes to continue the flow downstream. Figure 5 illustrates this "fill-and-spill" concept. At t_1, rain begins and the valley water table is below the topographic surface. By t_2, lateral inflow has entered from bedrock uplands, prompting a water table rise above ground to generate saturation overland flow at the valley sides. Storage demands downslope interrupt this overland flow before it can reach the valley outlet. As event inputs continue, the storage requirements in the valley are met so that by t_3 an ephemeral stream of saturation overland flow reaches the outlet and discharges runoff.

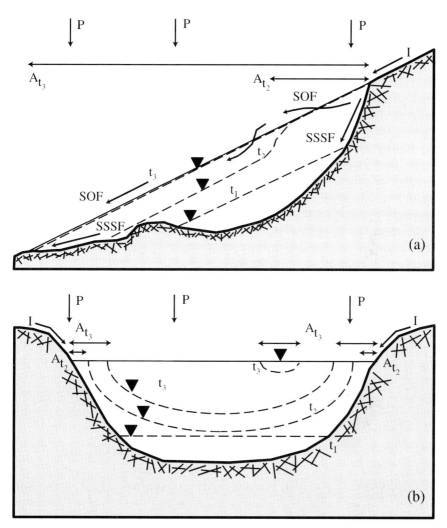

Fig. 5. Conceptualization of "fill-and-spill" runoff generation mechanism. *a* is a longitudinal profile of a valley and *b* is its cross section. *P* is precipitation, *t* is time at steps *1*, *2* or *3*. *SSSF* is subsurface stormflow and *SOF* is saturation overland flow. *A* is the runoff contributing area at t_2 or t_3

4 Headwater Catchment Hydrology

Streamflow generation in many headwater catchments in the humid temperate latitudes follow the variable source area concept described by Dunne (1978), based on the work of Hewlett and Hibbert (1965). In this conceptualization, the portion of a watershed yielding surface flow shrinks and expands as subsurface flow from upslope exceeds the capacity of the soil to transmit it, the magnitude of which is controlled by rainfall and antecedent moisture. Normally, the runoff first occurs as saturation overland flow in the downslope area, and then expands upslope (Dunne 1978). In the Shield environment, however, Allan and Roulet (1994), Buttle and Sami (1992), and Spence and Woo (2003) revealed that the upslope areas are the first sources of runoff, and these contributing areas expand downslope, stopping only when runoff cannot satisfy subsurface moisture demands (Fig. 6). Several runoff processes can occur simultaneously within the same headwater basin, including Hortonian overland flow from

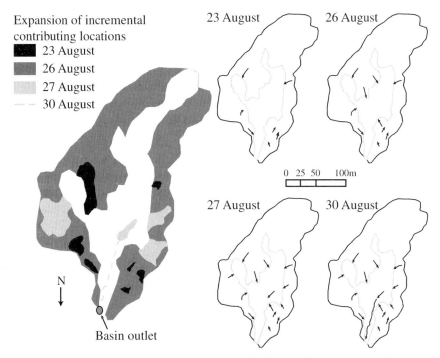

Fig. 6. A 5-ha headwater basin near Yellowknife: (left) expansion of the areas contributing to surface runoff, and (right) surface runoff linkages during an August 2000 event

exposed bedrock and saturation overland flow from soil filled areas, while infiltration and exfiltration are functions of the highly varied local topography and water table (perched or otherwise) position along the slopes. It is therefore useful to consider a headwater basin to comprise a number of hydrologic elements and introduce the element threshold concept to describe flow production in the Shield environment.

4.1 Element Threshold Concept

Runoff generation from catchments in the Canadian Shield should consider its topographic complexity in terms of landscape geometry and topology. Landscape geometry delineates the boundary of each physiographic unit and the variable flow contributing areas within it, thus allowing their relative size and shape to be discerned. Topology defines the arrangement of physiographic units and their locations relative to each other on the landscape. Taking into consideration the flow contributing areas and the hydrologic linkages between physiographic units in controlling flow production in a headwater catchment near Yellowknife, Spence and Woo (2006) proposed the "element threshold concept" which has the following attributes:

1. Canadian Shield catchments consist of a number of hydrologic elements.
2. In the context of runoff production and delivery, an element can perform one or more of the three functions of storing, contributing and transmitting water. The functional status of an element is determined by the water balance status relative to the thresholds that regulate runoff generation.
3. Differences in element physiography result in differences in storage capacity within a catchment. On the other hand, the status of available storage within each element is a result of the hydrologic processes within the element as well as its connections with the adjacent elements.
4. Landscape geometry and topology influence the behavior of elements because hydrologic connections influence the inputs to, and outputs from, the elements.
5. Hydrologic behavior is considered to be relatively uniform within elements and dissimilar among elements and this gives rise to variations in their runoff responses (uniformity is, of course, relative and scale dependent).

6. Runoff is released only when storage exceeds some threshold imposed by the physiography. Thus, flows along a drainage network can be disjointed during dry periods.

At a given time, an element performs one or more hydrologic functions of storing water, contributing runoff to another element, or providing a conduit that transmits water from and to its adjacent elements. Note that topological consideration dictates that only those elements connected to at least two other elements can perform the transmission function, though all elements have the potential to store and to contribute runoff. The water balance status of an element at a particular time and the thresholds that regulate runoff generation play the dominant role in determining whichever function(s) will be realized.

Elements with low thresholds relative to available water are the first to produce surface flow. The flow expands downslope but may be depleted to satisfy the storage requirements of the lower elements (Fig. 6). The storage capacity varies greatly among the elements, as does their antecedent moisture condition which changes in space and time. Thus, the areas that contribute to runoff can be highly disjointed as different elements generate flows at different times. Downstream passage of flow need not be continuous since the 'fill-and-spill' runoff mechanism (Spence and Woo 2003) requires the storage downstream to exceed their runoff thresholds before outflow occurs.

4.2 Examples

Field results from a 5-ha basin that drains into Pocket Lake near Yellowknife showed that bedrock uplands and soil-filled valleys behave differently over time. For example, runoff generation during minor rain events or at the beginning of large ones is restricted to the bedrock upland while storage demands in the valley curtail flow production (Fig. 6). At a longer time scale, the seasonal water budgets of each land cover type are different. In summer, bedrock upland yields larger runoff than the yield from the overall basin (Table 1). Uplands generate runoff more often while the valleys serve mainly to store or transmit water to the catchment outlet. This illustrates that basin runoff is not a simple additive function of runoff from each element, but their hydrologic connection is a major consideration in streamflow. To produce streamflow in a valley, the upland contributing area has to yield sufficient runoff to exceed valley storage demands.

Table 1. Magnitude of water budget components for a headwater basin that drains to Pocket Lake near Yellowknife, NWT, during the 2000 growing season and 2001 spring melt period.

	Summer 2000			Spring 2001		
	Upland	Valley	Basin	Upland	Valley	Basin
P+M	5,882±294	1,686±84	7,568±378	5,085±956	1,794±337	6,879±1293
ET	2,163±433	1,783±357	3,946±789	569±114	685±137	1,254±251
F	2,087±522			645±161		
I	1,684±118			4,212±815		
R		199±28	199±28		4,393±879	4,393±879
ΔS		1,127±282	1,127±282	607±152	598±150	-9±2
ΔS(calc)	-52	1,388		-341	928	

P rainfall, M snowmelt, ET evapotranspiration, F infiltration, I lateral runoff from upland to the valley, R outflow, ΔS and $\Delta S(calc)$ measured and calculated change in storage (all units in m^3).

The element threshold concept applies equally well to headwater basins with lakes. Storage demands in a lake near the outlet of a headwater basin allowed only 7% of snow meltwater to leave the catchment (Spence 2000). This lake initially behaved as a store (April 15–29), subsequently transmitted flow from upslope and contributed flow to the basin area below its outlet (April 30–May 8) once the lake level exceeded its outlet elevation. The dynamic aspect of the element threshold concept is highlighted by Mielko and Woo (2006) who observed that elements contribute water downslope with different time lags depending on their storage conditions and relative location in the basin. Furthermore, outflows from different lakes along a valley were independent of runoff generated at the lakes located above or below (Woo and Mielko 2007). Each lake contributed runoff at different times because of the particular amount of runoff produced from its direct catchment, and the antecedent storage status of the lake relative to its outflow threshold.

5 Basin Hydrology

The hydrology of a drainage basin may be distinguished into two components: land phase and channel phase. Land phase hydrology concerns land processes that generate, store and deliver water to the stream channels. Channel phase hydrology concerns the water in the streams and their riparian zones. In many hydrologic models, the former is often dealt with using land surface schemes while the latter is handled through routing procedures. This arrangement is unsatisfactory where the channels play major roles in flow production, storage and transmission, so that runoff received from the land phase is significantly modified before the water reaches the basin outlet. The Canadian Shield with its myriads lakes and wetlands along the stream channels is one such landscape.

Two situations favor the uncoupling of streamflow from land phase runoff in this landscape. First, an increase in channel surface area and volume enlarges vertical exchanges with the atmosphere and groundwater, and attenuates channel phase runoff. Second, storage demands by hillslopes and headwater lakes can physically disconnect source areas of land phase runoff from the channel. The mechanisms associated with the element threshold concept dictate whether outflow can be released from a particular segment of the drainage network. Briefly, a valley bottom (which may contain a lake or wetland) will receive water from adjacent elements and undergoes vertical water gains and losses, but will yield outflow only when the storage status exceeds its capacity. In the dry northwestern Canadian Shield, the only events capable of creating conditions that can exceed capacities are spring melt and large autumn rainfall events. After snowmelt, evaporation loss dominates and draws down valley storage. When the loss is not compensated by rainfall and lateral inflow, network runoff may be disrupted. Spence (2006) noted a shrinkage of the land areas contributing water to the channels of Baker Creek (basin area 137 km^2 at the outlet of Lower Martin Lake) during the dry summer of 2003 (Fig. 7). Only major rainstorms in this semi-arid environment yielded sufficient land phase runoff to replenish storage in a valley above its particular threshold to permit streamflow revival.

As basin size increases, discontinuity in streamflow becomes increasingly uncommon. The three catchments in Fig. 8 have similar lake coverage (~18%) and yet show three distinct stream discharge signals. The small Baker Creek basin (137 km^2) exhibits a subarctic nival streamflow regime (Church 1974) with high flows in the snowmelt season but zero flow in the winter. Among the three basins, it has the flashiest hydrograph, and is the only basin that shows flow response to late summer rainfall (Fig. 4 in Woo and Mielko 2007). The moderated and attenuated signal in streamflow is evidence of the lake storage effect in the larger Cameron (3630 km^2) and Camsell (32100 km^2) basins. These larger rivers exhibit a streamflow regime that is considered prolacustrine (Woo 2000) versus the nival regime of the smaller Baker Creek. It is not percentage of lake coverage, but the lake size and location in a basin (i.e., relative geometry and topology) that determine the degree of attenuation. A shift in regime occurs when the lakes in a basin become so large or so positioned that their storage and release functions overwhelm the seasonality of the land phase runoff. The result is a streamflow signal dominated by the hydraulic dynamics of the lakes.

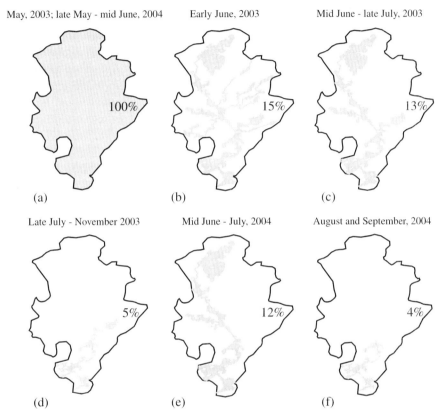

Fig. 7. Changing contributing areas to streamflow at the outlet of Lower Martin Lake, NWT, 2003 and 2004. The map denoting contributing area to the outlet of Lower Martin Lake from 1 May – 1 June 2003 and 25 May to mid June 2004 does not explicitly represent contributing area but rather event contributing area for events bounded by these dates. Percentage values are relative to the maximum contributing area denoted in (a)

6 Concluding Remarks

The key result of several years of field investigations in the northwestern Canadian Shield region is the understanding that the Shield is a landscape with spatially and temporally variable available storage, where the hydrologic connectivity that dictates runoff response is controlled in part by the geometry and topologic arrangement of landscape components (Spence and Woo 2006). Furthermore, it is in a semi-arid area where snowmelt is the primary source of water while summer rain may be insufficient to satisfy water demands to raise streamflow. The fill-and-spill principle and the element threshold concept explain the temporal and spatial variations in runoff. Flow linkage within the Shield drainage network can become discontinuous, and hydrologic connectivity significantly affects streamflow, especially in headwater Shield catchments. Connectivity has also been found to control runoff generation in the boreal plains (Quinton and Hayashi 2007), mountains (Carey and Woo 2001) and arctic tundra (Roulet and Woo 1988).

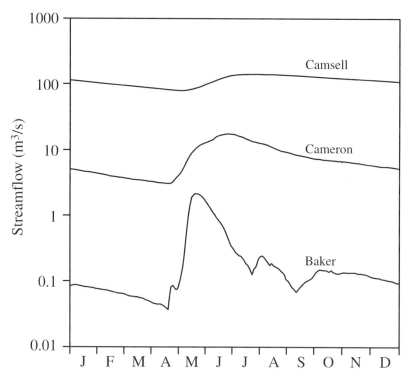

Fig. 8. Mean annual streamflow of Baker Creek (nival regime), Cameron River (transition between nival and prolacustrine regimes) and Camsell River (prolacustrine regime)

The Canadian Shield holds freshwater resource of immense quantity and it is of paramount importance to understand its hydrologic behavior. On a regional scale, several major rivers (e.g., Churchill in Labrador, la Grande in Quebec, Churchill and Nelson in Manitoba) have been harnessed to produce hydroelectricity. On a local scale, Shield terrain is considered to be a favorable repository site for mining and nuclear wastes with the assumption that the bedrock is restrictive to groundwater flow. Furthermore, there is great uncertainty over the impacts of climate variability and change (ACIA 2005) on the water resources of the Mackenzie Basin, a sparsely monitored environment because of its relative isolation and inaccessibility. Yet, our ability to model the Shield hydrology has yielded limited results (Pietroniro et al. 1998; Spence et al. 2005). Spence (2001) suggested that available models do not adequately simulate the subarctic Canadian Shield streamflow because they do not account for the relevant processes governing the lateral exchanges of energy and water. With an improved appreciation of the predominant processes, we are poised to incorporate them into numerical models.

Acknowledgements

Funding for this research was provided by Environment Canada, the Mackenzie GEWEX Study, Indian and Northern Affairs Canada through the Northern Student Development Program and the Northwest Territories Power Corporation. The authors thank Brian Yurris, Kerry Walsh, Doug Halliwell, Mark Dahl and Dave Fox of Environment Canada; Bob Reid, Denise Bicknell, Shawne Kokelj and Derek Faria of Indian and Northern Affairs Canada and Claire Oswald, Steve Kokelj, Iain Stewart, Devon Worth, Corrinne Mielko, Paul Saso and Jara Rausch for their assistance in the field. This project could not have been possible without Dan Grabke of the NWT Energy Corporation.

References

ACIA (2005) Arctic climate impact assessment. Cambridge University Press, Cambridge

Allan C, Roulet N (1994) Runoff generation in zero order precambrian shield catchments: the stormflow response of a heterogenous landscape. Hydrol Process 8:369–388

Bello R, Arama A (1980) Rainfall interception in lichen canopies. Climatol B 23:74–78
Bottomley DJ, Craig D, Johnston LM (1986) Oxygen-18 studies of snowmelt runoff in a small Precambrian shield watershed: implications for streamwater acidification in acid sensitive terrain. J Hydrol 88:213–234
Branfireun B, Heyes A, Roulet NT (1996) The hydrology and methylmercury dynamics of a Precambiran Shield headwater peatland. Water Resour Res 32:1785–1794
Branfireun B, Roulet NT (1998) The baseflow and storm flow hydrology of a Precambrian shield headwater peatland. Water Resour Res 32:1785–1794
Brown RJE (1978) Permafrost. In: Hydrological atlas of Canada, Plate 32. Dept Fish Environ
Buttle JM, Sami K (1992) Testing the groundwater ridging hypothesis of streamflow generation during snowmelt in a forested catchment. J Hydrol 135:53–72
Carey SK, Woo MK (2001) Spatial variability of hillslope water balance, Wolf Creek basin, subarctic Yukon. Hydrol Process 15:3113–3132
Church M (1974) Hydrology and permafrost with reference to North America. In: Permafrost hydrology. Proc Workshop seminar 1974, Can natl committee international hydrological decade, Ottawa, pp 7–20
Davison C (1984) Hydrogeological characterization at the site of Canada's underground research laboratory. In: Proc Internat symp on groundwater resources, utilization and contaminant hydrogeology, Montreal, Internat Assoc Hydrogeologists, pp 310–355
Devito K, Hill AR, Roulet N (1996) Groundwater–surface water interactions in headwater forested wetlands of the Canadian Shield. J Hydrol 181:127–147
Domenico PA, Schwartz W (1998) Physical and chemical hydrogeology. John Wiley and Sons, Toronto
Dunne T (1978) Field studies of hillslope flow processes. In: Kirkby MJ (ed) Hillslope hydrology, Wiley, Toronto, pp 227–293
Fracflow (1998) Preliminary hydrogeological, geochemical and isotopic investigations at the Giant Mine, Yellowknife, NWT. Report prepared for Water Resources Division, Indian and Northern Affairs Canada, Yellowknife, Northwest Territories
Hewlett J, Hibbert A (1965) Factors affecting the response of small watersheds to precipitation in humid areas. In: Sopper W, Lull H (eds) International symposium on forest hydrology, Pergamon Press, Oxford, UK, pp 275–290
Kochtubajda B, Flannigan MD, Gyakum JR, Stewart RE, Burrows WR, Way A, Richardson E, Stirling I (2007) The nature and impacts of thunderstorms in a northern climate. (Vol. I, this book)
Landals A, Gill D (1972) Differences in volume of surface runoff during the snowmelt period, Yellowknife, NWT. In: The role of snow and ice in hydrology, IAHS Publ no 107, pp 927–942
McDonnell J, Taylor C (1987) Surface and subsurface water contributions during snowmelt in a small Precambrian shield watershed, Muskoka, Ontario. Atmos Ocean 25:251–266

Metcalfe RA, Buttle JM (2001) Soil partitioning and surface store controls on spring runoff from a boreal forest peatland basin in north-central Manitoba. Hydrol Process 15:2305–2324

Mielko C, Woo MK (2006) Snowmelt runoff processes in a headwater lake and its catchment, subarctic Canadian Shield. Hydrol Process 20:987–1000

Peters D, Buttle JM, Taylor C, LaZerte B (1995) Runoff production in a forested shallow soil Canadian Shield basin. Water Resour Res 31:1291–1304

Pietroniro A, Martz L, Soulis ED, Kouwen N, Marsh P, Pomeroy JW, Spence C (1998) Hydrology for cold regions: GEWEX sub basin working group (ad hoc), paper presented at the 4th Sci workshop for the Mackenzie GEWEX Study, Montreal, November 1998, pp 16–18

Pomeroy JW, Gray DM, Marsh P (2007) Studies on snow redistribution by wind and forest, snow-covered area depletion, and frozen soil infiltration in northern and western Canada. (Vol. II, this book)

Quinton WM, Hayashi M (2007) Recent advances toward physically-based runoff modeling of the wetland-dominated central Mackenzie River Basin. (Vol. II, this book)

Raven K, Smedley J, Sweezey R, Novalowski K (1985) Field investigations of a small groundwater flow system in fractured monzonitic gneiss. In: Hydrogeology of rocks of low permeability. Proc Internat Assoc Hydrogeologists 17th Internat Congress, Jan. 7–12, 1985, Tucson, pp 72–85

Renzetti A, Taylor C, Buttle JM (1992) Subsurface flow in a shallow soil Canadian Shield watershed. Nord Hydrol 23:209–226

Roulet NT (1990) The hydrological role of peat covered wetlands. Can Geogr 34, 82–83

Roulet NT, MK Woo (1988) Runoff generation in a low arctic drainage basin. J Hydrol 101:213–226

Rouse WR, Mills P, Stewart R (1977) Evaporation in high latitudes. Water Resour Res 13:909–914

Spence C (2000) The effect of storage on runoff from a headwater subarctic Canadian Shield basin. Arctic 53:237–247

Spence C (2001) Sub-grid runoff processes and hydrological modeling in the subarctic Canadian Shield. In: Soil–vegetation–atmosphere transfer schemes and large scale hydrological models. Proc symp 6th IAHS Scientific assembly at Maastricht, The Netherlands, July 2001, IAHS Publ no 270, pp 113–116

Spence C (2006) Hydrological processes and streamflow in a lake dominated water course. Hydrol Process 20:3665–3681

Spence C, Dies K, Woo MK, Martz LW, Pietroniro A (2005) Incorporating new science into water management and forecasting tools for hydropower in the Northwest Territories, Canada. Proc Northern research basins 15th Int symposium and workshop, pp 205-214

Spence C, Rausch J (2005) Autumn synoptic conditions and rainfall in the subarctic Canadian Shield of the Northwest Territories, Canada. Int J Climatol 25:1493–5106

Spence C, Woo MK (2002) Hydrology of subarctic Canadian Shield: bedrock upland. J Hydrol 262:111–127

Spence C, Woo MK (2003) Hydrology of subarctic Canadian Shield: soil-filled valleys. J Hydrol 279:151–166

Spence C, Woo MK (2006) Hydrology of subarctic Canadian Shield: heterogeneous headwater basins. J Hydrol 317:138–154

Szeto K, Liu J, Wong A (2007) Precipitation recycling in the Mackenzie and three other major river basins. (Vol. I, this book)

Thorne G (1992) Soil moisture storage and groundwater flux in small Precambrian shield catchments. In: Proc 9th Int northern research basins symposium and workshop, Dawson City, Canada, pp 555–574

Thorne G, Laporte J, Clarke D (1994) Infiltration and recharge in granitic terrains of the Canadian Shield during winter periods. In: Proc 10th Int northern research basins symposium and workshop, Spitsbergen, Norway, pp 449–466

Thorne G, Laporte J, Clarke D, Davison C (1999) Water budget of an upland outcrop recharge area in granitic rock terrain of southeastern Manitoba, Canada. In: Proc 12th Int northern research basins symposium and workshop, Reykjavik, Iceland, pp 317–330.

Wells C, Taylor C, Cornett R, LaZerte B (1991) Streamflow generation in a headwater basin on the Precambrian shield. Hydrol Process 5:185–199

Wolfe SA (1998) Living with frozen ground: a field guide to permafrost in Yellowknife, Northwest Territories. Geol Surv Can misc report 64

Woo MK (2000) Permafrost and hydrology. In: Nuttall M, Callaghan TV (eds) The Arctic: environment, people, policy. Harwood Academic Pub, Amsterdam, The Netherlands, pp 57–96

Woo MK, Mielko C (2007) Flow connectivity of a lake–stream system in a semiarid Precambrian Shield environment. (Vol. II, this book)

Chapter 14

Recent Advances Toward Physically-based Runoff Modeling of the Wetland-dominated Central Mackenzie River Basin

William L. Quinton and Masaki Hayashi

Abstract Field studies were initiated in 1999 at Scotty Creek in central Mackenzie River Basin to improve understanding and model-representation of the major water flux and storage processes within a wetland-dominated zone of the discontinuous permafrost region. Four main topics were covered: (1) the major peatland types and their influence on basin runoff, (2) the physical processes governing runoff generation, (3) how runoff processes observed at the hillslope scale relate to basin-scale runoff, and (4) the water balance of Scotty Creek and its adjacent basins. A conceptual model of runoff generation was developed that recognizes distinct hydrologic roles among the major peatland types of flat bog, channel fen and peat plateau. This model contributes to resolving some of the difficult issues in the hydrologic modeling in this region, especially in relation to the storage and routing functions of wetlands-dominated basins underlain by discontinuous permafrost.

1 Introduction

Wetland-dominated terrain underlain by discontinuous permafrost covers extensive parts of northern North America and Eurasia. The hydrologic response of these areas is poorly understood, in large part due to the lack of study on the hydrologic functioning of the major wetland types, and the interaction among them. Near the center of the Mackenzie River Basin is an extensive area (ca. 53,000 km^2) of flat organic terrain with a high density of open water and wetlands, in the continental boreal region and the zone of discontinuous permafrost (Hegginbottom and Radburn 1992). With a limited understanding of the processes governing the cycling and storage of water in this region, attempts to model basin runoff have been met with limited success (Stewart et al. 1998). Discontinuous permafrost terrain is particularly sensitive to the effects of climatic warming, and pronounced changes in water storage and runoff pathways are expected with only small additional ground heat (Rouse 2000). Improved process understanding and description will reduce the uncertainty regarding future runoff production from wetland-dominated basins in the zone of discontinuous permafrost.

2 Methods

2.1 Field Studies

Wetlands occupy approximately 125×10^3 km^2 or 7% of the Mackenzie River Basin and are concentrated mainly in the Peace-Athabasca lowlands, the Mackenzie River delta and the lower Liard River valley. Scotty Creek (61°18' N; 121°18' W) lies in the lower Liard River valley, 50 km south of Fort Simpson (Fig. 1a) in discontinuous permafrost. The wetlands of Scotty Creek are typical of the 'continental high boreal' wetland region (NWWG 1988).

The Fort Simpson region is characterised by a dry continental climate, with short summers and long cold winters. It has an average (1971–2000) annual air temperature of -3.2°C, and receives 369 mm of precipitation annually, of which 46% is snow (MSC 2002). Snowmelt usually commences in the second half of March and continues throughout most of April so that by May, only small amounts of snow remain (Hamlin et al. 1998). Field measurements were taken at four gauged basins (Fig. 1b). At the Jean-Marie, Blackstone and Birch Rivers, measurements were limited to discharge at the basin outlets, aerial reconnaissance and ground verification surveys. Most fieldwork was conducted at Scotty Creek because it contains the major ground cover types found in the region, and was logistically manageable given its relatively small (152 km^2) size and proximity to Fort Simpson.

The stratigraphy in this region includes an organic layer of up to 8 m in thickness overlying a silt-sand layer, below which lies a thick clay to silt-clay deposit of low permeability (Aylesworth and Kettles 2000). Field reconnaissance at Scotty Creek revealed three major peatland types: peat plateaus, ombrotrophic flat bogs and channel fens (Quinton et al. 2003; Robinson and Moore 2000). These peatlands support a diverse vegetation community that includes four tree species (*Picea mariana, Larix laricina, Pinus contorta, Betula papyrifera*), fifteen shrub species (predominantly *Betula, Ledum, Kalmia and Salix*), sixteen species of lichen (predominantly *Cladina*), thirteen species of bryophytes (predominantly *Sphagnum*), in addition to species of vine, club-moss, fungi, liverwort, sedges, grasses, aquatic plants, horsetails and wild flowers. Peat plateaus are underlain by permafrost, and their surfaces rise 1 to 2 m above the surround-

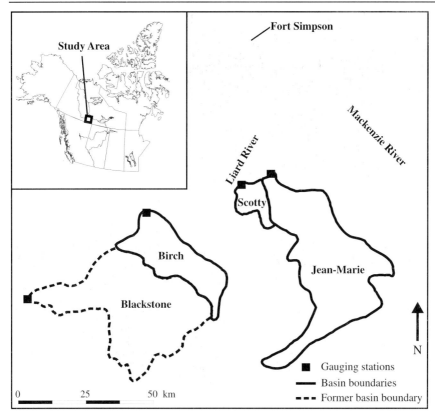

Fig. 1. Inset shows location of study area in Central Mackenzie Basin. Birch and Jean-Marie Rivers, and Scotty Creek study basins (boundaries shown in solid black lines) in the lower Liard River valley near Fort Simpson, Northwest Territories. Boundary of Blackstone River basin (dashed line) is based on published estimate but has now been revised (but not yet published by the Water Survey of Canada). Black squares indicate gauging stations operated by the Water Survey

ing bogs and fens. *Picea mariana* is the principal tree species, but *Pinus contorta* and *Betula papyrifera* are also present. Shrubs are most abundant on peat plateaus and lichen species dominate the ground cover, though patches of bryophytes also occur. Underlying the vegetation is sylvic peat containing dark, woody material

Channel fens take the form of broad, 50 to >100 m wide channels. Their surface is composed of a floating peat mat of sedge (*Carex* sp.) origin, approximately 0.5 to 1.0 m in thickness that supports sedges, grasses, herbs, shrubs, aquatic and plants including *Typha latifolia*, *Equisetum fluviatile* and *Menyanthes trifoliate*. Dense patches of Tamarack (*Larix laricina*) also occur in the fens. Flat bog surfaces are relatively fixed, and are covered with *Sphagnum sp.* overlying yellowish peat with well-defined sphagnum remains (Zoltai and Vitt 1995). The club-moss, fungi and liverwort species are most prevalent in the bogs. Most flat bogs are small features that occur within peat plateaus. As a result, they appear to be internally drained and hydrologically isolated from the basin drainage system. However, other flat bogs are connected to channel fens. These are relatively large bog complexes that often contain numerous peat plateaus. Surface drainage from connected flat bogs to channel fens has been observed during the spring freshet and in response to large, late-summer rain events.

2.2 Satellite Image Analysis

Two multi-spectral images were acquired for ground cover analyses, including (1) a 4×4 m resolution IKONOS image covering 90 km^2 of the 150 km^2 Scotty Creek basin, and (2) a 30×30 m resolution Landsat image covering a 32,400 km^2 area of the lower Liard River valley that includes four study basins (viz., Birch, Blackstone, Jean-Marie and Scotty). Both images were classified using the maximum likelihood method (Arai 1992; Richards 1984; Yamagata 1997) with training sites (Lillesand and Kiefer 1994) obtained from homogeneous areas, including flat bogs, channel fens and peat plateaus. The three major peatland types were readily identified on the basis of their contrast in surface characteristics. For example, because saturated surfaces absorb infra red light, the channel fens appear relatively dark compared with the surrounding bogs and peat plateaus (Fig. 2). This contrast is enhanced by the relatively high photosynthetic activity of the drier surfaces away from the channel fens. Since the reflection of red light increases with decreasing photosynthesis (Lillesand and Kiefer 1994) these drier areas are represented by bright surfaces on the image. Additional data layers containing topographic information, the location of drainage networks and basin boundaries, were included and used for computations of drainage area, drainage density, and average slope. The average slope was computed from the difference between the maximum elevation and elevation of the basin outlet, divided by the distance measured along the drainage way between these two points.

From the IKONOS image, a 22 km^2 area of interest that includes the main locations where field measurements were made, was chosen for the purpose of obtaining a detailed and accurate ground-cover classification from field knowledge. All peat plateaus, channel fen and flat bogs within

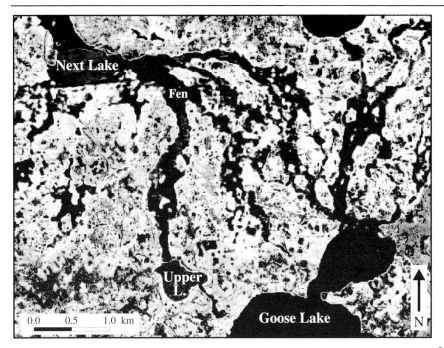

Fig. 2. Sample of high-resolution (4 m × 4 m) IKONOS image showing a 22 km^2 section in the southern part of Scotty Creek basin where field studies are concentrated. The unclassified image has been converted from false-color to a grey scale. Channel fens appear relatively dark compared with the surrounding areas composed of flat bogs and peat plateaus

the area of interest were digitised. Image analysis software was used to compute the proportion of this area occupied by each cover type, as well as the area and perimeter of individual peat plateaus, channel fens and flat bogs. Peat plateaus occupy the largest proportion (43%) of the target area (Table 1). Since permafrost occurs only beneath the peat plateaus, the analysis suggests permafrost occupies less than half of the area shown in Fig. 3. Despite the large number of isolated flat bogs, they account for less than 5% of the area, but the area covered by connected flat bogs is more than five times larger that of the isolated flat bogs, with a total area roughly equivalent to the area occupied by channel fens (Table 1). It is difficult to differentiate among individual connected flat bogs. The channel fens appear to separate much of the remaining landscape into distinct bog–peat plateau complexes, the size of which depends largely on the spacing of the channel fens. There also appears to be separation of connected flat bogs by large peat plateaus that extend between adjacent channel fens.

Table 1. Selected results of detailed ground cover classification of sub-section of the IKONOS image of Scotty Creek, representing an area of ca. 22 km^2 on the ground. N is the number of samples of each cover type. Deriving the number of connected flat bogs was not attempted

Cover type	N	Area [km^2]	Area [%]
Peat plateaus	609	9.52	43.0
Flat bogs (isolated)	999	0.89	4.0
Flat bogs (connected)	–	5.03	22.7
Channel fens	2	4.65	21.0
Lakes	4	2.06	9.3

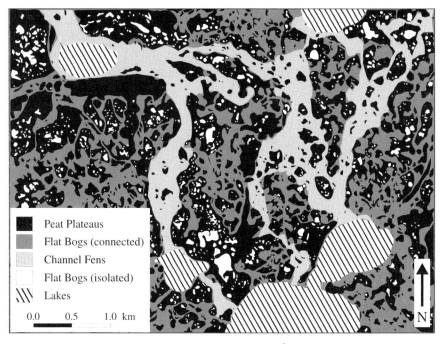

Fig. 3. Major ground-cover types in the same 22 km^2 area of the Scotty Creek basin presented in Fig. 2

3 Influence of Peat Bogs and Channel Fens on Basin Runoff

The arrangement of channel fens on the landscape, and observations of flow over their surfaces suggests that their hydrologic function is primarily one of lateral flow conveyance. Bogs are either surrounded by peat plateaus and therefore internally drained, as in the case of the isolated flat bogs, or have only ephemeral, tortuous surface flow routes to the channel fens, as in the case of the connected flat bogs. Flat bogs primarily serve the function of water storage rather than conveyance. This contrast between the channel fens and flat bogs suggests that the relative proportion of these two peatland types has implications for basin runoff. For example, a basin with a relatively high proportion of flat bogs should generate less runoff than a basin with a lower coverage of flat bogs. Figure 4 indicates that annual runoff was positively correlated with the percentage cover of channel fens, and negatively correlated with the percentage cover of flat bogs. The associations between channel fen coverage and runoff, and between bog coverage and runoff, are correlated in opposite directions because of the difference in the main hydrologic function of these two wetland types. Annual runoff correlates positively with both drainage density and the square root of basin slope, suggesting that the basins with more efficient drainage mechanisms have higher annual runoff.

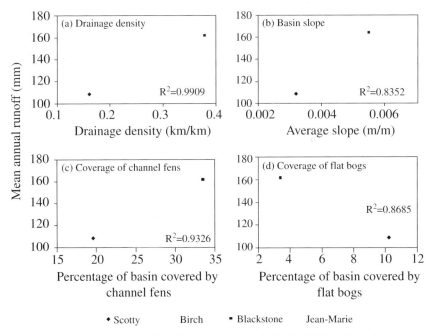

Fig. 4. Mean annual runoff of four-year period (1997–2000) plotted against (a) drainage density; (b) square root of the average basin slope; and percentage of the basin covered by (c) channel fens, and (d) flat bogs

The Scotty and Jean-Marie basins have relatively low average annual runoff values (Table 2), as these basins posses the characteristics that would diminish and delay runoff production: a relatively low average slope and drainage density, a low proportion of channel fens, but high coverage of flat bogs. The hydrographs of these basins are more delayed and have lower peaks than those of the Blackstone and Birch River basins (Fig. 5). However, Scotty and Jean-Marie differ in the timing of their runoff. On average, by the beginning of June, 41% of the annual runoff had drained from Scotty Creek, while at Jean-Marie, only 29% of the annual runoff had occurred. The greater basin lag of Jean-Marie reflects the fact that this river drains an area approximately 8.5 times larger than that of Scotty Creek, and as a result, the average flow distance to the basin outlet and the residence time are both longer at Jean-Marie.

Table 2. Size and percentage cover of major terrain types of the four study basins, derived from Landsat imagery. For each basin, the drainage density, average slope, and the average annual, spring (April–May) and summer (June–August) runoff for the four years (1997–2000) of observation are shown. The bog class refers to the sum of both isolated and connected flat bogs. The wooded class includes peat plateaus and wooded uplands in the northern parts of each basin

	Blackstone	Jean-Marie	Birch	Scotty
Area [km^2]	1910	1310	542	152
Average annual runoff [mm]	161.8	127.4	155.0	108.8
Average spring runoff [mm]	56.8	34.5	59.9	43.3
Average summer runoff [mm]	105.0	92.9	95.0	65.5
Wooded [% of basin]	66.8	65.6	64.9	63.2
Fens [% of basin]	33.5	27.4	30.7	19.6
Bogs [% of basin]	3.4	7.6	6.5	10.2
Drainage density [$km\ km^{-2}$]	0.378	0.237	0.373	0.161
Average slope [$m\ m^{-1}$]	0.0055	0.0034	0.0063	0.0032

Blackstone and Birch River basins both possess characteristics associated with higher runoff production, namely a relatively high average slope and drainage density, a high proportion of channel fens and a low coverage of flat bogs (Table 2). Consequently these two basins produce the highest average annual runoff (Fig. 4). The Birch River has a relatively small drainage area, and therefore would also have a relatively small average stream flow distance to the basin outlet. This could account for the slightly larger average runoff from this basin compared with the Blackstone during the April–May period (Table 2). Among the four basins studied, the Birch River basin was the first to commence runoff in each of the four study years. In three of these years, Scotty Creek, the other relatively small basin, was the second to respond.

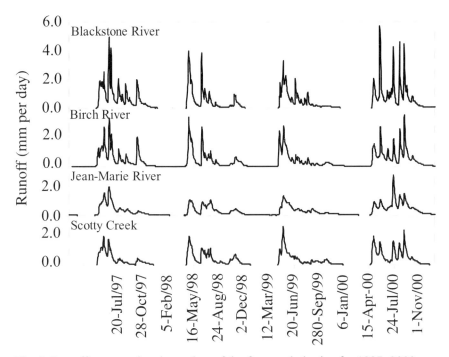

Fig. 5. Runoff measured at the outlets of the four study basins for 1997–2000

4 Influence of Peat Plateaus on Basin Runoff

Several physical attributes suggest that peat plateaus perform an important role in basin runoff generation. Annual late winter snow surveys over the period 1993–2005 indicated that a large amount of snow was stored beneath the tree canopy by late winter. The relatively high topographic position of peat plateaus produces a hydraulic gradient that is an order of magnitude larger than the adjacent flat bogs and channel fens. In addition, the presence of frozen, saturated soil close to the peat plateau surface severely restricts their capacity to store snowmelt and rainfall inputs so that much of the water received is shed laterally through their active layer (Fig. 6). However, the rate of subsurface drainage from peat plateaus to their adjacent bogs and fens strongly depends on the depth of ground thaw, as the frozen soil is relatively impermeable. Horizontal transmission also depends on properties of the active layer which vary sharply with depth (Quinton et al. 2000).

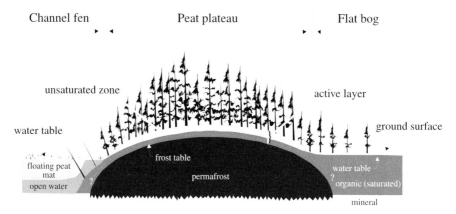

Fig. 6. Schematic cross-section of a peat plateau flanked by a channel fen on one side, and a flat bog on the other

4.1 Soil Conductance and Hydraulic Response

Similar to many other northern organic-covered terrains, such as arctic and alpine tundra, the soil profile on a peat plateau contains an upper, lightly decomposed layer, underlain by a darker layer in a more advanced state of decomposition (Fig. 7a), though the thickness of the upper layer can be highly variable over short (<1 m) distances. Because the degree of decomposition increases with depth below ground, the bulk density generally increases with increasing depth (Fig. 7b), while the porosity generally decreases (Fig. 7c). Peat development on the peat plateaus is derived mainly from sphagnum moss which, under similar environmental conditions, can form organic soils with a similar range of inter-particle pore diameters, regardless of the geographic setting. This is important hydrologically since pore size controls both the flux and the storage of water in the active layer. Detailed microscopic analysis (Quinton et al. 2000) of soils sampled from arctic tundra (Fig. 8a) indicates that the lower layer contains a larger proportion of small-diameter pores, with the consequence that both the hydraulic conductivity (Fig. 8b) and drainable porosity (Fig. 8c) are substantially reduced in the lower peat layer (Quinton and Gray 2003). Here, drainable porosity refers to the amount of water drained from a unit volume of sample, when a 0.4 m long core was placed vertically and allowed to drain freely for 24 hours.

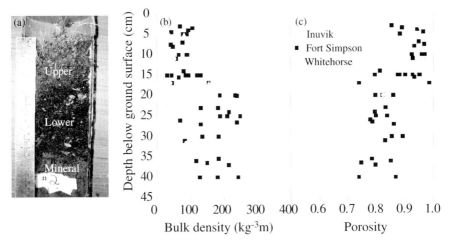

Fig. 7. (a) Typical organic soil profile in the study area showing upper and lower organic layers; variation in (b) bulk density and (c) total porosity with depth below the ground surface of a peat plateau at Scotty Creek. In (b) and (c), comparison is made with other organic cover types in northwestern Canada

For a peat plateau in Scotty Creek basin, Fig. 9 illustrates the sequence of events that produced a rapid runoff response to 23 mm rainfall. Prior to this rain, the water table was 0.43 m below the ground surface. The relatively large hydraulic conductivity near the surface (Fig. 8b) allowed the infiltrating rainwater to reach quickly the zone of high moisture content above the water table, thereby inducing an abrupt 0.13 m water table rise. The magnitude of this rise indicated a field-based drainable porosity of 18 %, which is consistent with the drainable porosity measured in the laboratory (Fig. 8c) on soils sampled from 0.3–0.4 m below the surface. Figure 8b indicates that this water table would rise into a zone where the hydraulic conductivity is one to two orders of magnitude higher, thereby allowing efficient lateral drainage of subsurface water.

4.2 Soil Water Storage

When saturated, the volumetric moisture content of the organic soil typically exceeds 80% (Fig. 10a). Laboratory data show that saturated peat from the upper layer can drain to a residual value of about 20%. Under

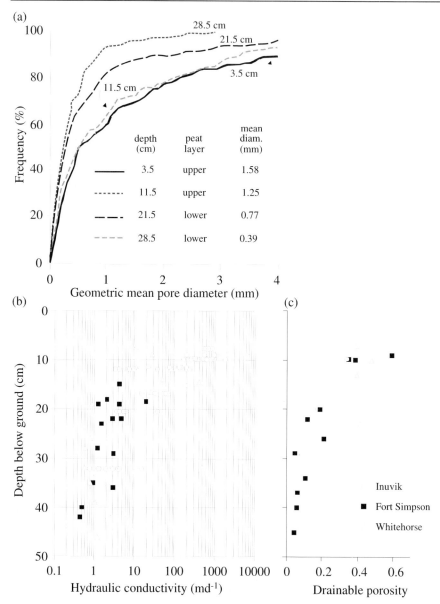

Fig. 8. (a) Cumulative frequency distribution of pore diameters for pores <4 mm diameter at four depths below the ground surface. The upper two depths (3.5 and 11.5 cm) are in the upper peat layer, and the lower two (21.5 and 28.5 cm) are in the lower peat layer; variation in (b) hydraulic conductivity and (c) drainable porosity with depth below ground of a peat plateau at Scotty Creek. In (b) and (c), comparison is made with other organic cover types in northwestern Canada

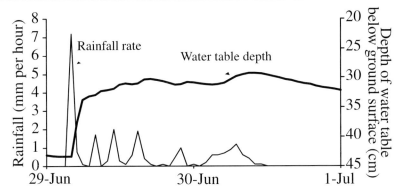

Fig. 9. Water table response in a peat plateau at Scotty Creek to a 23 mm rain event that occurred between 04:00 on 29 June and 11:00 on 30 June, 1999

field situations, this may occur during periods of relatively high soil tension, such as in late-summer and during soil freezing. The annual minimum unfrozen moisture content is around 20% at all depths but the annual maximum unfrozen moisture content increases with depth. The consequence is that the annual range of unfrozen moisture content changes with depth. For example, at 0.3 m below the ground surface, the annual range in the daily average unfrozen volumetric moisture content is about 50% (Fig. 10b), but decreases to 40% (20% to 60%) and 15% (20% to 35%) at 0.2 m and 0.1 m depths respectively.

Regardless of the moisture content at various depths prior to freezing, the unfrozen moisture content converges to about 20% throughout the active layer during soil freezing (Fig. 11). However, this does not suggest that the total moisture content remains at a constant value throughout winter. At freeze-up, the water table is typically deeper than 0.5 m below ground, while at the onset of spring melt, the upper surface of the frozen, saturated soil is typically about 0.1 m below the ground surface, within the zone of high hydraulic conductivity (Fig. 8b). How this condition develops during the winter period remains unclear. Recent field investigation suggests that the amount of water supplied to the soil during the spring melt event in addition to the cumulative amount of meltwater supplied during the preceding over-winter melt events, is sufficient to saturate the ~0.4 m thick soil zone between the water table position at the time of freeze-up and the frost table position at the end of winter. This is supported by recent (2002–03) measurements of liquid moisture using water content reflecto-

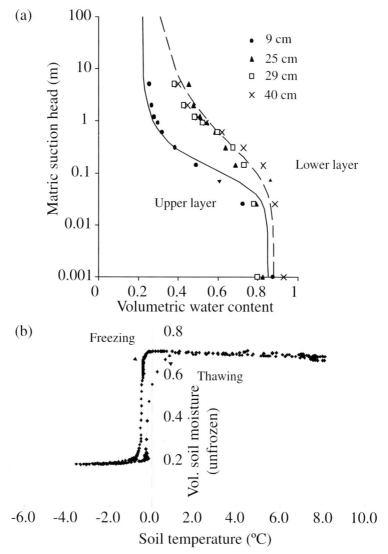

Fig. 10. (a) Variation in soil moisture with soil tension for samples from different depth positions in a soil pit on a peat plateau at Scotty Creek; and (b) unfrozen volumetric moisture content of a saturated sample from a pit at a depth of 0.3 m, plotted against soil temperature

meters (Campbell Scientific, CS 615) in a soil pit, and from measurements of total soil moisture (frozen and unfrozen) in two soil cores extracted in late winter (prior to snowmelt) near the pit (Fig. 12). Between freeze-up and late-winter, the total (frozen and unfrozen) soil moisture increased throughout the core profiles but was greatest near the ground surface, notably at the 0.1 m level. In both cores, the total soil moisture below 0.3 m was 5–10% below porosity, which was close to the saturated state. It was also indicated by the measured values in the soil pits that the unfrozen moisture content was only about 20% on the day when the two cores were extracted (Fig. 12).

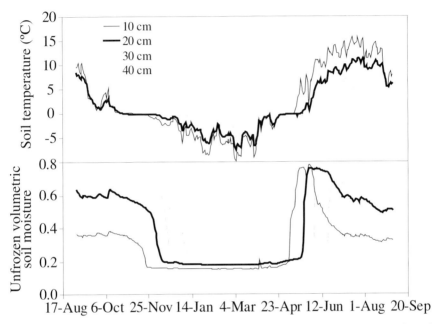

Fig. 11. Soil temperature and unfrozen volumetric moisture content at selected depths below ground surface of a peat plateau at Scotty Creek

5 Toward Basin Runoff Computation

The field and laboratory studies have produced an understanding of the key factors controlling subsurface runoff from a peat plateau, a critical step toward modeling the magnitude and timing of subsurface input from the peat plateaus to the basin drainage network. Depth to the frost table is a critical consideration for subsurface drainage but recent field measurements indicate that variations in the mean frost table depth and in the hydraulic gradient are small among peat plateaus, suggesting that the subsur-

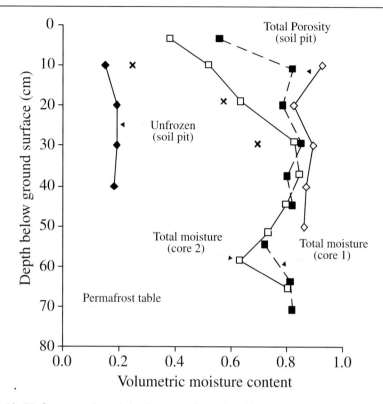

Fig. 12. Unfrozen and total (unfrozen + frozen) soil moisture content at a Scotty Creek site in late winter. Total soil moisture is obtained by gravimetric measurements of two soil cores extracted from snow-cover ground on April 6, 2003. Unfrozen moisture was measured on the same day as the soil pit measurements. Soil porosity was obtained from samples taken from the soil pit in August 2001 when sensors were installed. Crosses indicate soil moisture immediately before freeze-up, at 10, 20 and 30 cm depth increments

face drainage rate would not vary appreciably among plateaus. However, individual peat plateaus vary widely in size (Fig. 3) and so would the subsurface flowpath length and therefore the timing of subsurface drainage from plateaus. The hydraulic radius R_h provides a reasonable approximation of the average flow length to the edge of a peat plateau. It can be estimated by $R_h = 2A / P$, where P is the plateau perimeter and A is its area, both easily obtainable using image analysis software. Based on 609 peat plateaus identified (Fig. 3), R_h appears to follow a log-normal distribution (Fig. 13). Current research is focussed on applying the Cold Regions Hydrological Model (CRHM; Quinton et al. 2004) to derive a composite subsurface drainage hydrograph for the overall cover of peat plateaus for the range of flowpath lengths defined by the frequency distribution of R_h, using representative values of frost table depth and hydraulic gradient. This composite hydrograph represents the 'hillslope' input from the peat plateaus to the adjacent bogs and fens, including the basin drainage network.

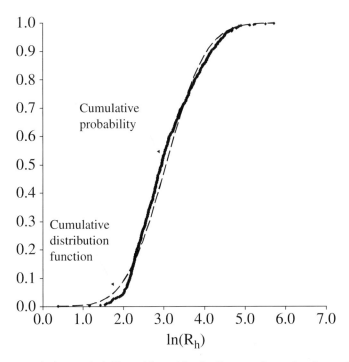

Fig. 13. Cumulative probability of logarithmically-transformed values of the hydraulic radius (R_h) of the 609 peat plateaus identified in Fig. 3. Dashed line indicates the lognormal distribution computed from the mean and standard deviation of the R_h values

Computing a composite hydrograph of drainage from the peat plateau land cover type, such as from all the peat plateaus shown in Fig. 3, is a first step toward computing the basin hydrograph for Scotty Creek and other basins in this region. The next step is to route the water from the peat plateaus, through the channel fens and connected flat bogs to the basin outlet, but there is little research on the hydrologic functioning of these two land-cover types. Water level recorded at several nodes within the Scotty Creek basin (Quinton et al. 2003) showed that following storm events, drainage water concentrates in the channel fens and moves toward the outlet at an average flood-wave velocity of 0.23 km h^{-1}. By tracking flood waves as they moved through the basin, it is evident that channel fens are an integral component of the overall drainage system, which also includes intervening lakes and open stream channels. The flood-wave velocity appears to be controlled by channel slope and hydraulic roughness in a manner consistent with the Manning formula, suggesting that a roughness-based routing algorithm might be useful. Although the apparent continuity of channel fens is clearly identified in satellite images (Fig. 2), Hayashi et al. (2004) demonstrated that the actual hydraulic connection varies over the snow-free period, and may depend on the water level in the fens. With a large supply of water such as during spring runoff, connected flat bogs often convey surface drainage along their perimeters and into channel fens. However, the spatial and temporal variation of this hydrologic connectivity and its role on basin drainage, are poorly understood. Furthermore, the possibility of deep subsurface flow below the peat plateaus cannot be ruled out. Such groundwater flow connections are well established in temperate wetlands (e.g., Siegel and Glaser 1987) but poorly documented in the discontinuous permafrost region.

6 Basin Water Balance

6.1 Evaporation

Over a four-year period (1999–2002) the cumulative precipitation was 1683 mm and only 593 mm discharged from Scotty Creek. Assuming that their difference was lost to evapotranspiration, the average annual evapotranspiration of this period was 273 mm yr^{-1}. Claassen and Halm (1996) showed that a chloride mass balance can be used to estimate the basin-scale evapotranspiration when the lithologic source of chloride is negligible. Scotty Creek basin is underlain by mineral sediments derived mainly from clay-rich glacial till with low hydraulic conductivity. Active flow of groundwater in such glacial till in western Canada is limited to a shallow (<10 m) local system (Hayashi et al. 1998a) in which the pre-Holocene chloride has been flushed out (Hayashi et al. 1998b). Scotty Creek therefore offers conditions suitable for applying the chloride method, where chloride enters the system predominantly through precipitation, and is lost mainly through stream flow. Using this method, evapotranspiration Et is given by

$$Et = P\,(Cs - Cp)\,/\,Cs \qquad (1)$$

where P is annual precipitation, and Cs and Cp are the volume-weighted average chloride concentration in stream water and precipitation, respectively. Forty-three water samples were collected at the outlet between March and December during 1999–2002 and analyzed for chloride (Hayashi et al. 2004). The volume-weighted average concentration was calculated by summing the product of the chloride concentration and the stream discharge at the time of sample collection, and dividing the total by the sum of all discharge values. The average Cs for the four-year period was 0.151 mg L^{-1}. The average Cp (0.044 mg L^{-1}) is given by the 10-year mean (1992–2001) of chloride in precipitation reported in the NatChem database (MSC 2002) at Snare Rapids, located 400-km northeast of Scotty Creek. This value is similar to the NatChem data from other stations in the interior western Canada (0.04 mg L^{-1}) presented by Hayashi et al. (1998b). The average precipitation for the water years 1999–2002 was 421 mm yr^{-1} (Table 3). Thus, Eq. (1) yields $Et = 298$ mm yr^{-1} which agrees with the hydrometric estimate of 273 mm yr^{-1}. A simple arithmetic average concentration of the 43 samples was 0.133 mg L^{-1}. Using this value for Cs in Eq. (1) gives $Et = 282$ mm yr^{-1} which is also in agreement with the hydrometric estimate. These results suggest that the chloride method has a great potential as a tool for estimating basin-scale evapotranspiration in ungauged basins.

Table 3. Fort Simpson annual and summer (May–September) precipitation (P) for each water year (October 1 to September 30), average snow water equivalent (SWE) in late March from snow survey data, and total annual runoff of Scotty Creek. All values in mm; n/a indicates the data not available.

	Normal	1999	2000	2001	2002
Period	1971–2000	10/98–09/99	10/99–09/00	10/00–09/01	10/01–09/02
Total P	369	409	431	431	412
May–Sep P	221	238	296	316	269
SWE	n/a	90	101	n/a	142
Total runoff	n/a	96	139	161	197

6.2 Precipitation and Runoff

For long term water balance estimation, the discharge record for the Birch (1974–2000), Blackstone (1991–2000), Jean-Marie (1972–2000) and Scotty (1995–2000) Creeks (Fig. 1b) were compiled. Annual runoff from these basins was generally below 200 mm (Fig. 14). The average annual runoff ratio (annual runoff expressed as a percentage of the annual precipi-

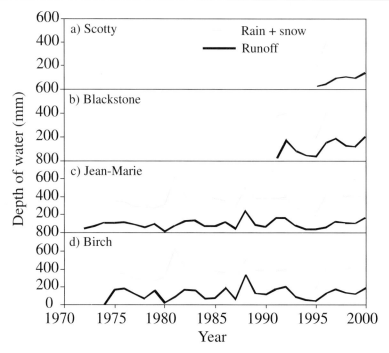

Fig. 14. Annual runoff measured at the outlets of: (a) Scotty Creek, (b) Blackstone River, (c) Jean-Marie River and (d) Birch River for the entire gauging period at these stations. Total Annual precipitation (rain and snow) measured at Fort Simpson is plotted for the same period

tation) ranged between 21% (Scotty) and 35% (Birch), indicating that ET is the dominant mechanism of water loss, with annual average rates of 297 mm (Scotty), 271 mm (Jean-Marie), 245 mm (Blackstone) and 241 mm (Birch).

7 Discussion

In the wetland-dominated central Mackenzie River Basin, peat plateaus are areas of saturated permafrost that support a tree canopy and rise above the surrounding terrain. This enables them to effectively retain water in isolated flat bogs, while re-directing runoff produced by snow melt and rainfall to the channel fens and connected flat bogs. Water entering channel fens is conveyed directly to the basin outlet, whereas water entering connected flat bog reaches the basin outlet via a channel fen. Runoff-generation algorithms in hydrologic models must account for the storage capacity of the isolated and flat bogs. Similarly, routing algorithms in distributed hydrologic models need to incorporate the network of connected flat bogs and channel fens. Preliminary studies indicate that surface roughness and channel slope may be the essential factors controlling the surface flow in channel fens.

Some major challenges remain before the conceptual model can be successfully implemented numerically. The apparent continuity of channel fens is clearly identified in satellite images, but their actual hydraulic connection likely depends on the water level. Further development of conceptual and numerical models requires the understanding of these subsurface and surface processes and their temporal and spatial variability. Likewise, the surface and subsurface hydrologic connection of flat bogs to channel fens has not been investigated. The exchange of mass and energy among the major peatland forms, and between them and the overlying atmosphere is poorly understood. Upon melt, large volumes of water are released, dramatically altering the heat exchanges (Marsh et al. 2007) and creates a mosaic of snow, bare ground and standing water for several weeks (Bowling et al. 2003). The infiltration, storage and redistribution of water within the active layer in organic terrain is exceptionally complex due to phase changes, abrupt depth-variations in soil transmission properties, and spatial and temporal configuration of the frost table. An improved understanding of the mass and energy exchanges among the peatland types, as well as the subsurface–surface–vegetation–atmosphere exchanges within each form, will permit proper modeling of the wetland-dominated drainage system in the subarctic region.

Acknowledgements

The authors acknowledge the financial support of the Natural Sciences and Engineering Research Council and the Canadian Foundation for Climate and Atmospheric Sciences. The authors also acknowledge the logistical support provided by the National Water Research Institute (Saskatoon) and by Mr. Gerry Wright and Mr. Roger Pilling of the Water Survey of Canada (Fort Simpson). We acknowledge the Aurora Research Institute for their assistance in obtaining a research license, and thank the Denedeh Resources Committee, Deh Cho First Nation, Fort Simpson Métis Local #52, Liidlii Kue First Nation and the Village of Fort Simpson for their support of this project. We also thank all those who have assisted in the field and laboratory research: John Bastien, Kelly Best, Tom Carter, Neil Goeller, Nicole Hopkins, Greg Langston, Cuyler Onclin, Jaqueline Schmidt, Jacek Scibek, Mike Toews, Jessika Toyra and Nicole Wright. Tom Brown's development of the computer code for CRHM is also gratefully acknowledged. The NatChem data were provided by the Meteorological Service of Canada and the Government of the Northwest Territories, Environmental Protection Service.

References

Arai K (1992) Maximum likelihood TM classification taking the effect of pixel to pixel correlation into account. Geocarto International 7:33–39

Aylsworth JM, Kettles IM (2000) Distribution of peatlands. In: Dyke LD, Brooks GR (eds) The physical environment of the Mackenzie Valley, Northwest Territories: a base line for the assessment of environmental change. Geol Surv Can B 547, pp 49–55

Bowling LC, Kane DL, Gieck RE, Hinzman LD, Lettenmaier DP (2003) The role of surface storage in a low-gradient Arctic watershed. Water Resour Res 39:1087

Claassen HC, Halm DR (1996) Estimates of evapotranspiration or effective moisture in Rocky Mountain watersheds from chloride ion concentrations in stream baseflow. Water Resour Res 32:363–372

Hamlin L, Pietroniro A, Prowse T, Soulis R, Kouwen N (1998) Application of indexed snowmelt algorithms in a northern wetland regime. Hydrol Process 12:1641–1657

Hayashi M, Quinton WL, Pietroniro A, Gibson JJ (2004). Hydrologic functions of wetlands in a discontinuous permafrost basin indicated by isotopic and chemical signatures. J Hydrol 296:81–97

Hayashi M, van der Kamp G, Rudolph DL (1998a) Water and solute transfer between a prairie wetland and adjacent uplands, 1. Water balance. J Hydrol 207:42–55

Hayashi M, van der Kamp G, Rudolph DL (1998b) Water and solute transfer between a prairie wetland and adjacent uplands, 2. Chloride cycle. J Hydrol 207:56–67

Hegginbottom JA, Radburn LK (1992) Permafrost and ground ice conditions of Northwestern Canada. Geol Surv Can, Map 1691A, scale 1:1 000 000

Lillesand TM, Kiefer RW (1994) Remote sensing and image interpretation. Wiley

Marsh P, Pomeroy J, Pohl S, Quinton W, Onclin C, Russell M, Neumann N, Pietroniro A, Davison B, McCartney S (2007) Snow melt processes and runoff at the arctic treeline. (Vol. II, this book)

MSC (Meteorological Service of Canada) (2002) National climate data archive of Canada. Environment Canada, Dorval, Quebec, Canada

NWWG (National Wetlands Working Group) (1988) Wetlands of Canada: ecological land classification series, no 24. Sustainable Development Branch, Environment Canada, Ottawa, Ontario, and Polyscience Publications Inc, Montreal, Quebec

Quinton WL, Carey SK, Goeller NT (2004) Snowmelt runoff from northern alpine tundra hillslopes: major processes and methods of simulation. Hydrol Earth Syst Sci 8:877–890

Quinton WL, Gray DM (2003) Subsurface drainage from organic soils in permafrost terrain: the major factors to be represented in a runoff model. Refereed Proc 8th International Conference on Permafrost, Davos, Switzerland

Quinton WL, Gray DM, Marsh P (2000) Subsurface drainage from hummock covered hillslopes in the Arctic tundra. J Hydrol 237:113–125

Quinton W, Hayashi M, Pietroniro A (2003) Connectivity and storage functions of channel fens and flat bogs in northern basins. Hydrol Process 17:3665–3684

Richards JA (1984) Thematic mapping from multitemporal image data using the principal components transformation. Remote Sens Environ 16:35–46

Robinson SD, Moore TR (2000) The influence of permafrost and fire upon carbon accumulation in high boreal peatlands, Northwest Territories, Canada. Arct Antarct Alp Res 32:155–166

Rouse WR (2000) Progress in hydrological research in the Mackenzie GEWEX Study. Hydrol Process 14:1667–1685

Siegel DI, Glaser PH (1987) Groundwater flow in a bog-fen complex, Lost River peatland, northern Minnesota. J Ecol 75:743–754

Stewart RE, Leighton HG, Marsh P, Moore GWK, Ritchie H, Rouse WR, Soulis ED, Strong GS, Crawford RW, Kochtubajda B (1998) The Mackenzie GEWEX Study: the water and energy cycles of a major North American river basin. B Am Meteorol Soc 79:2665–2683

Yamagata Y (1997) Advanced remote sensing techniques for monitoring complex ecosystems: spectral indices, unmixing, and classification of wetlands. Ph.D. thesis, University of Tokyo

Zoltai SC, Vitt D (1995) Canadian wetlands – environmental gradients and classification. Vegetation 118:131–137

Chapter 15

River Ice

Faye Hicks and Spyros Beltaos

Abstract River ice processes have an important influence on winter hydrology of cold regions. During freeze-up excessive frazil ice production can obstruct water intakes to constrain hydro-power production, and frazil accumulations can be detrimental to fish habitat. Frazil problems may persist through winter or, alternatively, mid-winter thaws may lead to premature breakup and possible ice jam flooding. The river ice breakup period may be characterized by severe ice runs associated with ice jam formation and release, with potential impacts on infrastructure, and a high risk of flooding. An overview of river ice research undertaken in the past decade is presented, including investigations into the potential impacts of climate change on rivers in the Mackenzie Basin, observations of dynamic river ice processes such as ice jam formation and release, application of satellite remote sensing techniques for river ice characterization and the development of new hydraulic and logic based models for ice jam flood forecasting.

1 Introduction

One unique aspect of cold region hydrology is the influence of winter on streamflow behavior. Most Canadian rivers experience some ice effects each year, and in many cases the runoff events associated with river ice have produced the most extreme and dangerous flood events on record. This is because breaking river ice forms ice jams that obstruct the passage of runoff and can raise water levels far higher than those experienced for the same flows under open water conditions. Therefore, despite the fact that river ice processes tend to occur on relatively small scales (in the order of tens of kilometers) they can significantly affect basin hydrologic response in term of channel routing efficiency, with the consequent influence felt over hundreds of kilometers.

River ice can also be beneficial. For example, in many areas of northern Canada, ice bridges across rivers provide access to remote communities, and many rely on these winter crossings for essential transport of supplies and people. Even more populated communities take advantage of river ice crossings for more convenient public transportation, or for industrial transport to and from mining or lumber operations. River ice covers have also been used as convenient platforms for bridge construction or bridge foundation testing. In less populated areas, river ice jams can actually be beneficial in creating water levels sufficiently high to replenish shallow lakes and wetlands, as in the case of the Peace-Athabasca Delta (Beltaos et al. 2006a).

Recent Canadian experience suggests that climate change has already begun to influence the winter regime of northern rivers. Many northern communities are experiencing warmer weather, which in turns limits the viability of some ice roads and crossings. Climate warming may also have the potential to increase the frequency and severity of ice jam related flooding in certain Canadian regions (Beltaos 2002). It is critically important to realize that river ice processes are not only affected by basin hydrology, they can affect basin hydrology. Consequently, realistic models of the impact of climate warming on basin hydrologic response in northern regions must have a deterministic component that considers the interaction of climate, hydrology and river ice hydraulics simultaneously.

This chapter provides an overview of the unique nature of river ice processes from freeze-up, through the winter and during breakup, including a discussion of the effects of streamflow regulation on the winter regime of rivers. Emphasis is placed on breakup and attendant ice jams, which are the processes that have the most serious ecological and socio-economic impacts. A synopsis of river ice research areas undertaken as part of the Mackenzie GEWEX Study (MAGS) is also presented, including investigations of: new streamflow monitoring and remote sensing techniques; novel ice jam flood forecasting methods; climate and regulation impacts on ice-jam flooding of northern rivers; and the development of numerical models of river ice processes. A summary discussion of the potential impacts of climate change on the winter regime of rivers is also presented.

2 Overview of River Ice Processes

2.1 Freeze-up

The first stage in river ice cover development on northern rivers is water cooling. The primary source of heat transfer is convective heat loss from the water surface to the colder overlying air. Solar radiation in the daytime contributes small amounts of heat to the overall energy budget; but in the Mackenzie River Basin daily heat gain from solar radiation during the freeze-up period is nearly balanced by daily heat loss due to long-wave radiation emissions, and thus the two can generally be neglected. As the cooling period progresses, water temperatures eventually reach 0°C. However, further cooling of the water to at least a few hundredths of a degree below 0°C is necessary before the first ice formation can practically occur. This is known as "supercooling".

The onset of freeze-up begins with the development of frazil particles (small discs of ice 1 to 3 mm in diameter) that form in the supercooled water. In the slower flow near the banks (e.g., less than about 0.1 m s^{-1} velocity), ice particles develop near the surface and accumulate to form a continuous layer of skim ice on the water surface. This skim ice effectively prevents further supercooling, and subsequent ice growth is thermal in nature. The resulting ice cover is typically termed "border ice". Because ice formed by thermal heat exchange across the ice layer usually results in crystal growth in the vertical direction, a characteristic of thermal ice is its columnar crystal structure, easily recognizable in the "candles" of ice seen as this type of ice melts.

Frazil particles also form in the faster moving portions of the flow (away from the banks). Figure 1 traces the formation of ice in the main-flow zone of a river, starting with ice-free, above-freezing water and ending with zero-degree water carrying large ice floes. Due to turbulence, water cooled at the surface is mixed through the flow and leads to an apparent spontaneous generation of frazil particles, occurring throughout the depth (once the water temperature cools below 0°C). Individual frazil particles tend to behave in a highly adhesive fashion while in supercooled water. Adhesion is generated when the particles melt briefly as they collide with other ice particles or objects, due to the small amount of heat produced by the collision, but refreeze readily in the supercooled water. This adhesive nature of the frazil particles causes them to accumulate, forming "frazil slush" (also known as "frazil flocs"). These frazil flocs eventually reach a size at which buoyant forces overcome the ability of the flow turbulence to maintain the flocs in suspension, and they float to the water surface. Here, they fuse with other flocs to form larger elements whose unsubmerged portion freezes into the familiar "pancake ice" (also known as "frazil pans"). Some of the frazil particles or pans may also collect along the border ice. This increases the border ice encroachment on the channel, and is termed "buttering".

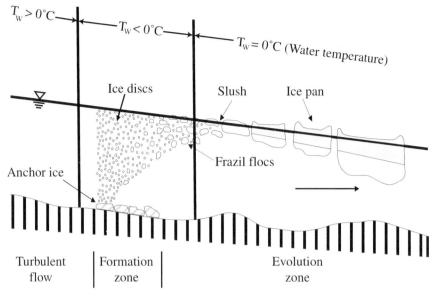

Fig. 1. Stages in river ice cover development

Turbulence also causes some frazil particles to impact the bed, and pick up small sediment particles before accumulating into large enough flocs to float to the surface. When this happens, the frazil slush layer underlying the pans may contain sediment particles. When the frazil particles adhere to very large gravel or boulders they can remain on the bed forming an ice accumulation known as "anchor ice", which releases and floats when the water is no longer supercooled.

Frazil pans float downstream on the water surface. As surface concentrations increase (both in time and in the downstream direction) the individual pans may ride up on, or freeze against other pans forming 'rafts'. When the concentration reaches about 80–90%, "bridging" often occurs. This involves a congestion of ice floes and a subsequent cessation of their movement at a site along the river. Once bridging is established, the incoming ice floes may lead to an upstream progression of the ice front by "juxtapositioning" with ice floes accumulating edge to edge on the water surface. However, if flow velocities are high enough, it is also possible that surface ice floes arriving at the ice front may be swept under the ice front and then deposited on the underside of the cover. This process is known as "hydraulic thickening". The increased thickness results in an increase in water level and a corresponding decrease in flow velocity. With a sufficient reduction in flow velocity, ice floes are no longer swept under the ice cover and the ice front can continue its upstream progression. In extreme cases, velocities may be high enough that the entire ice cover formed at the bridging site maybe swept downstream, after which bridging must again initiate before frontal progression of the ice cover can occur. Any one of these three scenarios may be observed at a given site at different times. Which of the three is to be expected at any given time is a function of both meteorological and hydraulic conditions.

As the ice front progresses upstream, either by juxtapositioning or by hydraulic thickening, the forces acting on the ice accumulation increase. These forces include the downslope component of ice weight within the ice accumulation, and the flow drag along the underside of the ice cover. These forces are resisted by the internal strength of the accumulation which, for freeze-up accumulations, is often enhanced by freezing between the individual ice floes. The forces acting on the ice cover increase as it lengthens, and when the magnitude of these forces approaches the internal strength of the ice accumulation, the ice cover is prone to collapse, or "shove", and thickens substantially as the ice front progresses upstream. The increased thickness and roughness of the ice cover after such a collapse is usually reflected in a dramatic increase in water levels. The resulting accumulation is termed a "freeze-up ice jam" or "hummocky ice cover". Normally once the accumulation has stabilized, the water between the ice floes freezes and gives strength to the accumulation, thereby inhibiting further consolidation.

2.2 Winter

Once a stable ice cover is established and cold weather persists, the solid-ice layer thickens by freezing at the water-ice interface. Where there is no slush deposit under the solid-ice layer, the original crystals grow vertically downward forming clear, columnar ice that is commonly called blue or black ice. If a slush deposit is present, the thickening process will be accelerated by the fact that a certain fraction of ice is already present and it is only the interstitial water that needs to freeze. Freezing causes expulsion of impurities, which tend to concentrate at the crystal boundaries where they can play an important role in the decay of the ice cover during the breakup period. The snow cover insulates the ice sheet and retards growth but it can also enhance growth via formation of snow ice. This is a relatively opaque layer that forms by freezing of overflow water in cases where the phreatic water surface is above the top of the ice sheet. For temperate lakes, Adams and Prowse (1981) found that the decrease in black ice growth due to insulation can be offset in the long term by the additional ice thickness created by snow-ice.

The growth of the solid-ice layer during the winter slows down as the layer becomes thicker. In many applications, the well-known Stefan formula, which is indexed by the accumulated freezing degree-days, provides a simple means for approximate prediction of the solid-ice thickness (Michel 1971). More sophisticated approaches explicitly account for various factors other than air temperature, such as solar radiation, wind speed, relative humidity, cloud cover, as well as snow depth and density (Menard et al. 2002). Values of the average maximum thicknesses of solid ice in Canadian rivers range from less than 0.3 m in the more temperate regions of Southern Canada to over 1.7 m in the Arctic (Prowse 1995). The dates on which the various maxima are attained vary significantly because of climatic differences. For example, river ice continues to grow well into April in the Mackenzie River delta (Sherstone et al. 1986), long after breakup has occurred in southern Canada.

Much higher growth rates and extreme thicknesses can be expected where thin layers of slowly moving water are continuously exposed to the atmosphere, forming aufeis, also known as icings or naleds. These are accumulations of solid ice produced by the seepage of water onto existing ice covers, and are most commonly encountered on arctic and sub-arctic rivers. During periods of runoff, icings determine the channel routing and can act as major flow restrictions. They are of special concern where flow is routed through narrow channels or culverts (Prowse 1995).

2.3 Breakup

The breakup of a river ice cover is triggered by mild weather and encompasses a variety of processes associated with thermal deterioration: initial fracture, movement, fragmentation, transport, jamming, wave motion and ice runs, and final ice clearance. Although several or all of these processes may occur simultaneously within a given reach, it is convenient to visualize the breakup period as a succession of distinct phases such as pre-breakup, onset, drive, and wash. During the pre-breakup phase, the ice cover becomes more susceptible to fracture and movement via thermally induced reductions in thickness and strength. At the same time, the warming weather brings about increased flows, due to snowmelt or rainfall or both. The increasing hydrodynamic forces fracture the ice cover, while the rising water levels reduce its attachment to the riverbanks. Eventually, large segments of the now fragmented cover are dislodged and set in motion by the flow. This is the onset of breakup, and is followed by the drive, that is, the transport and further breakdown of large ice sheets into smaller blocks and rubble. The onset is governed by many factors, including channel morphology, which is highly variable along the river. It is thus common to find reaches where breakup has started, alternating with reaches where the winter ice cover has not yet moved.

Invariably, this situation leads to jamming because ice blocks moving down the river in one reach encounter stationary ice cover in another reach and begin to pile up behind it, initiating a jam (Fig. 2). Ice jams can stay in place for a few minutes or for many days; they can be a few hundred meters or many kilometers long. Ice jams can cause much higher water levels than are possible under open-water conditions with the same river discharge, owing to their considerable thickness and roughness (Beltaos 1995). On many Canadian rivers, the highest water levels result from ice jams rather than from open-water floods. The wave and ice run that follow the release of a jam often dislodge and break up long sections of intact ice that they encounter; on other occasions, the stationary ice is too strong or the wave is too attenuated, and the ice run is arrested, forming a new jam. In this manner, more and more ice is broken up and carried down the river, until the final jam releases. This is the start of the wash or final clearance of ice.

Depending on hydro-meteorologic conditions, the severity of a breakup event can vary between two extremes, those of the thermal or overmature breakup and the premature breakup. The former type occurs when mild weather is accompanied by low runoff, due to slow melt and lack of rain. The ice cover deteriorates in place and eventually disintegrates under the limited forces applied by the modest current. Ice jamming is minimal, if any, and water levels remain low. Premature breakup on the other hand, is associated with rapid runoff, usually due to a combination of rapid melt and heavy rain. The hydrodynamic forces are sufficient to lift and break segments of the ice cover before significant thermal deterioration can occur. Ice jams are now the most persistent because they are held in place by sheet ice that retains its strength and thickness. This is aggravated by the prevailing high river flows, so that premature events are the most severe in terms of flooding and damages. Usually, a breakup event falls somewhere between these two extremes, and involves a combination of thermal effects and mechanical fracture of the ice. Herein, the term mechanical breakup is used to denote all non-thermal events because they are, at least in part, governed by the mechanical properties of the ice cover.

Fig. 2. Aerial and ground views of ice jams: (top) Peace River in Wood Buffalo National Park, Alberta (Photo: S. Beltaos), and (bottom) Hay River near the town of Hay River, Northwest Territories, ice floe thickness about 7 m. (Photos: F. Hicks)

In the colder continental parts of Canada such as the Prairies or the Territories, we are most familiar with a single event, the spring breakup, which is triggered by snowmelt. In more temperate regions, however, such as parts of Atlantic Canada, Quebec, Ontario and British Columbia, events called mid-winter thaws are common. Usually occurring in January and February, they consist of a few days of mild weather and typically come with significant rainfall. River flows may rise very rapidly and sufficiently to trigger breakup on many local rivers. This is the mid-winter breakup which can be much more severe than a spring event, because of the sharp rise in flow that results from the rain-snowmelt combination. The premature nature of a mid-winter breakup event almost ensures the occurrence of major jams. Moreover, dealing with the aftermath of flooding is hampered by the cold weather that resumes in a few days, while many mid-winter jams do not release but freeze in place, posing an additional threat during subsequent runoff events.

2.4 Breakup Regime of Northern Rivers

The breakup of northern rivers is triggered by spring melt and typically occurs after several days of bright sunshine that helps reduce ice thickness and strength. Mid-winter thaws and runoff events are either rare or completely unknown. Breakup is thus more predictable than on more southern rivers, and often can be anticipated days or weeks in advance, based on flow hydrographs occurring at upstream hydrometric stations. Maximum winter ice thickness is considerable, ranging from about 0.6 m on the upper Peace and Athabasca Rivers, to 1 m in the lower Peace and Slave, to over 1.5 m in the Mackenzie Delta channels.

Low water surface slope is another characteristic feature of the main rivers of the Mackenzie Basin. For instance, the slope of Peace River downstream of Carcajou drops below 0.1 m km^{-1} (Kellerhals et al. 1972). In the main channels of the Mackenzie Delta, the water surface slope is about 0.01 m km^{-1}. It is doubtful whether such flat reaches can generate sufficient driving forces to dislodge the thick ice cover under snowmelt-runoff conditions. Mechanical breakup is more likely to be generated by waves that result from the releases of upstream ice jams. Such waves can greatly amplify the driving forces (Beltaos and Burrell 2005b) and cause breakup over extensive river lengths (Gerard et al. 1984). Each wave is accompanied by an ice run that is eventually arrested by competent ice cover downstream to form a new jam. Essentially, mechanical breakup process consists of a sequence of jams and ice runs. Where this process is stalled for prolonged periods of time, thermal effects cause severe decay of the winter ice cover, resulting in a thermal event.

A key factor affecting many of the large rivers in the Mackenzie River Basin is that they are generally north flowing (e.g., Peace, Athabasca, Slave, and Mackenzie). Thus they tend to break up first in the headwaters, which are located in the most southerly portions of their respective basins. As they advance downstream, the dynamic ice runs discussed above are thus more likely to encounter a strong (undeteriorated) ice cover, since this cover is located further north. This sequence of events enhances the potential for ice jam formation as compared to rivers flowing in other directions.

2.5 Effects of Streamflow Regulation

On regulated rivers both water storage in the reservoir and flow release patterns have the potential to significantly affect river ice processes. Reservoir storage is important because it raises winter water temperatures in the downstream reach. This occurs because of the unique density characteristics of water (Ashton 1986). As with other fluids, water density varies with temperature. However, water density is maximum at 4°C, and decreases with temperatures both below and above 4°C. In deep reservoirs containing water at temperatures in excess of 4°C, the cooler, denser water is found at greater depths than the warmer, less dense water. This vertical stratification is stable because further heating of the surface layers of water only leads to reduced density in these upper layers. However, as water in a reservoir cools below 4°C, it develops an inverse temperature gradient. Initially, surface heat loss lowers the water temperature in the upper layers towards 4°C and this denser water then sinks to the lower levels. As the temperature cools the water further, the water density decreases and the colder (but less dense) water remains nearer the surface. The resulting profile is at 0°C near the surface and 4°C at the reservoir bed. Further heat loss through the winter season has the potential to cool the water through the entire depth. However, there will still be a temperature gradient until all of the water is cooled to 0°C. The temperature gradient generally persists throughout the winter, as the formation of an ice cover insulates the water from cold air temperatures. Snow accumulations on the ice cover enhance this insulating effect. Since water is typically drawn from the bottom of the reservoir, flow releases from dams are usually above 0°C throughout the winter. Note that the vertical temperature gradient does not persist once the water enters the river, due to the turbulent nature of the flow. Consequently, temperature in the river is expected to be homogeneous throughout the flow depth.

The release of warm water from a reservoir can affect a river's ice regime in three ways (see also Starosolszky 1990). First, it can inhibit the early formation of an ice cover in the upper reach of the river (near the reservoir outlet). This generally leads to a prolonged and relatively unstable freeze-up period (i.e., one prone to ice consolidation events). On rivers regulated for hydro-power production which are subject to hydro-peaking operations (highly variable discharge conditions), this unstable period may be prolonged. Second, it can limit the thermal growth of ice in the downstream channel. Therefore, thinner ice covers might be expected as compared to the pre-regulation period. Third, it could lead to the early melt of river ice in the spring and a greater tendency for thermal breakup events.

Regulation may also alter the flow hydrograph, and thence the ice regime, of the river downstream of the regulation site. For example, hydro-power generation by means of large reservoirs greatly augments fall and winter flows, resulting in thick ice covers and high freeze-up water levels, which can modify the frequency and severity of spring breakup jamming (Beltaos 1997; Beltaos et al. 2006a) or promote mid-winter breakup and jams (Andres et al. 2003). Breakup flows may be either augmented or reduced, and this effect can also modify the jamming regime.

Upstream effects arise mainly from a change in the water surface profile of the river. Relatively thick ice covers form over the surface of the reservoir, and appear earlier than under natural conditions (Starosolszky 1990). The thicker reservoir ice and the low water surface slope of a reservoir promote ice jam formation during breakup, not only within the main river but also at the mouths of tributaries located within the reservoir reach.

3 Monitoring River Ice Processes

3.1 Overview

Determining the areal extent and nature of the ice cover is a key objective of many river ice monitoring programs. Particularly during breakup, monitoring the development of open water leads, major ice movements, ice runs, and ice jam formation and release is often done with the aid of aerial reconnaissance flights, documenting with digital video or still camera. Water level monitoring is also generally undertaken, using both manual measurements and automated monitoring networks (e.g., Robichaud and Hicks 2001) with a key objective being to measure the water surface profiles associated with ice jam formation, or the propagating waves associated with ice jam release (Kowalczyk and Hicks 2003). A key variable to river ice studies is streamflow, as the variable ice conditions occurring in the freeze-up, winter, and breakup periods pose a challenge to conventional measuring techniques.

3.2 Discharge Determination under Ice Affected Conditions

Currently the only reliable method for determining discharge under ice-affected conditions is to conduct direct measurements. This involves the use of a current meter to obtain point velocity measurements at (typically) two points in the flow depth, at more than 20 vertical panels across a channel. These point measurements are then integrated over the flow area to determine the total discharge. Pelletier (1989) provides a detailed description of typical practices for streamflow gauging under ice-affected conditions in both Canada and the USA.

Because of the cost and logistical difficulties associated with direct measurement, winter discharge estimates may be inferred from as few as two direct measurements over a six-month winter period (Moore et al. 2002). Indirect determination of the streamflow is a relatively straightforward procedure under open water conditions. Data collected with automated water level recorders can be readily converted to discharge using established stage-discharge relationships (i.e., rating curves), developed by conducting simultaneous water level and direct discharge measurements over a wide range of stream flows. Whenever possible, gauging sites are placed in reaches relatively unaffected by backwater and drawdown, so that there is little scatter in the stage-discharge relationship. When this is achieved, the stage-discharge relationship is quite adequately defined by a simple uniform flow approximation (e.g., Manning's equation).

Although the stage-discharge relationship is a well researched problem in open channel flow situations, and is even amenable to prediction through hydraulic analyses for highly dynamic flow situations (where a looped rating curve occurs), at present this relationship is poorly defined under ice conditions. However, it is known to be highly variable, particularly when partial ice covers and/or ice jams are involved. The primary reason for this is that, under ice conditions, it is not possible to develop a unique relationship between stage and discharge, as variations in ice conditions result in a non-unique rating curve. Despite this variability, it is possible to quantify the relationship between stage and discharge at a given river section under simple ice-covered conditions, providing ice characteristics are known and gradually varied flow hydraulics are considered (Hicks and Healy 2003). Nevertheless, direct discharge measurement remains the principal method in determining winter discharge.

Recently, there has been interest in developing faster and more efficient methods for conducting direct discharge measurements in winter, necessary not only due to cost and access, but also because of the hazards to the operational staff. It is difficult to determine when the ice cover is safe, and it will become a greater concern if climate change results in thinner and more intermittent ice covers. Water Survey of Canada (WSC) staff began researching the suitability and accuracy of ultrasonic (acoustic) flow metering devices in the early 1970s, and have been pursuing the goal of automated discharge measurement using these devices in earnest since 1985 (Wiebe et al. 1993). The advantage of acoustic flow metering is that it can potentially eliminate the need for streamflow gauging using conventional current metering techniques, replacing it with an automated measuring system. Healy and Hicks (2004) explored the applicability of empirical index velocity measurements which have the potential to speed up the measurement procedure, by providing accurate relationships between a single point velocity (or single vertical velocity profile) and the channel mean velocity. Morse (2005) has been investigating analytical relationships that have the potential to expand the applicability of this approach.

3.3 Remote Sensing Techniques

The use of Synthetic Aperture Radar (SAR) imagery as acquired by the Canadian RADARSAT and other satellites for sea ice monitoring has reached the operational stage and indications are that it has similar potential in the field of river ice monitoring. Weber et al. (2003) in a study on the Peace River, employed a visual analysis combined with a semi-automated classification system to categorize seven different "ice types" ranging from open water and frazil pan ice covers to several levels of juxtaposed and consolidated ice covers. Further research has revealed that depending upon the scale of the river and the resolution of the satellite image, it may also be possible to identify the extent of ice in a navigable waterway. For example, a variant texture analysis procedure has been successfully applied to both the Mississippi and Missouri Rivers to delineate specific regions of brash ice, sheet ice and open water on satellite images (Tracy and Daly 2003).

As part of the MAGS study, Hicks et al. (2006) investigated the potential for characterizing river ice using satellite SAR. Field data collected on the Athabasca River at Fort McMurray, Alberta, in winter 2003 lends credence to the suggestion that it may be possible to infer, indirectly, information on relative ice thickness from RADARSAT-1 imagery. In addition, heavily consolidated ice covers were observed to exhibit a higher backscatter than less consolidated adjacent areas. During the 2003 breakup period it was possible to distinguish areas of open water, intact ice, ice jams and running ice in three satellite images taken from April 18 to April 22 (Fig. 9 in Woo and Rouse 2007). This shows promise as a cost-effective means of monitoring river breakup remotely. Further research investigating the protocols to be used for image analysis is underway, and this will aid in the analysis of images taken with different modes and incidence angles. In addition, further work investigating the potential for river ice characterization in freeze-up and winter is also underway.

4 Ice Jam Flood Forecasting

4.1 Overview of Ice Jam Formation and Release

While ice jams are formed during freeze-up, it is the breakup jams that have by far the greater flood damage potential, owing to the much higher flow that accompany the breakup events. Breakup jams are initiated where moving ice blocks encounter a river reach where the winter ice cover has not as yet been set in motion. Jam formation is thus linked to the mechanism of dislodgment of the winter ice cover and the onset of breakup (Beltaos 1997). In natural streams, this mechanism involves channel planform, bathymetry, and slope, as well as ice thickness and strength. Sharp bends, sudden reductions in slope, constrictions, or islands are frequent jamming sites, along with areas where the ice cover is relatively thick and strong.

Once primed, jams propagate upstream at a speed dictated by ice supply, flow and channel conditions, as well as internal strength of the accumulated ice blocks that comprise the rubble. During this time, local flow and ice conditions can be highly dynamic. Eventually, an approximation to a steady state may be established so that jam and flow properties change little with time. Given sufficient supply of ice blocks, a jam may be many kilometers long, and attain its full potential thickness and water level, becoming an "equilibrium" jam (Fig. 3). This appellation derives from the development of a reach with relatively constant flow depth and thickness, termed the equilibrium thickness (Uzuner and Kennedy 1976). Downstream of the equilibrium reach, the water surface slope increases rapidly so that the water level profile can meet the much lower stage that prevails at the toe (downstream end). Here, the thickness of the jam increases, and often leads to "grounding", particularly in steep and wide reaches (Beltaos 1995).

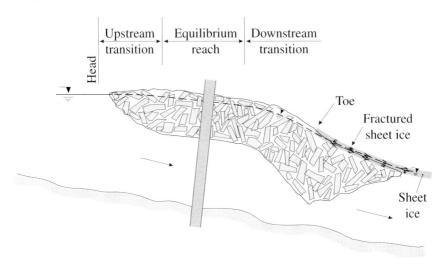

Fig. 3. Schematic illustration of an equilibrium jam. (From Beltaos (2001) with permission from ASCE)

Increasing flow and advancing thermal decay at the toe of an ice jam eventually cause it to release. Usually, the release is an abrupt event that generates a steep wave, as a large volume of water upstream of the toe can now move without obstruction other than bed friction. Release waves are characterized by rapid water-level rise (up to 0.8 m min^{-1} has been measured; Beltaos and Burrell 2005a; Kowalczyk and Hicks 2003), and very high water and ice velocities (several meters per second is not uncommon). Such waves have great damage potential, including the loss of human life. As already mentioned, release waves can also cause breakup of the winter ice cover over extended reaches, and are the main breakup-driving mechanism in flat rivers.

As part of the MAGS study, the hydro-climatic conditions leading to major ice-jam flooding of the Peace-Athabasca Delta have been identified (Beltaos et al. 2006a). Such flooding is essential for the replenishment of Delta lakes and ponds, especially those situated at higher elevations (perched basins). It was noted that significant jamming can only occur during a mechanical breakup event, and methodology to quantify the threshold between mechanical and thermal breakup was developed (Beltaos 2003a).

4.2 Empirical, Analytical, and Modeling Methods

It is not yet possible to predict the time and location of an ice jam, aside from likelihood statements in specific river reaches based on past experience. Ice-jam prediction is thus limited to thickness and attendant water levels, using analytical or numerical approaches. A key assumption is that the rubble in the jam behaves as a floating granular mass that obeys the Mohr-Coulomb failure criterion. The strength of the rubble is essentially supplied by the lateral confinement and by the effective upward stress, which is generated by the buoyancy of the ice in the rubble. When the jam forms, the rubble attains a thickness that is just enough to resist the internal stresses generated by the external forces (Pariset et al. 1966). Guided by this formulation, Beltaos (1983) obtained a simple functional relationship for equilibrium ice jams, and calibrated it with field data from Canadian rivers, spanning several orders of flow magnitude. The water depth within the equilibrium reach of a jam (Fig. 3) is primarily determined by river discharge, width, and slope.

The equilibrium water depth is a conservative estimate, because many jams do not attain the equilibrium condition. This is usually caused by limited supply of ice blocks and can be an important design factor in many practical situations. Steady-state, non-equilibrium jams are adequately predicted via one-dimensional numerical models, such as ICEJAM (Flato and Gerard 1986), RIVJAM (Beltaos 1993), and HEC-RAS (http://www.hec.usace.army.mil/software/hec-ras/hecras-download.html). These are public-domain models, but there are also proprietary models, such as ICEPRO and ICESIM (Carson et al. 2001). The proprietary DYNARICE model is both dynamic and two-dimensional (Liu and Shen 2000), and has been applied to field conditions. A radically different type of model is the DEM (discrete element model), which does not need to invoke the concept of a granular continuum. Instead, the motion of each block within the jam during small time steps is predicted by computing the forces applied on each block by the water and by the surrounding blocks. This approach provides important insights as to both the development and the final configuration of an ice jam and enables prediction of the forces exerted by jams on structures (Daly and Hopkins 1998; Hopkins and Tuhkuri 1999). As part of the MAGS study, the model RIVJAM was applied to the lower Peace River to determine the river discharge that is required for flooding of the Peace-Athabasca Delta when ice jams are present (Beltaos 2003b).

An important question pertaining to ice-jam flooding applications is how to determine the design stage for a variety of structures, including homes in residential subdivisions. For instance, when designing a bridge, the superstructure must be placed high enough to minimize the probability that it will find itself in the path of moving or jammed ice during the life of the structure. Such practical considerations underline the need for rational design criteria, based on stage-frequency relationships for ice-influenced conditions. Such relationships can be developed from historical records (empirical) or synthesized from local conditions (analytical).

4.2.1 Empirical Flood Frequency Estimates

The peak stages that can occur during breakup are strongly site-specific. Therefore, existing data should pertain to the site of interest or to its immediate vicinity. Transpositions and extrapolations should, as a rule, be avoided. Historical water level data may be available from various sources (e.g., hydrometric gauging stations, local residents, archives, photos, etc.) Ice scars on nearby trees also provide an indication of stages that occurred in the past and the year of occurrence (Gerard 1981). A method for performing a probability analysis on data deriving from such diverse sources as above is described by Gerard and Karpuk (1979). It is based on the "perception stage" concept (i.e., the stage below which any particular source would not have perceived, and recorded, the peak water level).

4.2.2 Analytical Flood Frequency Estimates

Crude indications of ice-caused flood frequencies can be obtained by a simple analysis based on the frequencies of the flow magnitudes occurring during the breakup period. A lower bound is obtained for the situation where no jams form near the site of interest. Then the peak stage can be calculated as a function of discharge using estimated values of ice cover thickness and hydraulic roughness. Similarly an upper bound may be established using the equilibrium-jam stage or by applying numerical models. For both the lower and upper bounds, the frequency of a given stage is that of the discharge associated with that stage. The actual frequency distribution will be somewhere between the two bounding distributions, and can be calculated if the probability of ice jam formation, $P(J)$, near the site of interest is known (Gerard and Calkins 1984). For high flows, the ice-jam stage may be limited by the configuration and elevation of the river floodplain, while further flow increases may result in jam release ("ice-clearing" flow) and open-water stages (Tuthill et al. 1996). In general, $P(J)$ could decrease with increasing flow (Grover et al. 1999), becoming zero when the flow exceeds the ice-clearing value.

4.3 Hydraulic Methods: Release Wave Modeling

Because of the dynamic nature of river ice jam release events and the significant flood risk they pose, it is desirable to be able to predict the speed and magnitude of the resulting release waves. This is a highly complex problem involving not only dynamic flow hydrodynamics but the interaction of the ice and water. Attempts have been made to model the propagation of ice jam release waves, most using one-dimensional hydrodynamic models and neglecting ice effects on the propagating wave (e.g., Blackburn and Hicks 2003). Although reasonable approximations could be achieved, difficulties were encountered in fully matching the shapes of measured stage hydrographs, suggesting that ice effects cannot be neglected. Jasek (2003) conducted field investigations documenting ice jam release events and found that the release wave celerity was affected by different ice conditions. Liu and Shen (2004) further explored this issue by applying a two-dimensional coupled flow and ice dynamic model (DynaRICE) to investigate the ice resistance effects (both internal resistance and boundary friction resistance) on ice jam release wave propagation in an idealized channel. Comparisons between the simulation results obtained with and without inclusion of ice dynamics showed that the ice effects reduce the peak discharge and slow down the release processes.

As part of the MAGS study, a number of ice jam release waves were measured on the Athabasca River at Fort McMurray. The most significant of these was in 2002 (Kowalczyk and Hicks 2003) during which the water level immediately downstream of the releasing ice jam rose 4.4 m in just 15 minutes. She and Hicks (2005) successfully modeled this event incorporating an uncoupled ice mass conservation equation and empirical momentum effects in the River 1D hydrodynamic model.

4.4 Regression and Logic-Based Ice Jam Forecasting Methods

Research into forecasting river breakup began over 50 years ago with simple single variable models and quickly led to multivariate models. Early work in this field was led by Shulyakovskii (1963) who viewed river breakup to be the result of deteriorating ice thickness and increases in river flows. Although Shulyakovskii's theory of river break up was simplistic, he did recognize the importance of the energy cycle in river ice processes and atmospheric circulation. Shulyakovskii quantified the concept of forecasting maximum stage rise during river break up as a function of several variables, including: ice thickness in the contributing reach; snow depth on the ice cover prior to ice melt; change in water level between the commencement of snow melt and the time of ice jam formation; the onset of negative air temperature in the ice-break up period; and total heat input. Both threshold models and regression models have evolved from these basic functional relationships.

Multivariate threshold models have been applied with modest success (e.g., Galbraith 1981; Wuebben and Gagnon 2005). However, specific additional variables and weighting factors required made these models highly site specific. White and Daly (2002) used stepwise selection of meteorological and hydrologic parameters to identify statistically significant input variables and then applied discriminant function analysis to predict ice jam occurrence. Massie et al. (2001) developed an artificial neural network model to produce a daily forecast of jam/no jam that required 22 input variables. Probably the most significant limitation of all of these models is that, while they do provide an assessment of the potential for ice jam occurrence, they cannot provide a prediction of the anticipated flood levels that might accompany an ice jam occurrence. Furthermore, they all exhibit a high degree of false positive indications of ice jam occurrence.

Single and multiple regression analyses have been applied to the problem of breakup water level prediction with moderate success. As part of the MAGS study, Robichaud (2003) developed a short term multiple linear regression model for predicting the maximum breakup water level for the Athabasca River at Fort McMurray, based on late winter snow pack and ice thickness as well as meteorological conditions in the few days immediately before breakup. Mahabir et al. (2007) extended the forecast lead time for Fort McMurray by several weeks, employing Fuzzy Expert Systems, a modeling technique based on set theory that allows variables to be described in linguistic terms. Fuzzy Logic has been applied successfully in a variety of fields where the relationships between cause and effect (variables and results) are difficult to express numerically but are conceptually well defined. Fuzzy Expert Systems produce a result based on logical linguistic rules rather than historical data, which allows this type of modeling to be less dependent upon the volume of historical data than many statistical methods. For the MAGS study, using only antecedent basin moisture, late winter snowpack conditions, and late winter ice thickness data, Mahabir et al. (2007) developed a Fuzzy Expert System that can forecast the potential for high water levels at breakup. Forecasts based on conditions known on April 1 (typically 3 weeks in advance of breakup) identified five years (of 22) for potentially high water levels at breakup, including all four actual high water events.

5 Climate Change Impacts on River Ice Processes

Numerous studies have examined the effects of climate on the timing of the ice season, using readily available data from hydrometric and related archives. It has been found that the general warming experienced in northern parts of the globe during the last 50 to 100 years is in step with a reduction in the duration of the winter ice cover and earlier occurrence of breakup on rivers (e.g., see review articles by Beltaos and Burrell 2003; Beltaos and Prowse 2001; Prowse and Beltaos 2002). The converse is true in isolated instances where the change involves cooling (Brimley and Freeman 1997). Andrishak and Hicks (2007) investigated the potential impacts of climate change on the thermal regime of the Peace River, as part of the MAGS study.

Such findings do not address the question of how climate change may alter the frequency and severity of extreme ice jam events. Relevant data are not as easy to find as freeze-up and breakup times, and their interpretation may be complicated by local precipitation and hydrograph characteristics. Consequently, prediction of breakup-related events would have to be based on a case-by-case analysis of current conditions and output from GCM/RCM-generated scenarios. A wide-ranging change that can be anticipated at present is the increased incidence of mid-winter breakups in parts of Atlantic Canada, Quebec, Ontario and British Columbia. This trend is already evident in parts of Atlantic Canada (Beltaos 2002, 2004; Beltaos et al. 2003). Mid-winter breakups are also expected to appear in certain regions that do not presently experience such events, such as the Prairies and northern Ontario and Quebec. For North America, Prowse and Bonsal (2004) applied simple but quantitative criteria to delineate present and future zones where mid-winter breakup events are likely to occur ("temperate region"; Fig. 4).

In certain rivers, mid-winter thaws can deplete the snowpack that is available for spring runoff, and thence reduce the severity of ice jamming during the spring breakup. This eventuality has been identified for the Peace River, as part of the MAGS study. Under future climate scenarios (2080s), depleted snowpacks are expected to severely inhibit ice-jam flooding in the lower Peace River, a necessary agent of replenishment of the lakes and ponds of the Peace-Athabasca Delta (Beltaos et al. 2006b, 2007).

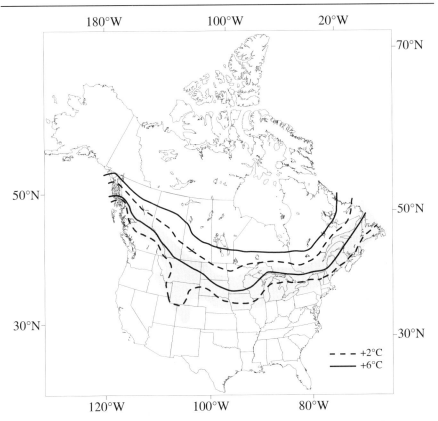

Fig. 4. Projected shift in the temperate region of North America based on mean winter temperature increases of 2 and 6°C. The shaded area is the current zone. (From Prowse and Bonsal (2004) with permission of the authors and of Environment Canada)

6 Summary

River ice plays a significant role in the lives of Canadians and other residents of high latitudes, providing a means of transport in the north, and threatening many communities across the country with flooding on an annual basis. River ice also plays a fundamental role in basin hydrology, and was a significant research component of the MAGS program. Investigations resulted in new methods for measuring streamflow under ice and the development of satellite remote sensing techniques for river ice characterization during breakup. Novel methods were developed for ice jam flood forecasting and to predict the flood waves associated with ice jam release. New, public domain, numerical models of river ice processes provide the means to assess the effects of climate change on the winter regime of rivers. Through a decade of field and modeling research, MAGS has substantially advanced the knowledge on river ice in cold regions.

References

Adams WP, Prowse TD (1981) Evolution and magnitude of spatial patterns in the winter cover of temperate lakes. Fennia 159:343–359

Andres D, Van der Vinne G, Johnson B, Fonstad G (2003) Ice consolidation on the Peace River: release patterns and downstream surge characteristics. Proc 12th Workshop on the hydraulics of ice covered rivers (CD format), CGU HS Committee on river ice processes and the environment, June 19–20, Edmonton, AB, pp 319–330

Andrishak R, Hicks F (2007) Impact of climate change on the Peace River thermal ice regime. (Vol. II, this book)

Ashton GD (1986) River and lake ice engineering. Water Resources Publications, Littleton, Colorado

Beltaos S (1983) River ice jams: theory, case studies and applications. J Hydr Eng Div–ASCE 109(HY10):1338–1359

Beltaos S (1993) Numerical computation of river ice jams. Can J Civil Eng 20:88–89

Beltaos S (1995) River ice jams. Water resources publications, Highlands Ranch, Co., USA.

Beltaos S (1997) Onset of river ice breakup. Cold Reg Sci Technol 25:183–196

Beltaos S (2001) Hydraulic roughness of breakup ice jams. J Hydr Eng Div–ASCE 127(HY8):650–656

Beltaos S (2002) Effects of climate on mid winter ice jams. Hydrol Process 16:789–804

Beltaos S (2003a) Threshold between mechanical and thermal breakup of river ice cover. Cold Reg Sci Technol 37:1–13

Beltaos S (2003b) Numerical modelling of ice-jam flooding on the Peace-Athabasca Delta. Hydrol Process 17:3685–3702

Beltaos S (2004) Climate impacts on the ice regime of an Atlantic river. Nordic Hydrol 35:81–99

Beltaos S, Burrell BC (2003) Climatic change and river ice breakup. Can J Civil Eng 30:145–155

Beltaos S, Burrell BC (2005a) Field measurements of ice-jam-release surges. Can J Civil Eng 32:699–711

Beltaos S, Burrell BC (2005b) Determining ice-jam surge characteristics from measured wave forms. Can J Civil Eng 32:687–698

Beltaos S, Ismail S, Burrell BC (2003) Mid-winter breakup and jamming on the upper Saint John River: a case study. Can J Civil Eng 30:77–88

Beltaos S, Prowse TD (2001) Climate impacts on extreme ice jam events in Canadian rivers. Hydrol Sci J 46:157–182

Beltaos S, Prowse TD, Carter T (2006a). Ice regime of the lower Peace River and ice-jam flooding of the Peace-Athabasca Delta. Hydrol Process 20:4009–4029

Beltaos S, Prowse TD, Bonsal B, MacKay R, Romolo L, Pietroniro A, Toth B (2006b) Climatic effects on ice-jam flooding of the Peace-Athabasca Delta. Hydrol Process 20:4031–4050

Beltaos S, Prowse TD, Bonsal B, MacKay R, Romolo L, Pietroniro A, Toth B (2007) Climate impacts on ice-jam floods in a regulated northern river. (Vol. II, this book)

Blackburn J, Hicks F (2003) Suitability of dynamic modeling for flood forecasting during ice jam release surge events. J Cold Reg Eng 17:18–36

Brimley W, Freeman C (1997) Trends in river ice cover in Atlantic Canada. Proc 9th Workshop on river ice, Fredericton, New Brunswick, pp 335–349

Carson RW, Andres D, Beltaos S, Groeneveld J, Healy D, Hicks F, Liu LW, Shen HT (2001) Tests of river ice jam models. Proc 11th Canadian river ice workshop, river ice processes within a changing environment, May 14–16, 2001, Ottawa

Daly SF, Hopkins MA (1998) Simulation of river ice jam formation. In: Ice in surface waters, Proc 14th Int symposium on ice, July 27–31, Potsdam, New York

Flato GM, Gerard R (1986) Calculation of ice jam profiles. Proc 4th Workshop on river ice, Paper C-3, Montreal, Canada

Galbraith PW (1981) On estimating the likelihood of ice jams in the Saint John River using meteorological variables. Proc 5th Can hydrotechnical conference May 26–27, 1981, Fredericton, New Brunswick, pp 219–237

Gerard R (1981) Ice scars: are they reliable indicators of past ice break-up water levels. Proc Int symposium on ice, Int Assoc for Hydraul Res, Quebec, Canada, pp 847–859

Gerard R, Karpuk EW (1979) Probability analysis of historical flood data. J Hydr Eng Div–ASCE 105(HY9):1153–1165

Gerard RL, Calkins DJ (1984) Ice-related flood frequency analysis: application of analytical estimates. Proc Cold regions specialty conference, Can Soc Civil Eng, April 4-6, 1984, Montreal, Quebec, pp 85–101

Grover P, Vrkljan C, Beltaos S, Andres D (1999) Prediction of ice jam water levels in a multi-channel river: Fort Albany, Ontario. 10th Workshop on river ice management with a changing climate: dealing with extreme events, Doering JC (ed) Winnipeg, MB, pp 15–29

Healy D, Hicks F (2004) Index velocity methods for winter discharge measurement. Can J Civil Eng 31:407–419

Hicks F, Healy D (2003) Determining winter discharge by hydraulic modelling, Can J Civil Eng 30:101–112

Hicks F, Pelletier K, Van der Sanden JJ (2006) Characterizing river ice using satellite synthetic aperture radar. Final report of the Mackenzie GEWEX Study (MAGS): Proc Final (11th) annual sci meeting, Nov 22–25, 2005, Ottawa, pp 490–508

Hopkins MA, Tuhkuri J (1999) Compression of floating ice fields. J Geophys Res 104(C7):15815–15825

Jasek M (2003) Ice jam release surges, ice runs, and breaking fronts: field measurements, physical descriptions, and research needs. Can J Civil Eng 30:113–127

Kellerhals R, Neill CR, Bray DI (1972) Hydraulic and geomorphic characteristics of rivers in Alberta. Alberta Research Council River Engineering and Surface Hydrology Report 72-1, Edmonton, Alberta

Kowalczyk T, Hicks F (2003) Observations of dynamic ice jam release on the Athabasca River at Fort McMurray, AB. Proc 12th Workshop on river ice. June 18-20, 2003, Can Geophys Union–Hydrol Sect, Committee on river ice processes and the environment, Edmonton, AB, pp 369–392

Liu L, Shen HT (2000) Numerical simulation of river ice control with booms. US army corps of engineers cold reg res eng lab, Tech Rep TR 00-10, Hanover, NH, USA

Liu LW, Shen HT (2004) Dynamics of ice jam release surges. Proc 17th Int symposium on ice, St. Petersburg, Russia

Mahabir C, Robichaud C, Hicks F, Fayek AR (2007) Regression and logic based ice jam flood forecasting. (Vol. II, this book)

Massie DD, White KD, Daly SF, Soofi A (2001) Predicting ice jams with neural networks. Proc 11th Workshop on river ice, Ottawa, pp 209–216

Menard P, Duguay CR, Flato GM, Rouse WR (2002) Simulation of ice phenology on a large lake in the Mackenzie River Basin (1960–2000). Proc 59th Eastern snow conference, Stowe, Vermont, USA, pp 3–12

Michel B (1971) Winter regime of rivers and lakes. Cold Reg Sci Eng Monograph III-B1a, Cold reg res eng lab, U.S. Army, Hanover, NH, USA

Moore RD, Hamilton AS, Scibek J (2002) Winter streamflow variability, Yukon Territory, Canada. Hydrol Process 16:763–778

Morse B (2005) River discharge measurement using the velocity index method. Proc 13th Workshop on river ice, Can Geophys Union–Hydrol Sect, Committee on river ice processes and the environment, Hanover, NH, pp 201–226

Pariset E, Hausser R, Gagnon A (1966) Formation of ice covers and ice jams in rivers. J Hydr Eng Div–ASCE 92 (HY6):1–24

Pelletier PM (1989) Uncertainties in streamflow measurement under winter ice conditions, a case study: the Red River at Emerson, Manitoba, Canada. Water Resour Res 25:1857–1867

Prowse TD (1995) River ice processes. Chap. 2, In: Beltaos S (ed) River ice jams, Water Resources Publications, Highlands Ranch, Co., USA

Prowse TD, Beltaos S (2002) Climatic control of river-ice hydrology: a review. Hydrol Process 16:805–822

Prowse TD, Bonsal BR (2004) Historical trends in river-ice break-up: a review. Nordic Hydrol 35:281–293

Robichaud C, Hicks F (2001) A remote water level network for breakup monitoring and flood forecasting. Proc 11th Workshop on river ice, Ottawa, May, pp 292–307

Robichaud C (2003) Hydrometeorological factors influencing breakup ice jam occurrence at Fort McMurray, Alberta. Report prepared in partial fulfillment of the requirements for the degree of M. Sc. in Water Resources Engineering (supervisor F. Hicks), University of Alberta, Edmonton, Alberta

She Y, Hicks F (2005) Incorporating ice effects in ice jam release surge models. Proc 13th Workshop on river ice, CGU–Hydrol Sect, Committee on river ice processes and the environment, Hanover, NH, pp 470–484

Sherstone DA, Prowse TD, Gross H (1986) The development and use of "hot-wire" and conductivity type ice measurement gauges for determination of ice thickness in arctic rivers. Cold Reg Hydrol Symposium, American Water Resources Association, Fairbanks, Alaska, U.S.A., pp 121–129

Shulyakovskii LG (1963) Manual of forecasting ice-formation for rivers and inland lakes. Manual of hydrological forecasting no. 4, Central Forecasting Institute of USSR, Translated from Russian, Israel Program for Scientific Translations, Jerusalem, Israel, 1966

Starosolszky Ö (1990) Effect of river barrages on ice regime. J Hydraul Res, IAHR, 28:711–718

Tracy BT, Daly FS (2003) River ice delineation with RADARSAT SAR. 12th Workshop on the hydraulics of ice covered rivers, CGU HS Committee on river ice processes and the environment, Edmonton, AB.

Tuthill AM, Wuebben JL, Daly SF, White KD (1996) Probability distributions for peak stage on rivers affected by ice jams. J Cold Reg Eng 10:36–57

Uzuner MS, Kennedy JF (1976) Theoretical model of river ice jams. J Hydr Eng Div–ASCE, 102 (HY9):1365–1383

White KD, Daly S (2002) Predicting ice jams with discriminate function analysis. Proc OMAE 2002: the 21st Int conf on offshore mechanics and arctic engineering, Session 7.2, June 23-28, 2002, Oslo, Norway

Weber F, Nixon D, Hurley J (2003) Semi-automated classification of river ice types on the Peace River using RADARSAT-1 synthetic aperture radar (SAR) imagery. Can J Civil Eng 30:11–27

Wiebe K, Fast EJ, Engel P (1993) Laboratory and field evaluation of the "AFFRA" acoustic flow meter. Joint report of operational technology sect., Monitoring and surveys div., EC, and Res. and Appl. Br., National Water Research Institute

Woo MK, Rouse WR (2007) MAGS contribution to hydrologic and surface process research. (Vol. II, this book)

Wuebben JL, Gagnon JJ (2005) Ice jam flooding on the Missouri River near Williston, North Dakota. CRREL Report 95–19

Chapter 16

Regression and Fuzzy Logic Based Ice Jam Flood Forecasting

Chandra Mahabir, Claudine Robichaud, Faye Hicks and Aminah Robinson Fayek

Abstract In Canada, ice jam events have frequently produced the most extreme and dangerous flood events on record, resulting in millions of dollars in associated damages. However, our ability to forecast such events remains quite limited. An example of this is the Athabasca River at Fort McMurray, Alberta, where severe ice jam events have been documented for over 100 years, and where breakup has been monitored intensively for the past 25 years. Despite these efforts, no reliable flood forecast model is yet available. Here, the use of Fuzzy Expert Systems is explored to examine their potential for developing long lead time ice jam risk forecasts for this site. The developed System identified seven out of twenty two years that had the potential for high water levels, including all four years where high water levels actually occurred. These preliminary results suggest that Fuzzy Expert Systems are promising tools for long range ice jam flood forecasting.

1 Introduction

Each spring, numerous rivers across Canada are watched closely because of the potential risk to property and lives should an ice jam form during the river breakup. Ice jams can produce rapid changes in water levels as river flow is impeded by the obstructive effects of ice accumulation, or as water is suddenly released from storage when an ice jam breaks. Unlike open water flood events that are preceded by heavy rains or snowmelt, at present spring ice jam events have no single generally identifiable predictable precursor. The lack of forecasting ability for such severe river conditions poses a serious threat to communities each year. For example, in 1997, ice jam related flooding caused $9 million in damages in the Alberta communities of Fort McMurray and Peace River.

Research into forecasting river breakup began over 50 years ago with simple single variable models and quickly led to multivariate models. Hicks and Beltaos (2007) provide a summary of the empirical, analytical, and modeling methods in river ice investigations. Single and multiple regression analyses have been applied to the problem of breakup water level prediction with moderate success. For example, Shulyakovskii (1963) used the freeze-up water level to forecast the water stage at the first ice movement on the Lena River at Krestoskaya, Russia. Beltaos (1984) used accumulated incoming heat to the ice cover to forecast the stage at breakup initiation, where the stage was a function of freeze-up levels and ice thickness. While most forecasting methods provide an assessment of the potential for ice jam occurrence, they do not offer a prediction of the anticipated flood levels that accompany an ice jam occurrence.

This study explores the applicability of Fuzzy Expert Systems for providing such a long lead time risk assessment tool, in terms of predicting in late winter whether major ice jam flooding events might be expected at breakup. Fuzzy Logic, pioneered in the 1960s by Zadeh (1965), has been applied successfully in a variety of fields where the relationships between cause and effect (variables and results) are difficult to express numerically but are conceptually well defined. See and Openshaw (1999) combined a fuzzy logic model with other methods of soft computing to enhance conventional flood forecasting techniques. Here, based on key factors previously identified as relevant for river breakup forecasting at this site, we illustrate the applicability of Fuzzy Expert Systems to ice jam flood risk forecasting for the Athabasca River at Fort McMurray, Alberta.

2 Study Area and Data

The Athabasca River originates in the eastern ranges of the Rocky Mountains and flows through Alberta in a north-easterly direction for hundreds of kilometers before reaching the city of Fort McMurray (Fig. 1). In the 80 km reach upstream of Fort McMurray, the river flows through an entrenched meandering channel, which is relatively steep and contains numerous rapids sections (Fig. 2). The Clearwater River joins with the Athabasca River at Fort McMurray, and the downtown area is built on a low floodplain at the confluence of the two rivers. Downstream of this confluence the slope of the Athabasca River is substantially reduced and the channel widens and contains numerous islands.

Every year the spring breakup of the Athabasca River causes concern for the residents of Fort McMurray and the community has a history of river ice jams resulting in flooding. Severe floods occurred most recently in 1977 and 1997, causing millions of dollars in damage. Minor flooding has also occurred several times in the last decade and although these small-

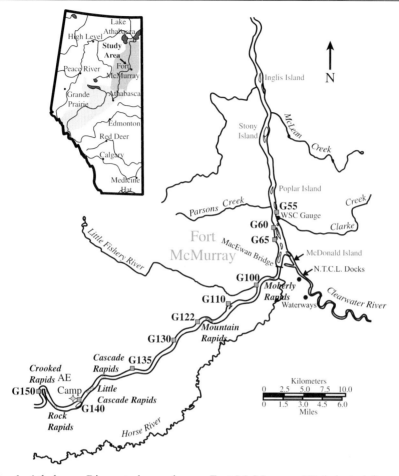

Fig. 1. Athabasca River study reach near Fort McMurray, AB (adapted from Robichaud 2003)

er floods resulted in only minimal damage to residential and commercial properties, expenses were still incurred by the government agencies responsible for flood monitoring and emergency preparedness.

The extensive monitoring over the past three decades has resulted in a significant volume of qualitative data (in the form of photographs and written descriptions of the river breakup processes) but not a rich set of quantitative data. The collection of quantitative data has been limited by an inability to predict the formation of ice jams, as their dynamic nature makes it extremely difficult to plan observations. Furthermore, critical reaches of this river are remote and relatively inaccessible. Most importantly, direct measurement of several of the key properties of an ice jam is logistically difficult, and in some cases impractical (e.g., porosity, internal strength, associated discharge, etc.). Consequently, many forecasting efforts have focused towards probabilistic models (empirical and statistical techniques).

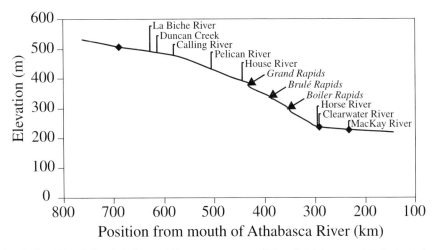

Fig. 2. Profile of the Athabasca River, upstream of Fort McMurray, AB (adapted from Robichaud 2003)

2.1 Nature of River Ice Breakup on the Athabasca River near Fort McMurray

River breakup generally occurs in the third week of April but has been documented to occur as early as March 8 and as late as May 11. Breakup is normally dynamic in nature, with numerous small ice accumulations toeing out over the shallow rapids in the reach upstream of town. Surges resulting from the release of ice accumulations in a reach extending hundreds of kilometers upstream of Fort McMurray appear to be responsible for the ice runs down through Fort McMurray. These ice runs frequently arrest, creating an ice jam in the vicinity of the Clearwater River confluence, due to the sudden marked drop in bed slope and the numerous islands obstructing the wide shallow flow downstream. Flooding in the town generally results when such ice jams cause water to back up the channel of the Clearwater River.

2.2 Historical Data

Due to the severity and frequency of ice jams in the vicinity of Fort McMurray, considerable time and effort have been invested to monitor and study river ice breakup on the Athabasca River there, particularly since 1977 when a major ice jam event occurred. From the late 1970s to the late 1980s, the Alberta Research Council (ARC) conducted a river ice breakup monitoring program through its Surface Water Engineering group. Unfortunately, the ARC group has since been disbanded due to government funding cutback. Alberta Environment continues to monitor the site to provide real-time flood warning, but conducts a limited data collection program. Since 1999, a comprehensive breakup monitoring program has been undertaken at this site by the University of Alberta (UofA), initially as a component of the Mackenzie GEWEX Study (MAGS) and more recently through research funding from Alberta Environment's Water Research Users Group. In addition, hydraulic and meteorological data for the Athabasca River Basin have been collected for many years by various government agencies, private companies and other interest groups. Thus, for forecasting purposes, there were several sources of data available at different time intervals. A hydrometeorological database for researching spring breakup on the Athabasca River at Fort McMurray, starting from the 1972 breakup, was compiled by Robichaud (2003).

2.2.1 Meteorological Data

The Meteorological Service of Canada (MSC) operates a station at the airport in Fort McMurray recording both hourly air temperature and precipitation. Also, as part of this research program, the UofA has operated a near real-time meteorological site in Fort McMurray since 2000, collecting solar radiation in addition to air temperature, precipitation, and barometric pressure data at 30 minute intervals (Robichaud and Hicks 2001).

A key variable for ice event forecasting is the solar radiation. Environment Canada provided data on the hours of bright sunshine measured with a sunshine ball at the Fort McMurray Airport from November 1, 1971 to March 31, 1996. As the UofA meteorological station (which uses a pyranometer for measuring incoming solar radiation) did not become operational until 2000, there was a large gap in the data series. Fortunately, solar radiation data were provided by Golder Associates from their Aurora station (located approximately 55 km north of Fort McMurray) for the years 1988, 1989, as well as 1995 to 2001. This, together with data from the UofA meteorological station from October 2000 to June 2001 allowed us to fill the gap (1996–2001) in the record and to evaluate the potential effects of change in station location (UofA station is near, but not at the airport). Hours of bright sunshine were also measured with a sunshine ball at the UofA station in 2001 to enable conversion of hours of bright sunshine to solar radiation. Details of this effort and the resulting homogeneous dataset are presented by Robichaud (2003).

2.2.2 Snow Course Data

Alberta Environment provided SWE data for 1972 to 2003 from 18 snow course stations in the Athabasca River drainage basin upstream of Fort McMurray. Using the Thiessen polygon method, data from these sites were converted into a single basin average for the Athabasca Basin upstream of Fort McMurray. Data obtained in late February (March 1 record) and late March (April 1 record), were used in the forecasting models. March 15 and April 15 snow course records were available, but generally incomplete and therefore were not considered.

2.2.3 Hydraulic Data

Water Survey of Canada (WSC) measures water levels at two locations in the Fort McMurray area. The station 'Athabasca River below Fort McMurray' is located 5.6 km downstream of the confluence with the Clearwater River. The 'Clearwater River at Draper' station is located approximately 13 kilometers upstream of the confluence with the Athabasca River.

The maximum water levels during breakup and the breakup dates at Fort McMurray were documented by various agencies over the years with the earliest breakup event documented in 1875 (Robichaud 2003). WSC provided the freeze-up water level at the gauge below Fort McMurray associated with the 1973 to 2003 breakup years. Several years of breakup water levels at the WSC gauge below Fort McMurray were documented by Doyle (1987).

3 Methodology

Robichaud (2003) examined a total of 16 parameters for the Athabasca River in an investigation of the relationship between individual variables and peak water levels during breakup, using single value threshold and linear regression models. Given the complexity of processes involved, no practical relationships were found. Robichaud (2003) then explored multiple linear regression models and obtained very good correlations between the forecasted peak breakup water level, H_B, at the Clearwater River confluence and the actual observed peaks water levels at the same site. Model results are shown in Fig. 3, where it is seen that extreme water levels (> 246.0 m) predicted by the model were within ±0.5 m of the observed values ($r^2 = 0.95$ overall). In the figure it should be noted that the one high water level associated with a non-ice jam event actually resulted from a large ice run that left remnant ice at the mouth of the Clearwater River, restricting its outflow.

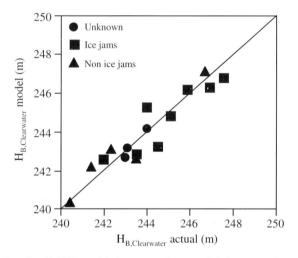

Fig. 3. Robichaud's (2003) multiple regression model forecasts in comparison to observed data (adapted from Robichaud 2003)

The parameters that were identified as most significant in Robichaud's (2003) relationship were:
- accumulated solar radiation prior to breakup,
- SWE measured in March,
- SWE measured in April,
- late winter ice thickness,
- late fall soil moisture (using an antecedent precipitation index), and
- the rate of rise in water level on the Athabasca River (measured at the WSC gauge site) just prior to breakup.

A validation test of this model was conducted in 2003 and successfully predicted the breakup water level within 0.5 m (Fig. 4).

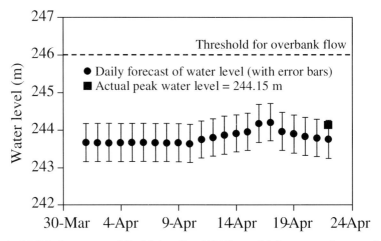

Fig. 4. Validation test of Robichaud's (2003) multiple regression model in breakup 2003 (adapted from Mahabir et al. 2002)

While the multiple regression technique used by Robichaud (2003) offers a short-term forecasting tool, it does not allow an assessment of the potential risk more than just a few days in advance of breakup. From an emergency preparedness perspective, it would be extremely useful to have even a qualitative assessment of flood risk a few weeks in advance of breakup. In this context, it is significant to note that three of the six variables determined to be significant in Robichaud's (2003) model could be classified as antecedent conditions, in that they are known in late winter, well before breakup. These variables are (1) basin average soil moisture, (2) basin-average snow water equivalent, and (3) ice thickness. It is heuristically known that if the values of all these variables are much lower than normal, the risk of ice jam flooding would be low. Conversely, if the values are much higher than normal, a higher risk would exist. Fuzzy Expert Systems capture this type of heuristic knowledge and allow for overlapping ranges of values of variables when boundaries are not clearly defined. For example, an expert may be able to forecast the risk if all the variables are "high" but may not be able to describe "high" as a single value. The purpose of this research was to develop a Fuzzy Expert System to recognize years with high risk of ice jam flooding based on these three antecedent conditions. If successful, the risk of severe flooding could be forecast weeks in advance.

3.1 Fuzzy Expert Systems

Fuzzy Logic is a modeling technique that allows variables to be described in linguistic terms. Pioneered by Zadeh (1965), it has been applied successfully in a variety of fields where the relationships between cause and effect (variables and results) are difficult to express numerically but are conceptually well defined. Fuzzy Expert Systems produce a result based on logical linguistic rules rather than historical data, which allows this type of modeling to be less dependent upon the volume of historical data than are many statistical methods. The Systems consist of four basic steps.

1. All variables (both dependent and independent) must be defined in terms of sets of linguistic classifiers. Membership functions are used to relate the degree to which a particular value of a variable is described by each linguistic term.
2. Membership functions of each independent (input) variable are related to the dependent (output) variable by defining statements or 'rules'. Normally rules are defined as a series of IF-THEN statements that relate the premise(s) to the conclusion.
3. The rules are mathematically evaluated and the results are combined through processes called implication and aggregation, respectively.
4. The resulting solution set of values is then transformed into a single number by a process called defuzzification.

In developing the Fuzzy Expert Systems, particularly the membership functions, extensive knowledge of the model subject is required. As a result, historical data and expert opinion are often combined in defining membership functions and rules. Bardossy and Duckstein (1995) provided detailed information for the development of Fuzzy Expert Systems.

3.1.1 Membership Functions

Each parameter in the model is generally described by a set of linguistic terms. For example, the value of the variable soil moisture could be described using the linguistic concepts of low, average and high. Membership functions are then used to relate the degree, μ, to which a particular value of soil moisture is described by each of these linguistic term. The value of μ ranges from 0 (not part of the set) to 1 (perfectly represents the linguistic concept). A value of a variable may belong to more than one membership function, with varying degrees of membership. As an example, consider an ice thickness of 0.85 m measured by WSC on the Athabasca River below Fort McMurray in late spring. While the ice is thicker than the average historical value of 0.75 m, it is not "thick" relative to the maximum historical ice thickness recorded (1.10 m). Through the use of membership functions, Fuzzy Logic is able to define variables and/or results as belonging to linguistic groupings, such as thick or thin, to varying degrees. In this case, an ice thickness of 0.85 m might belong to "thick" to a large degree (e.g., 0.8), but would also belong to "thin" to a much lesser degree.

One of the main challenges in developing a Fuzzy Expert System is establishing these membership functions and the number of linguistic groupings for each variable. In cases such as this, with small data sets, it is necessary to incorporate the judgment of experts to define membership functions based on experience, logic, and physical bounds. Membership functions must be defined over the entire range of possible values of the variable they describe and can take on a variety of shapes, depending on the philosophy behind the concept of the linguistic term, as described by Mahabir et al. (2002). Figure 5 illustrates the membership functions adopted for the maximum water level at breakup, as an example.

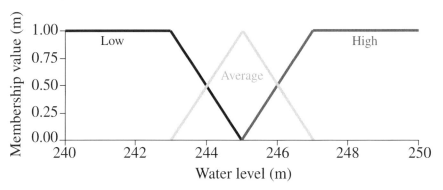

Fig. 5. Membership functions for the flood risk zones (adapted from Mahabir et al. 2002)

3.1.2 Rule Definition

The Fuzzy Expert System consists of IF...THEN rules relating the linguistic terms of the input variables to the linguistic terms of the output variable. Operators such as AND can be used to relate the input variables to each other to define the result as a combination of the input variables. The AND operator is mathematically applied as an intersection operator by either the Minimum or Product function. Minimum is commonly used when the input data are independent of each other, and Product is often applied if input variables are interdependent.

In a rule-based model, the relationship between the input variables and the results is easily understood by simply reading the rule. For example, one rule could be: IF the ice thickness (premise) is high (linguistic term represented by a membership function) THEN the risk of an ice jam (conclusion) is high (linguistic term represented by a membership function).

Rules are influential in selecting the number of variables and the membership functions to be used within the Fuzzy Expert System, since the number of rules required increases if the number of membership functions increases, and increases exponentially with an increase in the number of input variables. To avoid undue complexity, a minimum number of parameters and membership functions should be considered for a Fuzzy Expert System.

Each rule also has an associated weight or certainty grade. A certainty grade can be used to weigh rules between 0 and 1. As a starting point, certainty grades are normally set to 1 meaning all rules have equal weighting. Rule weights are often used to improve the model performance without modifying the membership functions of each linguistic term. Ishibuchi and Nakashima (2001) discussed the improvement of Fuzzy Rule-Based Systems by modification of certainty grades and contrasted this method with modification of membership functions.

3.1.3 Implication and Aggregation

Implication is a process that evaluates the portion of the membership function that is active for a particular rule. Depending on the method of implication, the active set area can be considered to be all of the values in the membership function that belong to the membership function to an equal or lesser degree (known as the Minimum Operation), or the active set could be the entire membership function scaled by the degree to which the variable belongs (Product Operation). Implication results in one set of values for each rule evaluated.

The sets from Implication are combined into a single set in a process called Aggregation. If the sets from Implication are summed together, the method of Aggregation is called Summation. If Aggregation of the sets occurs by combining the maximum values obtained for each output membership function after Implication, then the Maximum method has been used. No firm guidelines have been developed for applying various methods of Implication and Aggregation. Typically, a sensitivity analysis is performed to determine which methods perform best for a particular Fuzzy Logic model.

3.1.4 Defuzzification

Defuzzification is the process by which a Fuzzy Logic solution set is converted into a single crisp value. The solution set is in the form of a function, relating the value of the result to the degree of membership. The Center of Area (or Center of Gravity) method is one of the most common of the Defuzzification methods. The Bisector method produces a value that will split the area of the solution set in half. Three other Defuzzification methods focus on the maximum membership value attained by the solution set. Frequently, the maximum value of the solution set is a range of values rather than a point value. Smallest of Maxima selects the lowest value at which the highest membership value is attained. Similarly, Middle of Maxima and Largest of Maxima select the middle value of the maximum membership and the largest value at which the largest membership value occurs, respectively.

The method of Defuzzification is normally the most sensitive of the calculation parameters. For example, consider the case where a resultant set has a simple shape of 0.75 membership between A and C, otherwise 0 membership (shown in Fig. 6). Smallest of Maxima will produce the lowest value where the highest membership occurs, which corresponds to value A in Fig. 6. In contrast, the Bisector method will find the value that splits the resultant set in half by area and produce that value as the numerical result, such as Value B. By choosing the Largest of Maxima, the largest parameter value for which the largest membership is achieved will be selected. This corresponds to point C. The objective of the model will influence the selection of the Defuzzification method.

3.2 Fuzzy Expert System Design

A Fuzzy Expert System was created to evaluate the potential risk of ice jam flooding at breakup based on the three antecedent conditions found relevant in Robichaud's (2003) short-term forecasting model. Spring snowpack conditions in terms of basin averaged SWE, late fall soil moisture conditions, and ice thickness were obtained from the historical database for the Athabasca at Fort McMurray discussed earlier. In terms of 'consequence', flood occurrence was assessed based on the peak water levels attained at breakup, as compared to the various thresholds for flooding concerns at the town. Where possible, the recorded water levels were used. For those years where the peak water level had not been documented, it was estimated using the multiple linear regression model discussed above (Robichaud 2003). In the end, this facilitated a data set consisting of 22 points, or 22 breakup events ranging from 1977 to 1999.

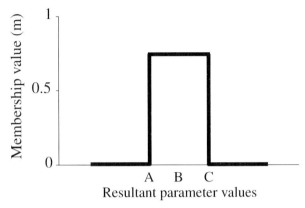

Fig. 6. Defuzzification of Resultant Set where A represents result of Smallest of Maxima, B represents the result of Centroid and C represents the result of Largest of Maxima (adapted from Mahabir et al. 2002)

3.2.1 Membership Functions

The ranges of each membership function were defined by the distribution of recorded values. Three membership functions were defined to describe Snowpack SWE and Antecedent Soil Moisture (viz., low, average, and high). In this study, "average" was considered to be the median value of the data set and was therefore a single point. A triangular membership function was defined by limiting the definition of average to values above the 25% quartile and below the 75% quartile. Values less than the 25% quartile were defined as belonging 100% to the definition of low, while data values between "low" and the median were defined as "low" to a lesser extent; this created a trapezoidal membership function. Similarly, data points above the 75% quartile to the maximum recorded value were defined as "high". Due to data limitations, it was not possible to define the transition from 100% membership to 0% membership based on physical evidence or inherent knowledge. For these reasons, straight lines were used.

Applying three linguistic terms to ice thickness generated a membership function for "average" that included a range of only 10 cm which, considering the natural variation of ice thickness, was too narrow a range to have any physical meaning. For this reason, it was decided that only two linguistic terms, thick and thin, would be used to describe the entire possible range of ice thicknesses.

The membership functions for the risk zones to be forecast were based on the known 'Alert' and 'Minor' flooding levels, specifically 244.0 m and 246.0 m, respectively (Fig. 5). If the forecast water level is higher than the Alert Level, then it belongs more to the Average membership function than to the Low membership function. If the water level is greater than the level at which minor flooding occurs, then it belongs to the High membership function to a larger degree than to the Average membership function.

3.2.2 Rule Definition

The rule base of the Fuzzy Expert System was defined based on historical data and heuristic knowledge. Ideally, historical data would be available to define each of the rules. Since there are three input parameters with a total of 8 membership functions (3 for SWE, 3 for soil moisture and 2 for ice thickness), the required number of rules is 18 (3 x 3 x 2). Three years of data representing low, average and high water levels at breakup were used to assist in defining the rulebase. Logical interpolation between rules was employed when little knowledge was available to define particular rules. While it is possible to define the outcome when extremes occur in all input variables (e.g., all variable are described as "low"), the three calibration years provided a basis for intermediate rules in the database. The 22 years of data used to develop the Fuzzy Expert System did not include occurrences of all 18 possible combinations of variables, which means that historical records were not available to validate each and every rule. Rules were determined based on the logical knowledge wherever possible.

3.2.3 Model Construction

The platform selected for the development of the Fuzzy Expert System was MatLab (Version 6.1.0.450, Release 12) and MatLab's Fuzzy Logic Toolbox (Version 2.1.1). The premises were combined using the concept of "AND" (minimum operator). Implication, aggregation and defuzzification were performed by the Minimum, Maximum and Centroid operators, respectively. A description of these operators and how they are applied are available in Klir et al. (1997). For the base model, all rules were weighted equally.

Using the base model configuration, two Fuzzy Expert Systems were created. The March Fuzzy Expert System included the parameter values that were typically available on March 1, specifically March 1 snow water equivalent (measured in late February), late fall soil moisture and winter ice thickness. The April Fuzzy Expert System was identical except that it used snow data collected during the last week of March (available on April 1). The following statement provides an example of the possible rules in the April Fuzzy Expert System database (note that one condition in italics would be selected for each variable for each rule).

If the Snowpack in April is (*low, average, high*) and the Ice Thickness is (*low, average, high*) and the Soil Moisture is (*low, average, high*) then the water levels during spring breakup will be (*low, average, high*).

For a complete rule base, all combinations of possible input variables must have a define rule or outcome.

3.3 Sensitivity Analysis on the Historical Record

While the Fuzzy Expert Systems were found to be less sensitive to the weightings of the rule base, the results of the model were found to be highly sensitive to the method of defuzzification. If the soil moisture or the SWE was described as low in the data set, then the maximum water level attained at breakup was either low or average. If the ice thickness was low, the maximum water level could be any of the three possible linguistic descriptors. This led to the hypothesis that ice thickness may be less important than the other two parameters. Two models were created with weightings assigned to the rules. In the first model, rules for a premise with high ice membership functions were given half the weight of the other rules in the rule base. In the second model, rules regarding the low membership functions for Soil Moisture or SWE were given a higher weight than other rules. Neither of these models produced a significant change in the assessed water level risk compared to the base model.

The method of defuzzification may hold the key for forecasting the occurrence of extremely high water levels at Fort McMurray caused by breakup ice jams. The Bisector method of Defuzzification produced similar, but slightly lower, forecasts of water levels compared to the base model, which applied the Center of Gravity for defuzzification. The performance of the base model was more accurate than that of Bisector Model. Smallest of Maxima, Mean of Maxima and Largest of Maxima produced forecasts similar to each other, but very different from the base model. The Maxima models selected 1977 and 1997 as the only years with the risk of high water levels, and these were indeed the years when major ice jam flooding occurred in Fort McMurray. However, these models did not forecast the occurrence of high water for years when high water levels were caused by factors other than ice jams. For the four years that were forecast as lower risk years by the Maxima models, the actual water levels remained below the defined average water level of 245.0 m. This represents only four of the actual fifteen years of record when water levels were below average.

4 Results

Using the base model configuration, two Fuzzy Expert Systems were created: the March 1 and the April 1 Fuzzy Expert Systems. While the April system was found to be more accurate, the March system allowed for a longer lead forecast time (Table 1). The March 1 Fuzzy Expert System identified seven years out of the twenty-two years of data as having the potential for high water levels at breakup. These seven years included all four years where ice jam flooding was actually experienced. The April 1 Fuzzy Expert System identified five years as potentially high water levels at breakup, including all four actual high water events. The April system produced fewer false-positive results, hence providing a better assessment of potential risk closer to the occurrence of river breakup.

5 Conclusions

Ice jams are frequent in many rivers in the cold region and it is highly desirable, from an emergency preparedness planning perspective, to be able to forecast the potential for severe ice jam flooding before the onset of breakup. This investigation explored the use of Fuzzy Expert Systems for long lead forecasts of flood risk related to ice jam breakup, by demonstrating its application to the Athabasca River at Fort McMurray, Alberta, Canada.

Both multiple linear regression and fuzzy logic have potential for river ice breakup forecasting. Regression models can produce a quantitative forecast of maximum water levels during spring breakup but require input

Table 1. Forecasting results from the Fuzzy Expert System

Year	SWE [mm] Mar.1	SWE [mm] Apr.1	Soil moisture index [mm]	Ice thickness [m]	Peak water level[a] [m]	Ice jam flooding observed	Forecasted flooding Mar.1	Forecasted flooding Apr.1
1977	69	67	438.1	0.88	247.6	yes	yes	yes
1978	77	27	280.0	0.88	242.0		yes	
1979	90	38	345.8	1.10	246.9	yes	yes	yes
1980	63	83	335.2	0.69	244.8*			
1981	54	11	380.1	0.75	244.1*			
1982	110	141	234.9	0.65	242.2			
1983	39	60	260.8	0.54	242.3			
1984	58	22	280.5	0.81	243.5			
1985	117	89	425.5	0.73	243.5		yes	yes
1986	50	27	262.0	1.05	244.0			
1987	56	63	258.0	0.87	245.1			
1988	42	16	249.9	0.66	244.5			
1989	68	83	347.5	0.62	243.1			
1990	61	36	382.9	0.63	243.0			
1991	73	83	289.0	0.77	245.8*			
1992	92	9	463.2	0.75	242.1*		yes	
1993	50	20	295.3	0.82	243.4*			
1994	129	112	299.1	0.68	241.2*			
1995	58	33	228.8	0.85	245.6*			
1996	107	81	365.0	0.73	248.7*	yes	yes	yes
1997	117	128	460.1	0.77	247.0	yes	yes	yes
1998	36	10	378.9	0.58	244.8*			
1999	108	86	162.9	0.81	240.4			

[a]asterisk (*) indicates water level deduced using Robichaud's (2003) regression.

variables measured at the time of breakup, thus limiting the lead time available to undertake mitigative measures. Fuzzy logic models are based on logic allowing more data for validation, and they can provide qualitative forecasts of water levels with information available weeks prior to river breakup. Using only antecedent basin moisture, late winter snowpack conditions, and late winter ice thickness data, we developed a Fuzzy Expert System that can forecast the potential for high water levels at breakup. Forecasts based on conditions known on April 1 (typically 3 weeks in advance of breakup) identified five years (of 22) for potentially high water levels at breakup, including all four actual high water events.

While further research is needed to refine the predictive capabilities of the model, our results show that it is possible to forecast flood risk, and possibly even major ice jams, weeks before the onset of river breakup.

Acknowledgements

The authors thank Alberta Environment for their support for this research. Support of Claudine Robichaud's MSc thesis project, through the NSERC scholarship program is also gratefully acknowledged. The authors thank Larry Garner of Alberta Environment, who aided in this research effort. Thanks are also extended to: Dr. Nathan Schmidt of Golder Associates, who kindly provided critical meteorological data from their Aurora station; Bob Kochtubajda, Meteorological Service of Canada for providing their data, and Kim Epp of WSC for facilitating access to the gauge data. Finally, the authors wish to acknowledge the support Fred Baehl, Director, Emergency Preparedness, Regional the Municipality of Wood Buffalo (RMWB) for his continued support and facilitation of this research effort as well as Gord Harlow and the other employees of the RMWB for the data provided.

References

Bardossy A, Duckstein L (1995) Fuzzy rule-based modeling with applications to geophysical, biological and engineering systems. Bahill AT (ed) CRC Press, Inc. Boca, Florida

Beltaos S (1984) Study of river ice breakup using hydrometric station records. Proc Workshop on Hydraulics of River Ice. Fredericton, Canada, pp 41–59

Doyle CJ (1987) Hydrometeorological aspects of ice jam formation at Fort McMurray, Alberta. M.Sc. thesis, University of Alberta, Edmonton, Alberta

Hicks F, Beltaos S (2007) River ice. (Vol. II, this book)

Ishibuchi H, Nakashima T (2001) Effect of rule weights in fuzzy rule-based classification systems. IEEE T Fuzzy Syst 9:506–515

Klir GJ, St Clair UT, Yuan B (1997) Fuzzy set theory foundations and applications. Prentice-Hall, Inc.

Mahabir C, Hicks F, Fayek AR (2002) Forecasting ice jam risk at Fort McMurray, AB, using fuzzy logic. Proc 16th IAHR International Symposium on Ice, International Association of Hydraulic Engineering and Research, New Zealand, December 2–6, pp 91–98

Robichaud C (2003) Hydrometeorological factors influencing breakup ice jam occurrence at Fort McMurray, Alberta. M.Sc. thesis, University of Alberta, Edmonton, Alberta

Robichaud C, Hicks F (2001) A remote water level network for breakup monitoring and flood forecasting. Proc. 11th Workshop on River Ice, Ottawa, May, pp 292–307

See L, Openshaw S (1999) Applying soft computing approaches to river level forecasting. Hydrol Sci J 44:763–776

Shulyakovskii LG (1963) Manual of ice-formation forecasting for rivers and inland lakes. Israel Program for Scientific Translations TT 66-51016, Jerusalem, Israel (1966)

Zadeh LA (1965) Fuzzy sets. Inform Control 8:338–353

Chapter 17

Impact of Climate Change on the Peace River Thermal Ice Regime

Robyn Andrishak and Faye Hicks

Abstract A one-dimensional hydrodynamic model that includes river ice formation and melting processes is developed and used to assess climate change impact on the ice regime of the Peace River in Alberta. The model employs an Eulerian frame of reference for both the flow hydrodynamics and the ice processes (ice cover formation and deterioration) and uses the characteristic-dissipative-Galerkin finite element method to solve the primary equations. Model calibration and validation results with historical data are presented; these indicate that the present model adequately simulates water temperature and ice front profiles. Higher air temperatures predicted by the CGCM2 climate model were used to generate future ice front profiles that correspond to the historical runs. This preliminary climate change impact analysis suggests that there is a significant potential for a shorter ice-covered season on the Peace River by the 21st century. At the Town of Peace River, the average total reduction in ice cover duration is 28 days (31%) under the scenario applied.

1 Introduction

The winter ice cover that forms on most northern rivers plays an important role in ecosystems and water quality (Prowse and Culp 2003) and in many cases is a significant factor in Canada's northern transportation network (Gerard et al. 1992; Kuryk and Domaratzki 1999). Climate change, in particular climate warming, has the potential to affect not only the duration and extent of ice cover on northern rivers but also, potentially, the frequency and/or severity of ice jam events (Beltaos and Prowse 2001; Beltaos et al. 2007). Clearly, it is important to be able to assess the impact of climate change and climate variability on the ice regime of rivers and to develop adaptive strategies to minimize the negative impacts of these changes.

This preliminary investigation explores the potential impacts of climate warming on the thermal regime of the Peace River for the reach extending from Hudson's Hope in British Columbia (BC) to Fort Vermilion in Alberta (AB) (Fig. 1 in Beltaos et al. 2007). This river is regulated by hydropower dams at its upstream end, and as a result thermal ice processes are a dominant feature of its winter regime. Building on the validated dynamic hydraulic model of the Peace River developed by Hicks (1996), and also applied by Peters and Prowse (2001), we employ a fully Eulerian framework to incorporate thermal ice formation and deterioration processes. The model is validated with available historical data and then applied to provide an evaluation of the potential magnitude and significance of climate change impacts on thermal ice processes.

2 Model Development

The model developed for this study is built upon the University of Alberta's public domain, dynamic river routing model, *River1D* (Hicks and Steffler 1992). *River1D* is a one-dimensional finite element-based numerical model that solves conservation of water mass and longitudinal momentum using the characteristic-dissipative-Galerkin (CDG) scheme. For this investigation, the rectangular channel version of the *River1D* model has been enhanced to incorporate thermal ice related processes including consideration of water temperature, suspended frazil ice, surface ice coverage, surface frazil ice, and solid surface ice. Ice front location is a supplementary solution variable that determines where the free-drift assumption is applied to surface ice and where surface ice velocity is zero.

Existing river ice models (e.g., Lal and Shen 1989; Shen et al. 1995) use an Eulerian-Lagrangian approach to model the governing equations. In contrast, the thermal process equations modeled in *River1D* have been developed from control volume principles in a completely Eulerian frame of reference. Each equation can be written in the form:

$$\frac{\partial}{\partial t}(\Phi) + \frac{\partial}{\partial x}(U\Phi) = \Sigma F \qquad (1)$$

where Φ represents the solution variable, U is the mean flow *or* surface ice velocity, ΣF is the sum of the mass or energy fluxes applying to the control volume, t is the temporal coordinate, and x is the longitudinal coordinate. Equations in this general form were subsequently developed into weak statement formulations based on the characteristic-dissipative-Galerkin finite element scheme, and then incorporated into *River1D*. For each time step in the transient solution, the modeling procedure involves a decoupled solution of the total (ice and water) mass and longitudinal momentum conservation equations, followed by solution of the water temperature equation, and finally the ice mass conservation equations.

Figure 1 illustrates the generalized river cross section, upon which the model is formulated, containing suspended frazil and a variable ice cover made up of slush and solid ice layers. The variable dimensions indicated on the figure are defined with the equations that follow. The modeling approach assumes that the drifting ice moves at the mean water velocity, with ice resistance effects only considered once the ice itself is arrested. This version of the model does not include consideration of dynamic ice jam formation or release processes.

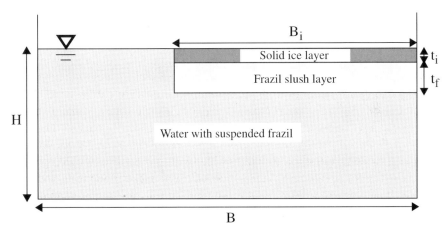

Fig. 1. Schematic cross section of a rectangular river channel with a variable ice cover and frazil ice in suspension

Conservation of thermal energy is used to derive the model for river water temperature:

$$\frac{\partial}{\partial t}(AC_pT_w)+\frac{\partial}{\partial x}(UAC_pT_w)=-\frac{(B-B_i)\phi_{wa}}{\rho}-\underbrace{\frac{B_i\phi_{ia}}{\rho}}_{\phi_{ia}>0}-\underbrace{\frac{B_i\phi_{iw}}{\rho}}_{\phi_{iw}>0} \quad (2)$$

where A is the liquid water flow area, C_p is the specific heat of water, T_w is the water temperature, B is the channel width, B_i surface ice coverage, ϕ_{wa}, ϕ_{ia}, ϕ_{iw} are the net rates of heat exchange per unit area between water and air, ice and air, and ice and water, respectively, and ρ is the density of water. The flux term containing ϕ_{ia} only applies if heat is being lost, as any heat gain over an ice-covered area is directed towards ice melt. Similarly with the third flux term on the right hand side of Eq. (2), which represents heat transfer associated with solid ice melt, there must be a positive heat loss from the water to the ice cover for this term to affect water temperature.

The water-air heat exchange rate is calculated using a typical linear heat transfer method, also employed by Lal and Shen (1989) for river ice modeling, whereby all of the temperature dependent components of the energy budget are consolidated into a single linear function of the form:

$$\phi_{wa} = h_{wa}(T_w - T_a) \qquad (3)$$

where h_{wa} is a calibration coefficient in the order of 15 W m^{-2} °C^{-1} for the Peace River (Andres 1996) and T_a is the ambient air temperature. Following Lal and Shen (1989), the present model simply adjusts the open water heat exchange rate calculated in Eq. (3) for the thermal conductivity of an ice cover, K_i, to evaluate the ice-air heat exchange:

$$\phi_{ia} = \phi_{wa} / \left(1 + \frac{h_{wa} t_i}{K_i}\right) \qquad (4)$$

where t_i is the thickness of solid ice. The method used to assess the ice-water heat exchange rate is adopted from the work of Ashton (1973), where the flow velocity and depth of water flow beneath the ice cover are used in a non-linear relationship to compute the heat transfer coefficient. This coefficient is then multiplied by T_w to calculate the rate of heat exchange.

Once a 0°C isotherm has developed within the simulated reach, the ice mass conservation component of the model is activated. The initial ice mass process being modeled is the generation of suspended frazil uniformly distributed throughout the flow, given by the concentration fraction C_f, where:

$$\frac{\partial AC_f}{\partial t} + \frac{\partial UAC_f}{\partial x} = \frac{1}{\rho_i}\left[\underbrace{\frac{\rho(B-B_i)\phi_{wa}}{\rho_i L_i}}_{\substack{\text{frazil formation}\\ \text{if } T_w = 0}} - \underbrace{\rho_i \eta C_f B}_{\text{frazil rise}}\right] \qquad (5)$$

and ρ_i is the density of ice, L_i is the latent heat of ice, and η is a calibration parameter that controls the rate at which frazil ice rises out of suspension due to its buoyancy to form surface ice pans. Note that the flux term associated with frazil formation requires the water temperature be 0°C at that location.

As suspended frazil ice rises, the surface ice coverage increases at a rate also influenced by an initial frazil floe thickness, t'_f, that is assumed constant and is specified by the modeler. This value represents the minimum pan thickness that can remain stable at the water surface without being overcome by turbulence and is necessary to numerically control the rate of change of surface ice concentration. Deviating from the two-layer surface ice structure proposed in Fig. 1, the change in surface ice coverage, B_i, can be evaluated by simply conserving all of the frazil slush and pore water added to the surface layer within a "virtual" surface ice layer t'_f thick:

$$\frac{\partial t'_f B_i}{\partial t} + \frac{\partial U_i t'_f B_i}{\partial x} = \frac{1}{\rho'} \left[\underbrace{\left(\rho_i + \rho \frac{e_f}{(1-e_f)}\right) \eta C_f (B - B_i)}_{\text{frazil and pore water deposition}} \right] \quad (6)$$

where U_i is the surface ice velocity, ρ' is the combined density of frazil slush and pore water, and e_f is the porosity of frazil slush. Pore water mass is captured within the frazil slush layer in proportion to e_f, as given by the term $\rho e_f /(1-e_f)$.

The thickness of frazil slush on the underside of pans and solid ice at the surface is derived from the following equations when frazil slush exists:

$$\frac{\partial B_i t_f}{\partial t} + \frac{\partial U_i B_i t_f}{\partial x} = \frac{1}{\rho'} \left[\underbrace{\left(\rho_i + \rho \frac{e_f}{(1-e_f)}\right) \eta C_f B}_{\text{frazil and pore water deposition}} - \underbrace{\frac{\rho' B_i \phi_{ia}}{\rho_i \ L_i}}_{\substack{\text{pore water freezing} \\ \text{if } \phi_{ia} > 0}} - \underbrace{\frac{\rho' B_i \phi_{iw}}{\rho_i \ L_i}}_{\substack{\text{slush melt} \\ \text{if } \phi_{iw} > 0}} \right] \quad (7)$$

$$\frac{\partial B_i t_i}{\partial t} + \frac{\partial U_i B_i t_i}{\partial x} = \frac{1}{\rho_i} \left[\underbrace{\frac{\rho' B_i \phi_{ia}}{\rho_i \ L_i}}_{\substack{\text{pore water freezing} \\ \text{if } \phi_{ia} > 0}} + \underbrace{\frac{\rho B_i \phi_{ia}}{\rho_i \ L_i}}_{\substack{\text{solid ice melt} \\ \text{if } \phi_{ia} < 0}} \right] \quad (8)$$

where t_f is the thickness of the frazil slush layer on the underside of ice at the surface. If this layer freezes completely (i.e., $t_f \rightarrow 0$), the flux associated with pore water freezing no longer applies to the solid ice equation, and new terms for growth of columnar ice and solid ice melt due to warm water replace the pore water freezing term in Eq. (8). In this case, the solid ice equation becomes:

$$\frac{\partial B_i t_i}{\partial t} + \frac{\partial U_i B_i t_i}{\partial x} = \frac{1}{\rho_i} \left[\underbrace{\frac{\rho \, B_i \phi_{ia}}{\rho_i \, L_i}}_{\substack{\text{growth of columnar ice} \\ \text{if } \phi_{ia} > 0 \\ \text{and } T_w = 0}} + \underbrace{\frac{\rho \, B_i \phi_{ia}}{\rho_i \, L_i}}_{\substack{\text{solid ice melt} \\ \text{if } \phi_{ia} < 0}} - \underbrace{\frac{\rho \, B_i \phi_{iw}}{\rho_i \, L_i}}_{\substack{\text{solid ice melt to warm water} \\ \text{if } \phi_{iw} > 0}} \right] \qquad (9)$$

In the present version of the model, the user must specify the time at which bridging occurs at the downstream boundary. When this time in the simulation is reached, the ice front location is set at the downstream boundary and the approach of ice from upstream leads to the upstream progression of the ice front. A simplified version of the *RICEN* model's (Shen et al. 1995) conservation of surface ice discharge method is used to determine the rate of ice cover progression. Surface ice entrainment at the leading edge of the ice cover and thickening processes, which both reduce the rate of ice cover advance, are approximated by the constant calibration factor, P_{jux}, in the following equation that tracks ice front location:

$$X_i^{t+\Delta t} = X_i^t - \frac{\frac{B_i}{B} U_i}{P_{jux}} \Delta t \qquad (10)$$

where $X_i^{t+\Delta t}$, X_i^t are the ice front locations at time $t + \Delta t$ and t, respectively, and Δt is the solution time step used to run the model.

Recession of the ice cover due to melt is handled more naturally by the model. The ice front location moves node by node downstream as the ice thickness at the ice front decreases towards zero. Intermediate ice front locations are not calculated during the melt process, as they are during upstream progression.

3 Model Calibration and Application to the Peace River

3.1 Data Requirements

The data required to run the *River1D* thermal river ice model consists of initial conditions and inflow time series for the hydraulic, water temperature, and ice conditions. A downstream water level time series is also required for the hydraulic modeling component. Finally, one or more air temperature time series must be specified to drive the thermal modeling components. Heat input from solar radiation can also be considered, but this feature was not employed in this case study application, as insufficient data were available. Downstream boundary conditions for water temperature and ice conditions are not required as the finite element method employed in *River1D* uses the applicable 'natural' boundary conditions.

Discharge records and tailrace water temperature for the Bennett/Peace Canyon Dam were made available by Alberta Environment and used to develop the upstream boundary conditions for the model. Ice inflow at the upstream boundary was set at zero for all simulations, which is consistent with the physical situation in this case. Extensive air temperature records at Fort St. John, BC, the Town of Peace River, AB, and at High Level, AB, were used to construct the air temperature time series for the simulations (station locations shown in Fig. 1 in Beltaos et al. 2007). For this study, the average mean daily air temperature at the first two stations was used to define the air temperature upstream of the Town of Peace River and the mean daily air temperature at High Level was applied to the lower reach. The remaining input parameter is the time of ice front initiation at the downstream boundary. This value was either estimated or, when known, taken directly from historical ice front observations provided by Alberta Environment.

3.2 Model Calibration and Validation

Calibration and validation of the thermal river ice model involved two phases: the first required calibration/validation of the air-water heat exchange coefficient to observed water temperature data; the second phase involved calibrating and validating the remaining set of parameters that dictate the simulated ice front profile.

3.2.1 Calibration/Validation Using Measured Water Temperature

Water temperature observations on the Peace River in Alberta are currently limited to two locations: the Water Survey of Canada (WSC) gauges at Alces River (164 km downstream of the Bennett Dam) and the Town of Peace River (396 km downstream of the Bennett Dam). As only two years of record were available, for the 2002/03 and 2003/04 ice seasons, the former season was used for calibration and the latter for validation.

For the calibration using the 2002/03 data, heat exchange coefficients of 15 and 20 W m^{-2} °C^{-1} were tested (chosen based on the results of previous studies on the Peace River (e.g., Andres 1993, 1996)). The resulting simulated water temperature profiles at the Alces and Peace River gauge sites are shown in Figs. 2a and 2b. As the figures illustrate, both values of the heat exchange coefficient tested appear to simulate reasonable water temperatures at the Alces and Peace River gauge sites. However, for this preliminary investigation, it was decided to proceed using a value of 15 W m^{-2} °C^{-1} for the heat exchange coefficient, as it provided the better representation of the date the zero degree isotherm arrived at the gauge sites.

Figures 2c and 2d compare the simulated water temperature profile (using 15 W m^{-2} °C^{-1} for the heat exchange coefficient) with the observed data at the two gauge sites for the 2003/04 (validation) season. As the figure illustrates, model results do not as closely match but are generally consistent with the measured data. In particular, the timing of the arrival of the zero degree isotherm is well represented by these simulation results.

Overall, the results of this preliminary calibration and validation to the water temperature data suggest that the model is producing reasonable results, though not perfectly capturing the water temperature behavior. A choice of 15 W m^{-2} °C^{-1} seems a reasonable compromise given the limited available data for calibration and validation. Clearly more data and additional modeling is required to refine the simulation results for water temperature over the course of the cooling period. The differences between simulated and recorded water temperature may simply be a reflection of the suitability of the air temperature data for that location or the quality of the water temperature data at that site or at the upstream boundary. In any case, the water temperatures were considered to be adequately modeled for the purposes of this preliminary investigation, on the basis of the consistent simulation of the timing of the zero degree isotherm at Alces and the Town of Peace River, which can be considered the most important prerequisite to accurate ice process modeling.

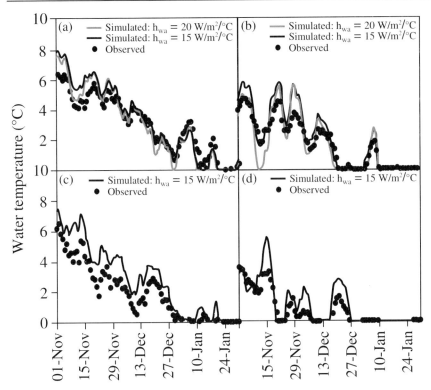

Fig. 2. Peace River water temperature calibration for 2002/03, using gauges at (a) Alces and (b) the Town of Peace River. Model validation for 2003/04, using gauges at (c) Alces and (d) the Town of Peace River

3.2.2 Calibration/Validation of the Ice Process Parameter Set

More than 20 years of historical records, including documentation of the ice front progression in each year, were supplied by Alberta Environment for calibration and validation of the ice process model components. Unfortunately, much of the record is sparse, most notably in terms of the water temperature information required for the inflow boundary condition. It was eventually decided that most of the empirical ice process parameters would be set to 'typical' values and only the juxtaposition parameter, P_{jux}, would be adjusted.

A value of P_{jux}= 2.5 was found to produce the best overall ice front results for the 20 years of historical record simulated, 2003/04 through 1984/85. The remaining parameters and their adopted values were:
- frazil floe porosity = 0.5;

- frazil rise parameter = 0.0001 m s^{-1};
- Manning's n for ice cover = 0.02; and
- ice-water heat exchange constant = 1187 W s$^{0.8}$ m$^{2.6}$ °C^{-1} (Ashton 1973).

Other variables can be used to calibrate the model parameters and to assess the quality of its performance: measured water levels, documented surface ice concentrations, and observed ice thicknesses. However, given that the objective of this study was to investigate climate warming influences on the extent and duration of ice cover, ice front location was considered the most relevant. Thus, validation consisted of comparing the modeled ice front locations to the observed data.

In comparing model results to measured ice front progression, it was found that the performance of the model varied from year to year. For example, the simulated profile for 2002/03 and 2003/04 extended considerably farther upstream than the observed profile in the reach upstream of the Town of Peace River (TPR) as shown in Fig. 3. However, in other years, such as 1995/96 and 1996/97, the model performed extremely well, as also shown in the figure. Among the remaining 16 years simulated, 14 were qualitatively considered to have 'good' ice front profile agreement with available observations, one 'very good' (1988/89), and one 'fair' (1985/86).

In addition to the necessary approximations regarding the inflow boundary water temperatures for those years when the data was suspect or missing, a key factor contributing to the variable accuracy of the ice front profile simulation is that at present, the *River1D* thermal ice model does not explicitly consider ice cover consolidation or hydraulic thickening, but rather approximates these effects with the constant calibration parameter, P_{jux}. These dynamic processes are known to occur on occasion along the Peace River, particularly during the freeze-up period; however, with a constant P_{jux} the variability of these effects from reach-to-reach or year-to-year cannot be considered. Therefore, it is not unexpected that the current version of the model would over-estimate the upstream progression of the ice cover in some of the historical simulations.

Despite these limitations in the model's capabilities, it still produces sufficiently reasonable results to be useful in conducting a preliminary assessment of the potential influences of climate change on the thermal ice regime of the Peace River. This is particularly true at the Town of Peace River where the timing of ice front arrival was simulated very well in the majority of years run, as opposed to the results in the reaches farther upstream where larger discrepancies were apparent. For example, based on the 20 years of historical record simulated, *River1D* predicted the average duration of ice cover and dates of freeze-up and breakup at the Town of Peace River to within two days of the observed. The maximum extent of ice simulated was, on average, 50 km farther upstream than the observed value.

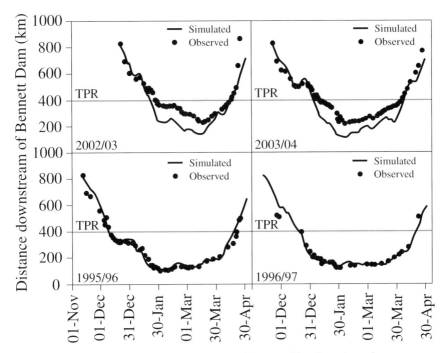

Fig. 3. Modeled and observed historical ice front profiles for selected years

4 Climate Change Analysis

The Canadian CGCM2 climate model was selected to assess the impact of climate change on the historical winter seasons modeled. Two standardized future climate scenarios were available from the Mackenzie GEWEX Study (MAGS) research network database; these are commonly referred to as the A2 and B2 scenarios. The A2 scenario, based on larger population growth and higher cumulative CO_2 emissions than the B2 scenario over the 1990–2100 period, was chosen for this study in order to examine the more severe climate prediction. The CGCM2 model provides mean monthly temperature change projections relative to 1961–90 for various locations in Canada. For this preliminary analysis, the mid-range projection for the year 2050 was selected over the two extremes of 2010 and 2080.

To assess the potential effects of climate change on the winter regime of the Peace River, it was necessary to assume that the mean monthly air temperature change projections from the climate change scenario could be applied directly to the mean daily historical values used as model input. Other compounding potential effects of climate warming, such warmer water temperatures in the hydropower dam reservoir and/or a delay in the timing of initial ice cover bridging at Fort Vermilion, could not be considered here but do warrant future investigation. Intuitive judgment suggests that neglecting these factors would mean that the results of this analysis would likely underestimate the potential impact of warming on the duration and extent of the river's ice cover. Discharge is another key factor in the winter regime of rivers, and although natural flow rates would be expected to change in a future climate scenario, information from BC Hydro indicates that the nature of the hydropower facilities controlling the headwaters of the Peace River is to service base power demand as opposed to long- or short-term growth in demand. As a result, flow releases at this site are not heavily influenced by changes in weather or population growth that would correspond with a future climate scenario. Therefore, an assumption of change in future river flows is not considered a significant uncertainty for the purposes of this investigation.

Examples of climate change ice front profiles compared with the historical simulations are presented in Fig. 4 for the same example years presented in the previous section. The predicted November, December, January, February, March, and April temperature increases for the southern region (including the Town of Peace River) of 0.37, 4.02, 5.11, 3.85, 4.10 and 1.85°C, and 0.30, 3.82, 5.67, 3.90, 4.05 and 1.70°C for the northern region, clearly have a significant impact on the overall ice front progression notwithstanding the fact that the bridging date has not been delayed.

In particular, the duration and maximum extent of ice cover are reduced. The maximum upstream extent of the ice cover would be expected to be consistently farther downstream of the Bennett Dam by an average of 60 km (Fig. 5a). In terms of the delay in the date of freeze-up at the Town of Peace River, an average of 13 days is indicated and for breakup, the average date is 15 days earlier. A 28 day reduction in the average duration of ice cover (Fig. 5b) represents a shortening of the ice-covered season by 31% compared to the historical observations from the 1961–90 period.

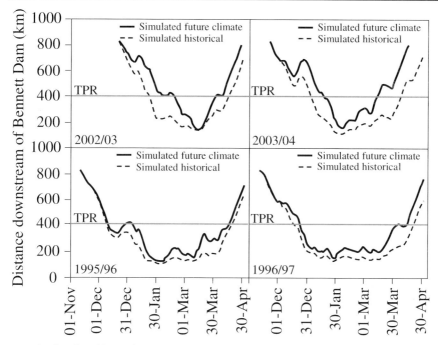

Fig. 4. Simulated historical and future climate ice front profiles for selected years

5 Discussion and Future Research

This investigation applied the *River1D* thermal ice model to assess the potential impacts of climate change on the thermal ice regime of Peace River. In general, the model produced reasonable predictions of ice cover progression for the validation period, but it cannot precisely capture ice cover progression in years where dynamic processes, such as secondary ice cover consolidation, significantly influence the location of the ice front. Nevertheless, the model's capability is considered adequate for the purposes of this preliminary study.

The model was used to explore the potential impacts of climate change on the Peace River thermal ice regime. Results indicate that both the duration and the maximum extent of the ice cover will be reduced under the CGCM2/A2 scenario. Given the limited input and validation data, the fact that the model only considers thermal ice processes at this time, the lack of consideration of the effects of climate change on reservoir outflow temperatures and ice cover initiation date, and uncertainties associated with

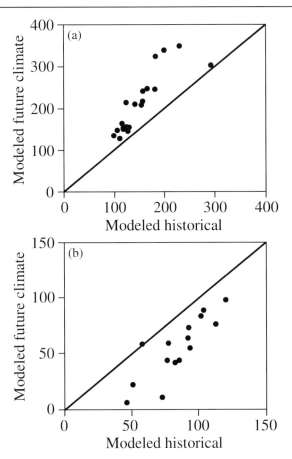

Fig. 5. Modeled historical versus future climate change (a) minimum ice front distance (in kilometers) from the Bennett Dam in British Columbia and (b) duration of ice cover (in days) at the Town of Peace River in Alberta

the meteorological climate change analysis itself (as well as its applicability for this particular period of record), these quantitative averages cannot be considered firm predictions. However, their magnitudes definitely suggest that there will be a measurable, and possibly even significant, impact attributable to climate change on the future ice regime of the Peace River. Therefore, it is important to start developing adaptive strategies as well as improved models and data archives, in order to gain a more reliable quantitative assessment of these impacts.

Opportunities exist to improve the current *River1D* thermal model. The most immediate need is to incorporate the physics of ice cover stability and mechanical thickening into the current version. This should greatly improve the model's consistency when simulating the ice front profile from year to year. Secondarily, computation of ice floe velocity can be advanced to include the effect of channel constriction on the passage of large concentrations of surface ice. In addition, consideration of natural channel geometry would facilitate validation of water levels, not just ice front progression.

The bridging phenomenon is still not completely understood and remains largely site specific. It would be extremely beneficial, particularly with respect to modeling climate change effects on river ice, to have a reliable bridging criterion built into the model. Future research, modeling, and field observation could reveal a great deal about this aspect of the river's ice regime. Other issues not currently included in the model such as lateral thermal inflow and snow cover could also be the focus of future work.

Continued and improved data collection is also critical to the quality and success of this and other river ice studies. For the Peace River, one or more additional water temperature monitoring sites downstream of the Town of Peace River would provide extremely useful calibration and validation data. The temperature of water discharged from the hydropower dam should continue to be measured and reservoir models should be developed to evaluate the effect of climate change on the seasonal water temperature boundary condition. In terms of modeling climate change impacts on the river ice, additional scenarios can be investigated using the current model to assess the importance of the inflow water temperature and date of bridging on the overall ice front simulation.

Acknowledgments

Funding for this study was provided by the Climate Change Research User's Group (CCRUG) at Alberta Environment as well as by the Natural Sciences and Engineering Research Council of Canada. The authors would also like to thank Kim Westcott and Chandra Mahabir of Alberta Environment for their support of this project. Thanks are also extended to Martin Jasek of BC Hydro for providing historical data for the study, and to BC Hydro/Glacier Power/Alberta Environment for their joint monitoring program on the Peace River which provided the detailed data for 2002/03 and 2003/04.

References

Andres DD (1993) Effects of climate change on the freeze-up regime of the Peace River: phase I ice production algorithm development and calibration. Report No. SWE 93/01, Environmental Research & Engineering Department, Alberta Research Council, Edmonton, Alberta

Andres DD (1996) The effects of flow regulation on freeze-up regime: Peace River, Taylor to the Slave River. Northern River Basins Study Project Report No. 122, Northern River Basins Study, Edmonton, Alberta

Ashton GD (1973) Heat transfer to river ice covers. Proc 30th Eastern Snow Conference, Amherst, Massachusetts, pp 125–135

Beltaos S, Prowse TD (2001) Climate impacts on extreme ice-jam events in Canadian Rivers. Hydrol Sci J 46(1):157–181

Beltaos S, Prowse TD, Bonsal B, Carter T, MacKay R, Romolo L, Pietroniro A, Toth B. (2007) Climate impacts on ice-jam floods in a regulated northern river. (Vol. II, this book)

Gerard R, Hicks FE, MacAlpine T, Chen X (1992) Severe Winter Ferry Operation: The Mackenzie River at Ft. Providence, NWT. Proc 11th International Assoc. for Hydraulic Research Ice Symposium, Banff, Alberta, June 1992, pp 503–514

Hicks FE, Steffler PM (1992) A characteristic-dissipative-Galerkin scheme for open channel flow. J Hydraul Eng-ASCE 118:337–352

Hicks FE (1996) Hydraulic flood routing with minimal channel data: Peace River, Canada. Can J Civil Eng 23:524–535

Kuryk D, Domaratzki M (1999) Construction and maintenance of winter roads in Manitoba. Proc 10th Workshop on the Hydraulics of Ice-covered Rivers, Winnipeg, pp 265–275

Lal AMW, Shen HT (1989) A mathematical model for river ice processes (RICE). Report No. 89-4, Department of Civil and Environmental Engineering, Clarkson University, Potsdam, New York

Peters DL, Prowse TD (2001) Regulation effects on the Lower Peace River, Canada. Hydrol Process 15:3181–3194

Prowse TD, Culp JM (2003) Ice breakup: a neglected factor in river ecology. Can J Civil Eng 30:1–17

Shen HT, Wang S, Lal AMW (1995) Numerical simulation of river ice processes. J Cold Reg Eng 9:107–118

List of Symbols

A	liquid water flow area [m^2]
B	top width of channel [m]
B_i	surface ice width or coverage [m]
C_f	suspended frazil ice concentration [dimensionless]
C_p	specific heat of water [J kg^{-1} °C^{-1}]
e_f	porosity of frazil slush [dimensionless]
ΣF	sum of mass or energy fluxes [units vary]
h_{wa}	linear heat transfer coefficient [W m^{-2} °C^{-1}]
K_i	thermal conductivity of ice [W m °C^{-1}]
L_i	latent heat of ice [J kg^{-1}]
P_{jux}	juxtaposition parameter [dimensionless]
t	time [s]
T_a	air temperature [°C]
t_f	thickness of frazil ice layer at the surface [m]
t'_f	initial frazil ice thickness [m]
t_i	thickness of solid ice layer at the surface [m]
T_w	water temperature [°C]
U	mean water velocity [m s^{-1}]
U_i	surface ice velocity [m s^{-1}]
x	longitudinal distance along channel centerline [m]
X_i	location of ice front / distance from upstream boundary [m]
Δt	solution time step [s]
η	frazil rise parameter [m s^{-1}]
ρ	density of water [kg m^{-3}]
ρ'	combined density of frazil slush and pore water [kg m^{-3}]
ρ_i	density of ice [kg m^{-3}]
Φ	solution variable of interest (units vary)
ϕ_{ia}	net rate of heat exchange per unit area between ice and air [W m^{-2}]
ϕ_{iw}	net rate of heat loss per unit area between ice and water [W m^{-2}]
ϕ_{wa}	net rate of heat loss per unit area between water and air [W m^{-2}]

Chapter 18

Climate Impacts on Ice-jam Floods in a Regulated Northern River

Spyros Beltaos, Terry Prowse, Barrie Bonsal, Tom Carter,
Ross MacKay, Luigi Romolo, Alain Pietroniro
and Brenda Toth

Abstract The Peace-Athabasca Delta (PAD) in northern Alberta is one of the world's largest inland freshwater deltas. Beginning in the mid-1970s, a scarcity of ice-jam flooding in the lower Peace River has resulted in prolonged dry periods and considerable reduction in the area covered by lakes and ponds that provide habitat for aquatic life in the PAD. Using archived hydrometric data and in situ observations, the ice regime of the lower Peace is quantified and ice-jam flooding is shown to depend on freeze-up stage and spring flow. The former has increased as a result of flow regulation; the latter has decreased due to recent climatic trends. This has contributed to less frequent ice-jam flooding. The frequency of ice-jam floods is further explored under "present" (1961–90) and "future" (2070–99) climatic conditions. The ice season duration is likely to be reduced by 2–4 weeks, while future ice covers would be slightly thinner than they are at present. More importantly, a large part of the Peace River basin is expected to experience frequent and sustained mid-winter thaws. These events are expected to cause significant melt and depleted snowpack in the spring, leading to severe reduction in the frequency of ice-jam flooding.

1 Introduction

Climate change is expected to modify streamflow (Burn and Hesch 2007) and thence, the ice regime of major rivers affected by hydro-electric regulation in Western Canada. Of particular interest are extreme flood events, including those caused by ice jams that form during the spring breakup of the ice cover (Hicks and Beltaos 2007). The focus of the present study is the Peace-Athabasca Delta (PAD) in northern Alberta (Fig. 1), one of the world's largest inland freshwater deltas and home to large populations of waterfowl, muskrat, beaver, and free-ranging wood bison. In the last half century, mean annual temperature over western Canada, including the

PAD region, has increased by 1° to 2°C with the majority of this warming occurring during winter and spring (Zhang et al. 2000). During roughly this same period, winter snow accumulation in the upper reaches of the PAD has significantly decreased (Keller 1997, Romolo et al. 2006).

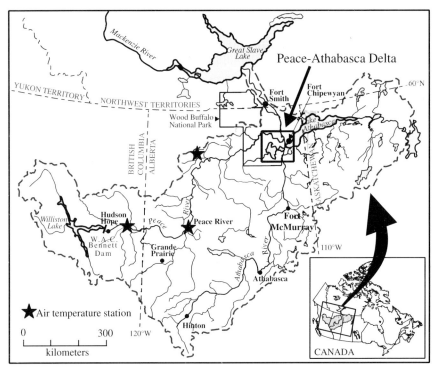

Fig. 1. Location of Peace-Athabasca Delta in the upper Mackenzie River Basin. (From Beltaos et al. 2006a, reproduced with permission from John Wiley & Sons Limited)

The Peace River rises in the Rocky Mountains of northern British Columbia. Since 1968, it has been regulated by the W.A.C. Bennett hydroelectric generation dam. The upstream limit of the PAD is located near Peace Point, some 1100 km downstream of the dam (Fig. 1). The PAD comprises many lakes and ponds that are connected to Lake Athabasca and the Peace/Slave rivers by scores of active and inactive channels. Flow is normally northward, but can reverse when the Peace is at high stage, thus providing the PAD ecosystem with crucial water supply and essential nutrient replenishment.

Beginning in the mid-1970s, this complex and dynamic region has experienced prolonged dry periods, leading to considerable reduction in the area covered by lakes and ponds that are habitat for aquatic life. A common perception during the 1970s and 1980s was that regulation by the Bennett dam reduced the frequency of large open-water flood events that were capable of inundating the entire PAD. More recent studies, however, revealed that open-water floods, including the historically high event of 1990, did not produce sufficiently high water levels to inundate the higher-elevation areas of the Delta, such as the perched basins (Demuth et al. 1996). It was only in the 1990s that the essential role played by spring ice jams in the flooding of such areas was demonstrated (Prowse and Conly 1998). Prowse et al. (1996) analyzed historical hydrometric records kept by the Water Survey of Canada (WSC) at the Peace Point gauge (Fig. 1). They found that the peak water levels that occurred during several breakup events exceeded that of the 1990 open-water flood, some by as much as 2 m. It is now known that the major flood events of recent decades (1963, 1965, 1972, 1974, 1996 and 1997) were all caused by ice jams whose occurrence is influenced by both regulation and climate (Beltaos et al. 2006a; Prowse and Conly 1998). Demuth et al. (1996) identified four likely ice-jam lodgment sites (or jam "toe" locations, see Fig. 1): at the downstream end of the first large island in Slave River; near the confluence of Peace and Slave Rivers, at the mouth of Riviere des Rochers; at the sharp bend near Rocky Point; and at the upstream end of Moose Island. Until this study was undertaken, there was hardly any information concerning the freeze-up and breakup processes within the delta reach of the Peace (Sweetgrass landing to mouth of Peace, Fig. 1). The objectives of this study, which commenced in 1999, were (1) to identify and quantify the hydro-climatic parameters that are conducive to ice-jam flooding, and (2) to predict changes to the frequency of ice-jam floods in response to anticipated climate change scenarios.

2 Data Collection and Analysis

The lower reach of Peace River is largely inaccessible by road. Consequently, very limited information exists about local ice processes in general, and ice breakup and jamming in particular. For the present study, the only source of comprehensive data is the hydrometric station record for the WSC gauge at Peace Point located well upstream of the delta reach (~110 km above the mouth of Peace) or some 60 km above Sweetgrass Landing.

However, the local channel configuration and slope are similar to those encountered within the delta reach. Hence Peace Point records was used as a surrogate for delta-reach processes, with the understanding that various details may not always be adequately represented. This approach was first adopted by Prowse et al. (1996) who examined the records up to 1993; subsequent years (to 2003) have been filled in as part of the present study.

The information gleaned from the historical record at Peace Point was corroborated with field measurements and observations in the entire study reach (from Peace River mouth to Peace Point and beyond), using winter-road and snowmobile access, fixed- or rotary-wing aircraft, boats, or special-purpose vehicles on loan from Parks Canada. Without the logistical assistance support from the Fort Smith and Fort Chipewyan offices of this agency, the field program would be limited to mainly qualitative aerial observation. Occasional logistical assistance was also provided by B.C. Hydro field crews. Important supplementary information and occasional field support were provided by the Yellowknife office of WSC.

Field data collection comprised the following components:
1. Measurements of channel bathymetry between Boyer Rapids and Sweetgrass Landing, to characterize channel hydraulics in this, not previously surveyed, reach. Particular attention was paid to a 5 km stretch centred at the Peace Point gauge in order to describe local hydraulics for use in slope-area and ice jam calculations.
2. Ice thickness measurements in the fall and winter, including early-spring data, intended to document decay processes during the pre-breakup period.
3. Continuous water-level monitoring at selected stations in the delta during the ice season.
4. In situ ice breakup observations and measurements during the spring events of 2000, 2002 and 2003; and freeze-up observations in late 2000.

To characterize relevant climatic variables, archived data at meteorological stations operated by the Meteorological Service of Canada (MSC) were used. Parks Canada reports on the 1996 and 1997 floods (Giroux 1997a, 1997b) provided helpful information on ice-jam locations and high water marks. On occasion, WSC hydrometric information at the town of Peace River and on major tributaries helped distinguish between natural trends and regulation effects. The numerical models RIVJAM (Beltaos 2003a) and WATFLOOD (Pietroniro et al. 2003) were applied to quantify ice-jam flood stages and climate impacts on spring flows, respectively.

3 Results

3.1 Ice Processes in Lower Peace River (Boyer Rapids to Slave River)

Within the study area, the Peace River is large but flat (width ~ 600 m, slope ~ 0.05 m/km), and is ice covered for nearly six months of each year (November to late April or early May). Freeze-up in a flat reach would be expected to occur by surface juxtaposition of slush ice floes because hydrodynamic forces would not be sufficient to overturn and submerge the floes when they come to rest at a constriction or against already formed ice cover. This process still occurs on occasion, but the hydrometric record suggests that there are years when the large, regulation-generated, flows during late fall produce a thickened ice cover. This was corroborated by in situ observations and stage records in November of 2000: a newly formed surface cover collapsed to form a thickened accumulation of ice fragments and floes that raised water levels to ice-jam values (Beltaos et al. 2006a). On the average, post-regulation freeze-up stages at Peace Point are 1.4 m higher than pre-regulation values.

Once a stable cover has formed, a layer of solid ice (sheet ice cover) grows vertically by freezing into the underlying water or ice-water mixture of a thickened porous accumulation. WSC archived data for the Peace Point gauge site indicate that the solid ice layer attains a maximum thickness of 0.7 to 1.2 m, with an average value of 0.9 m. Typically, this occurs some time between mid-March and mid-April. Measurements that started in 2000 reveal a longitudinal variation of ice thickness. Comparable values have been found at various locations downstream of Peace Point, though there is often a tendency towards thinner ice within the delta reach (Fig. 2).

After mid-March, ice thickness may begin to decrease in response to positive heat inputs to the ice cover, due to milder air temperatures and increased solar radiation. Thickness reduction has been empirically linked to degree-days of thaw above a base temperature of -5°C (Bilello 1980), though such a statistical relationship yielded considerable scatter when applied to the present day situation. On average, the implied rate of thinning is about 0.0015 m/(°C day) which is comparable to values obtained in more southern Canadian rivers (Beltaos et al. 2006a). In addition to thinning, the ice cover is subject to strength reductions resulting from penetration of short-wave radiation and preferential melting along ice crystal boundaries. This effect is difficult to measure but calculations by Prowse et al (1996)

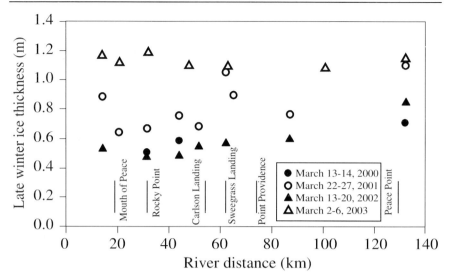

Fig. 2. Spatial variation of ice thickness along the lower Peace River, as revealed by measurements obtained in March of 2000–03. Peace-Point data were provided by WSC; 2002 data point is interpolated (modified after Beltaos et al. 2006a)

indicated significant strength loss during the pre-breakup period. Observations near Peace Point on April 29, 2003, two days prior to breakup initiation, support this prediction, albeit qualitatively. Holes drilled in the ice cover revealed that the lower 50 cm or so were completely saturated with water, suggesting the presence of a dense network of interconnected cracks and voids.

As the flow begins to increase, typically in mid-April, so does the river stage and the hydrodynamic forces exerted on the ice cover. Hinge cracks form first, running along the river and close to the banks to produce shore leads. This reduces the degree of ice cover attachment to the channel boundary and, with further increases in flow and stage, leads to formation of transverse cracks. Eventually, the stage and hydrodynamic forces may become sufficient to dislodge and transport the ice down river. Once in motion, large ice slabs collide with each other and with channel boundaries, generating rubble that accumulates to form ice jams where they are halted by a competent ice cover that has not yet been dislodged. The first sustained movement of the winter ice cover at a given site is defined as the *onset* or *initiation* of breakup (Beltaos 1997).

Depending on hydro-meteorological conditions, breakup events can be either thermal or mechanical. The former typically occurs when mild weather is accompanied by low runoff due to gradual melt and lack of rain. The ice cover decays in situ and eventually disintegrates or is sufficiently weakened that it can be dislodged by the mild current. Ice jamming is minimal, if any, and water levels remain low. Mechanical breakup occurs when the ice cover is dislodged while still retaining significant strength and thickness, and is typically triggered by rising flows. Major jams can now form because they are held in place by competent sheet ice that has not yet moved.

Field observations during April and May of 2000 and 2002 documented thermal breakup events, characterized by advanced decay of the ice cover prior to disintegration in situ, or dislodgement by the flow. In 2000, spring flows were relatively low (maximum = 1740 $m^3 s^{-1}$) and the ice cover gradually disappeared by thermal attrition. On the other hand, the higher flows of 2002 (maximum = 4010 $m^3 s^{-1}$) were insufficient to dislodge the ice cover while it remained strong and thick. At Peace Point, the ice did not move out until the spring flow recession, after sustaining considerable thermal decay. Neither of these events produced significant jamming.

The 2003 event, attended by high flows (maximum = 5770 $m^3 s^{-1}$), was partially of the mechanical type but did not result in ice-jam flooding of the delta reach. Essentially, breakup below the town of Peace River consisted of sequential releases and arrests of bank-to-bank rubble fields that attained lengths of tens of kilometers. When arrested, the rubble formed major jams. One of these jams caused flooding at Carcajou located several hundred kilometers above Peace Point, while another nearly flooded the community of Garden Creek some 120 km above Peace Point. The farthest downstream jam formed at Boyer Rapids on May 1; the wave generated by the preceding release dislodged the ice cover in the vicinity of Peace Point. This was as far as the mechanical type of breakup advanced. The Boyer Rapids jam gradually melted away, producing a minor wave at Peace Point upon its release on May 4. Downstream of Peace Point, the ice cover disintegrated by thermal decay and gradual movements of ice sheets. Only minor accumulations of rubble formed at a few locations in the delta reach.

3.2 Conditions for Ice-jam Flood Occurrence

A necessary but not sufficient condition for the occurrence of significant jams is a mechanical breakup event. Detailed analysis of hydrometric records at Peace Point yielded a methodology to characterize past events as either mechanical or thermal (Beltaos 2003b). The results are consistent with findings on other rivers and underscore the importance of spring flows and freeze-up levels. Within the period covered by hydrometric gauge archives (1960–present), there have been six major ice-jam floods:

under the natural-flow regime in 1963 and 1965; and under regulated-flow conditions in 1972, 1974, 1996, and 1997. There is also evidence that brief jamming and limited flooding occurred in 1994 when a jam was formed at Rocky Point following a release at Boyer Rapids (Kevin Timoney, pers. comm. 2005, quoting the local newspaper Slave River Journal, May 4 1994). All these events were associated with mechanical breakup at Peace Point, but there were many other mechanical events in the record that did not lead to flooding. Beltaos et al (2006a) reasoned that two additional conditions must also be fulfilled: an ice jam must form in the delta reach of the Peace, and the river flow must exceed a certain threshold before the water levels along the jam rise above the bank elevation to cause significant overbank flooding.

The flood-threshold flow was determined using the model RIVJAM (Beltaos 2003a), after calibration with high-water-mark data from the 1996 and 1997 flood events (Fig. 3). It was found that a minimum Peace-Point flow of approximately 4000 m^3 s^{-1} is necessary to initiate flooding when a major jam has formed in the delta reach. The extent of a flood and its eff-

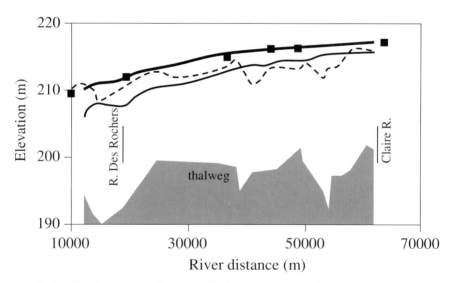

Fig. 3. RIVJAM prediction for April 29, 1997 and comparison with 1997 high water marks (square symbols). Continuous lines show predicted water surface and bottom of jam; dashed line indicates approximate elevation of south bank. R. des Rochers marks the mouth of Peace River and the head of Slave River. (From Beltaos 2003a, reproduced with permission from John Wiley & Sons Limited)

ectiveness in replenishing PAD habitat will also depend on the duration that high flows are maintained. This duration is governed by the shape of the flow hydrograph, but can be approximately indexed by the maximum flow during the event. As indicated in Table 1, all of the ice-jam flood events had maximum breakup flows well in excess of 4000 $m^3 s^{-1}$, hence there would have been ample time for sustained inundation of the delta.

The condition under which an ice jam is created in the delta reach of the Peace, cannot be quantified at present because it is not possible to predict whether and where an ice jam will form in a given reach even if a mechanical event occurs at Peace Point. Such uncertainty is aptly illustrated by the 1994 and 2003 events, both of which had comparable breakup flows, together with a jam forming at Boyer Rapids and a mechanical event occurring at Peace Point. In 1994, the release wave of the Boyer Rapids jam dislodged the downstream ice cover all the way to Rocky Point where a new jam formed. However, in 2003, the Boyer Rapids jam remained intact for several days and was thermally eroded to insignificance before it was released to produce a minor wave. The result was a thermal event in the delta reach.

Table 1. Peace-Point water levels and flows associated with ice-jam flood events (modified after Beltaos et al. 2006a)

Year	H_F^a [m]	H_m^b [m]	Q_m^b [$m^3 s^{-1}$]	Q_{max}^c [$m^3 s^{-1}$]
1963	213.32	220.55	4000	9000
1965	212.16	219.02	4400	7000
1972	213.01	220.30	4200	5600
1974	212.71	220.17	6500	8700
1994[d]	213.57	220.64	5310	5310
1996	212.38	220.28	4800	5800
1997	214.22	220.75	9000	9160

[a]Freeze-up stage.
[b]Peak breakup stage and concomitant flow.
[c]Maximum flow obtained during the spring breakup event.
[d]Not a major event – limited flooding.

3.3 Significance of Freeze-up Stage and Breakup Flow

Consistent with results on other rivers, Beltaos et al. (2006a) found a strong positive correlation between the breakup onset stage, H_B, and the freeze-up stage H_F. Other things being equal, high values of H_F are likely to be associated with more frequent thermal events, as indicated in a plot of Q_{max} (maximum flow during breakup event) vs. H_F (Fig. 4). Thermal events tend to cluster towards the higher freeze-up levels and mechanical events are far more frequent at the lower levels. This result can be deduced from the equations presented by Beltaos (1997) based on extensive data on Canadian rivers, and is consistent with findings in the United States (Kathleen White, pers. comm., 2005). On the other hand, the inhibiting effect of high freeze-up stages can be offset by large spring discharge, as was demonstrated by the 1997 event. Table 1 indicates that this event was a case when H_F was well above the values of the other five events, but the concurrent flow (Q_{max} = 9160 m^3 s^{-1}) was the highest on record for breakup conditions. Thus, other things being equal, higher flows enhance the probability of flooding and the magnitude of the flood, which is also consistent with current understanding of breakup and jamming processes (Beltaos 1995, 1997).

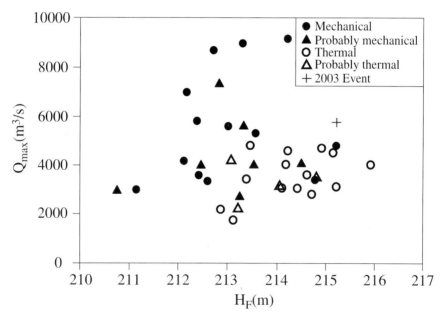

Fig. 4. Influence of freeze-up stage and maximum breakup discharge on the type of breakup event. Major flood events are identified by year of occurrence. The qualifier "probably" is used where the event designation is uncertain, but there are strong indications in favor of one or the other type (modified after Beltaos et al. 2006a).

Particularly pertinent to this discussion is the 2003 breakup event which was attended by high flows but did not generate ice jams and flooding in the delta reach. The value of Q_{max} is estimated by WSC as 5770 m^3 s^{-1} (and as 6100 m^3 s^{-1} by Beltaos et al. 2006a), values that are comparable to that of the 1996 flood event (Table 1). However, the freeze-up stage (H_F) was 215.2 m, some 3 m higher than that of the 1996 event.

Beltaos et al. (2006a) found positive and negative temporal trends for H_F and Q_{max}, respectively, for the post-regulation period (1971–2003; the reservoir-filling years 1968–1971 were excluded). The increase in freeze-up stage (significance probability = 0.004) is consistent with the higher fall and winter flows that have resulted from regulation. The decrease in breakup discharge (significance probability = 0.20) was attributed to climatic effects, such as a decline in winter precipitation in the upper Peace River basin (Keller 1997, Romolo et al. 2006). However, the short pre-regulation record at Peace Point precludes assessment of the relative effects of these two factors (H_F and Q_{max}) on the frequency of ice-jam flooding during that period (Beltaos et al. 2006a).

3.4 Climatic Effects and Future Ice Regime

Given the sensitivity of ice-jam processes to climatic inputs (Beltaos 2002, 2004; Beltaos and Prowse 2001), the issue of climate change is a matter of increasing concern to the long-term health and survival of the PAD ecosystem. Beltaos et al. (2006b) examined how the frequency of ice-jam flooding may be altered under projected climate scenarios using available hydro-meteorological data and recent Global Climate Model (GCM) output. The meteorological station at Fort Smith was used to represent climatic conditions applicable to the ice cover of the lower Peace River. On the other hand, breakup flows in the lower Peace are largely generated by snowmelt, which is linked to winter snow accumulation and spring melt in the upper basin. Thus, climate records from the Grande Prairie station can be used to characterize runoff conditions in this region (Prowse et al. 1996). In fact, Beltaos et al. (2006b) found that the total winter precipitation (Nov. 1 to Mar. 31) at Grande Prairie can be used as an empirical predictor of ice-jam floods: the probability of occurrence of an ice-jam flood exceeds 0.5 when winter snow accumulation exceeds 150 mm (snow water equivalent is calculated as one-tenth of snowfall amount).

Future climatic conditions at these two stations were projected based on GCM output from the Canadian Climate Centre (CGCM2) for the 30-year period centered on the 2080s (i.e., 2070–2099). This period was chosen to maximize climate change impacts which are projected to be greatest by the end of the 21st century. The choice of CGCM2 among the different international GCMs is discussed by Beltaos et al. (2006b). Model projections are based on the A2 and B2 greenhouse gas/sulfate emission scenarios (Nakicenovic et al. 2000). The A2 scenario is associated with a more fossil-fuel intensive world (projected mean global temperature change of 2.5°–4.5°C by 2100), and the B2 with smaller greenhouse-gas emissions (1.5°–3.0°C by 2100).

For each scenario, the potential for ice-jam flooding in the PAD was examined using November to March projected temperature and precipitation changes at Grande Prairie, and temperature changes at Fort Smith. First, monthly GCM temperature and precipitation changes were spatially interpolated to determine regional changes at the Grande Prairie and Fort Smith climate stations. Future daily temperature and precipitation changes were then generated by linearly interpolating the monthly changes to the daily scale. These changes were superimposed on the observed 1961–90 time series to yield future daily temperature and precipitation series for the years 2070–2099. Since this procedure only incorporates projected changes to the mean climate, the variability in the future series remains the same as that of 1961–90.

The future series indicate expected changes, such as shorter ice seasons (by 2 to 4 weeks) and thinner ice covers (by 13 to 20%). More importantly, both the A2 and B2 scenarios exhibit frequent occurrences of mid-winter thaws (MWTs) at Grande Prairie, with temperatures sometimes attaining 10°C and lasting for several weeks. Defining mid-winter as the period between Jan. 1 and March 15, the snowmelt potential (indexed by the total number of positive degree-days) of Grande Prairie MWTs under present and future climatic conditions is illustrated in Fig. 5. Such thaws could severely deplete the snowpack and minimize the probability of ice-jam flooding during the spring breakup. Taking potential snowmelt into account, future values of the winter precipitation index were computed and shown to be considerably smaller than the present values. Exceedance of 150 mm will be far less frequent in the future, suggesting a four-fold reduction in the frequency of ice-jam floods during the spring breakup (Beltaos et al. 2006b). At the same time, the future temperature series for Fort Smith indicate that MWTs are far less frequent and much shorter than at Grande Prairie. Thus, flow generated by a MWT in the upper basin would encounter a strong and thick ice cover upon arrival at the delta reach. A breakup event would be unlikely.

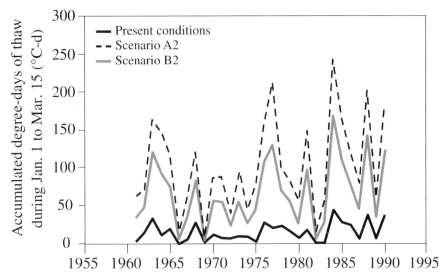

Fig. 5. Melting potential of mid-winter thaw events occurring at Grande Prairie under present and future climatic conditions. (From Beltaos et al. 2006b, reproduced with permission from John Wiley & Sons Limited)

These predictions were further corroborated by applications of the WATFLOOD hydrologic model. WATFLOOD has been calibrated for the Peace-Athabasca basin and used in conjunction with the ONE-D hydrodynamic model to simulate daily flow hydrographs at various hydrometric gauging stations over the period Oct. 1 1965 to Sept. 30 1989 (Pietroniro et al. 2003). Though the model tends to overestimate spring flows at Peace Point, comparisons of model-generated hydrographs for corresponding present and future years are revealing. For instance, the model predicts slightly earlier and higher flows in the future year corresponding to the ice-jam flood year 1974 (Fig. 6). In the following two seasons, the model indicates substantial winter runoff in response to extensive MWTs at Grande Prairie, but the resulting flows are not large enough to cause mid-winter ice jam flooding. Furthermore, as a result of partially depleted snowpacks, spring flows are reduced considerably, thus lessening the likelihood of an ice-jam flood.

4 Discussion and Summary

The ice regime of a regulated northern river, which is closely linked to the maintenance of a major delta ecosystem, has been studied using hydromet-

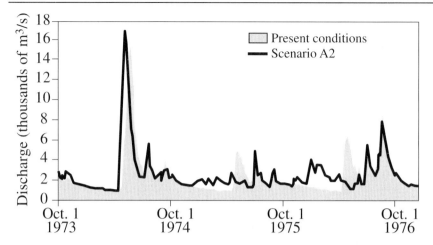

Fig. 6. Typical comparison between present and future Peace Point flows, as generated by WATFLOOD. (Modified from Beltaos et al. 2006b, reproduced with permission from John Wiley & Sons Limited)

ric records, numerical modeling, and in situ field observations. Observations indicate that mechanical breakup occurs as a sequence of jams and releases. Each release generates a wave that dislodges the downstream ice cover for a certain distance. The rubble is eventually arrested and a new jam forms. This process may advance all the way to the mouth of Peace River, or it may stall upstream of the delta reach, as occurred in 2003. Thus there can be years for which the hydrometric record at Peace Point indicates a mechanical breakup event, while the ice cover further downstream disintegrates thermally. It is unlikely that the reverse can occur: a combination of thermal breakup at Peace Point and a mechanical one in the very flat delta reach would require dislodgement of the ice cover without the large hydrodynamic and body forces generated by the release wave. This aspect is presently explored using recently developed analytical methodology (Beltaos and Burrell 2005; Beltaos 2007).

Empirical evidence (Fig. 4) suggests that mechanical breakup and ice-jam flooding are promoted by low freeze-up stages and high spring flows and vice versa. This is in full agreement with data and physically-based relationships from other Canadian rivers and points to potential flood-enhancing reservoir operation strategies (Beltaos et al. 2006a). For instance, a timely flow release in the spring of 1996 helped increase the flood stage (Prowse et al. 2002).

Climatic impacts on ice-jam flood frequency were studied for two 30-year periods, 1961–90 (present) and 2070–99 (future), using CGCM2 output for two greenhouse-gas emission scenarios. Results showed that sustained and frequent mid-winter thaws that appear in the future period over the upper portion of the basin will deplete spring snowpacks and result in reduced spring flows. Consequently, the frequency of ice-jam flooding will be reduced considerably. To prevent this outcome, costly measures would have to be implemented, such as modified regulation procedures or erection of flow and ice control structures. Adaptation measures could be considered as an alternative (Beltaos et al. 2006b).

Though the present study focuses on the PAD, an additional concern pertains to more southern parts of the basin. Here, the local ice cover will decay in response to the mild temperatures associated with a mid-winter thaw and could be dislodged and broken up by the attendant increase in flows. Several communities can be affected, especially those located along the upper Peace and its major tributaries.

Acknowledgements

Funding for this study has been provided by the Program on Energy Research and Development, the Northern Rivers Ecosystem Initiative, the Climate Change Action Fund and by the National Water Research Institute (NWRI). The authors acknowledge the technical field support by Earl Walker and Cuyler Onclin of NWRI, and by the Parks Canada offices at Fort Smith and Fort Chipewyan (Jonah Mitchell, Charlie Risteau, and colleagues). The authors also thank Alberta Environment (Gordon Fonstad) for information on ice conditions in upper reaches of Peace River during the spring runoff, and BC Hydro (Jay Joiner, Martin Jasek, and colleagues) for similar information and logistical field support. Hydrometric data and occasional field support provided by the Yellowknife office of Water Survey of Canada (Dan Dube, Murray Jones, Dale Ross, and colleagues) are greatly appreciated. Acknowledged with thanks are advice on the hydrologic modeling that has been kindly provided by Prof. Nicholas Kouwen, University of Waterloo, and insightful discussion on snowmelt with Stefan Pohl, NWRI at Saskatoon.

References

Beltaos S (1995) River ice jams. Water resource publications, Highlands Ranch, Colorado, USA

Beltaos S (1997) Onset of river ice breakup. Cold Reg Sci Technol 25:183–196

Beltaos S (2002) Effects of climate on mid-winter ice jams. Hydrol Process 16:789–804

Beltaos S (2003a) Numerical modelling of ice-jam flooding on the Peace-Athabasca Delta. Hydrol Process 17:3685–3702

Beltaos S (2003b) Threshold between mechanical and thermal breakup of river ice cover. Cold Reg Sci Technol 37:1–13

Beltaos S (2004) Climate impacts on the ice regime of an Atlantic river. Nord Hydrol 35:81–99

Beltaos S (2007) The role of waves in ice-jam flooding of the Peace-Athabasca Delta. Hydrol Process (in press)

Beltaos S, Burrell BC (2005) Determining ice-jam surge characteristics from measured wave forms. Can J Civil Eng 32:687–698

Beltaos S, Prowse TD (2001) Climate impacts on extreme ice jam events in Canadian rivers. Hydrol Sci J 46:157–182

Beltaos S, Prowse TD, Carter T (2006a) Ice regime of the lower Peace River and ice-jam flooding of the Peace-Athabasca Delta. Hydrol Process 20:4009–4029

Beltaos S, Prowse T, Bonsal B, MacKay R, Romolo L, Pietroniro A, Toth B. (2006b) Climatic effects on ice-jam flooding of the Peace-Athabasca Delta. Hydrol Process 20:4031–4050

Bilello MA (1980) Maximum thickness and subsequent decay of lake, river and fast sea ice in Canada and Alaska. U.S. Army CRREL Report 80-6, Hanover, NH, USA

Burn DH, Hesch N (2007) Trends in Mackenzie River Basin streamflows. (Vol. II, this book)

Demuth MN, Hicks FE, Prowse TD, McKay K (1996) A numerical modelling analysis of ice jam flooding on the Peace/Slave River, Peace-Athabasca Delta. Peace-Athabasca Delta technical studies – PADJAM, Sub-component of Task F.2: Ice studies. National Hydrology Research Institute Contrib. Series CS-96016, Saskatoon, Canada. In: Peace-Athabasca Delta technical studies appendices: I, Understanding the ecosystem, Task Reports

Giroux S (1997a) 1996 Peace-Athabasca Delta flood report. Wood Buffalo National Park, Fort Chipewyan, Alberta, Canada

Giroux S (1997b) 1997 Peace-Athabasca Delta flood report. Wood Buffalo National Park, Fort Chipewyan, Alberta, Canada

Hicks F, Beltaos S (2007) River ice. (Vol. II, this book)

Keller R (1997) Variability in spring snowpack and winter atmospheric circulation pattern frequencies in the Peace River Basin. M.Sc. thesis, University of Saskatchewan, Saskatoon, Canada

Nakicenovic N, Alcamo J, Davis G, de Vries B, Fenhann J, Gaffin S, Gregory K, Grübler A, Jung TY, Kram T, La Rovere EL, Michaelis L, Mori S, Morita T, Pepper W, Pitcher H, Price L, Raihi K, Roehrl A, Rogner HH, Sankovski A, Schlesinger M, Shukla P, Smith S, Swart R, van Rooijen S, Victor N, Dadi Z (2000) IPCC special report on emissions scenarios, Cambridge University Press, Cambridge and New York

Pietroniro A, Conly M, Toth B, Leconte R, Kouwen N, Peters D, Prowse T (2003) Modeling climate change impacts on water availability in the Peace and Athabasca Delta and catchment. NREI (Northern River Basins Initiative) Final project report, draft, June 2003, Saskatoon, Canada

Prowse TD, Conly M (1998) Impacts of climatic variability and flow regulation on ice jam flooding of a northern Delta. Hydrol Process 12:1589–1610

Prowse TD, Conly M, Lalonde V (1996) Hydrometeorological conditions controlling ice-jam floods, Peace River near the Peace-Athabasca Delta. Northern river basins study, Project report no. 103, NRBS, Edmonton, Canada

Prowse TD, Peters D, Beltaos S, Pietroniro A, Romolo L, Toyra J, Leconte R (2002) Restoring ice-jam floodwater to a drying delta ecosystem. Water Int 27:58–69

Romolo LA, Prowse TD, Blair D, Bonsal B, Marsh P, Martz LW (2006) The synoptic climate controls on hydrology in the upper reaches of the Peace River Basin. Part I: snow accumulation. Hydrol Process 20:4097–4111

Zhang X, Vincent LA, Hogg WD, Niitsoo A (2000) Temperature and precipitation trends in Canada during the 20th century. Atmos Ocean 38:395–429

Chapter 19

Trends in Mackenzie River Basin Streamflows

Donald H. Burn and Nicole Hesch

Abstract Trends in the hydrologic regime were analyzed for three major rivers in the Mackenzie River Basin: the Athabasca, Peace, and Liard Rivers. Monthly and annual trends were identified using the Mann-Kendall test with an approach that corrects for serial correlation. The global (or field) significance of the results for each watershed was evaluated using a bootstrap resampling technique. The results reveal more trends in some hydrologic variables than are expected to occur by chance. There are both similarities and differences between the trend characteristics for the three watersheds investigated.

1 Introduction

This paper explores and compares hydrologic trends for the Athabasca, Peace, and Liard River basins in the Canadian north. A limited study of other sites within the Mackenzie River Basin has also been conducted. Climate change modeling studies have hypothesized that northern basins will be particularly sensitive to the impacts of climatic change. However, comparatively little research has been conducted on trends and variability in northern basins (Spence 2002; Woo and Thorne 2003) in part because of the lack of data for unregulated rivers in the remote areas that characterize much of the far north. Often, anthropogenic effects, such as the construction of large reservoirs or changes in land use, can hinder the ability to understand the impact climate change may have on water resource systems. The Athabasca, Peace, and Liard River basins were chosen for this study because they have very little natural storage (no large lakes) and have not been subject to any major water diversions. Note that while the Peace River flow is impacted by the WAC Bennett dam, this study only considers portions of the Peace River basin that are not affected by this regulation.

The research described herein explores the trend characteristics of monthly and annual streamflow within the three subject basins to better understand the potential impacts of climate change on the northern environment. Future work will explore the trend characteristics of modeled runoff data with the aim of comparing the modeled and observed data for a common time period.

2 Study Area

The study area consists of streamflow gauging stations from the Athabasca, Peace, and Liard River basins. A total of 47 stations were examined, including 16 stations for the Athabasca, 16 for the Peace, and 15 stations for the Liard (Table 1). Monthly and annual streamflow were examined for three analysis periods: 1965–2004, 1969–2004, and 1975–2004, giving record lengths of 40, 35, and 30 years, respectively. Longer record lengths result in greater power for the statistical tests used but shorter record lengths allow greater spatial coverage through a larger number of stations having sufficient data for analysis during the period. For example, although there are 47 stations for the 30 year analysis period, the available number of stations drops to 46 for the 35 year analysis period and to 37 for the 40 year analysis period.

The stations examined include both traditional streamflow gauging stations as well as artificial stations, created as the difference in streamflow between an upstream and a downstream gauging station. The latter type of station provides a means to examine the streamflow characteristics for ungauged tributary areas of a watershed. This was particularly useful for the Peace River and allowed the estimation of inflow to the mainstem of the Peace River through the calculation of differences in flows between two gauged stations on the (regulated) mainstem of the river. The focus in this work was on moderate to mid-sized stations with drainage areas ranging from around 500 km^2 to approximately 100 000 km^2 (Table 1).

3 Methods

The trends and variability in hydrological variables are assessed using the Mann-Kendall non-parametric trend test. The version used incorporates a correction for serial correlation in the data (Yue et al. 2002). The global (or field) significance of these results for each watershed was next evaluated using a bootstrap resampling technique (Burn and Hag Elnur 2002) to determine whether the observed number of trends exceeds what is expected to occur by chance. The determination of field significance reflects

Table 1. Summary of stations included in the analysis of trends

Station ID	Drainage area [km^2]	Latitude [°N]	Longitude [°W]
Athabasca			
07ag003	829	53.6	116.27
07ca006	1,110	55.2	112.47
07bf002	1,160	55.45	116.5
07bj001	1,900	55.32	115.41
07af002	2,560	53.47	116.63
07dd002	2,700	58.36	111.24
07aa002	3,880	52.91	118.06
07bb002	4,420	53.61	115
07ad002-07aa002	5,900	53.42	117.57
07bc002-07bb002	8,680	54.45	113.99
07ad002	9,780	53.42	117.57
07bc002	13,100	54.45	113.99
07cd001	30,800	56.69	111.25
07da001-07be001	58,400	56.78	111.4
07be001-07ad002	64,820	54.72	113.29
07be001	74,600	54.72	113.29
Peace			
07hf002	667	57.74	117.62
07fc003	1,750	56.68	121.22
07ha003	1,960	56.06	117.13
07hc001	4,660	56.92	117.62
07ee007	4,900	55.08	122.9
07fd002-07ef001	9,200	56.14	120.67
07ke001	9,860	58.32	113.07
07gh002	11,100	55.46	117.16
07ge001	11,300	55.07	118.8
07fb001	12,100	55.72	121.21
07fc001	15,600	56.28	120.7
07jd002	35,800	57.88	115.39
07gj001	50,300	55.72	117.62
07fd003-07fd002	50,900	55.92	118.61
07ha001-07fd003	56,000	56.24	117.31
07kc001-07ha001	107,000	59.11	112.43

Table 1. Summary of stations included in the analysis (continued)

Station ID	Drainage area [km²]	Latitude [°N]	Longitude[°W]
Liard			
10ed003	542	61.34	122.09
10ac005	888	59.12	129.82
10be007	1,190	59.33	125.94
10fb005	1,310	61.45	121.24
10cb001	2,160	57.23	122.69
10be004	2,570	58.86	125.38
10ab001	12,800	60.47	129.12
10eb001	14,600	61.64	125.8
10cd001-10cb001	18,140	58.79	122.66
10cd001	20,300	58.79	122.66
10aa001	33,400	60.05	128.9
10ed002-10ed001	53,000	61.75	121.22
10be001-10aa001	70,600	59.41	126.1
10be001	104,000	59.41	126.1
10ed001-10be001	118,000	60.24	123.48

the spatial correlation structure of the data set. The methodology used is described in greater detail in Burn et al. (2004a, b).

4 Results

Although results were evaluated for the three analysis periods noted above, the focus will be on the shortest and longest analysis periods which provide 30 (1975–2004) and 40 (1965–2004) years of record, respectively. Table 2 summarizes the results for each river basin separately and for the three watersheds combined. Shown in the table is the percentage of stations with significant trends, at the 5% significance level, for each of the variables. Analysis performed at the 10% significance level leads to similar results. The results give both the percentage of stations with an increasing trend and the percentage of stations with a decreasing trend. Results in bold indicate a variable for which the results are field significant, implying that the number of trends observed for that variable exceeds the number expected to occur by chance.

Table 2 reveals that the number of trends exhibited in several months is larger than what would be expected to occur by chance. January to March

Table 2. Percentage of stations with a significant trend (5% level) for the 30 and 40 year analysis periods

Time Period	Athabasca River		Peace River		Liard River		Combined	
	30 years	40 years	30 years	40 years	30 years	40 years	30 years	40 years
January	+0/-43	+0/-20	+0/-25	+0/-20	+47/-0	+55/-0	+17/-22	*+19/-13*
February	+7/-50	+0/-40	+0/-17	+0/-20	+47/-0	+64/-0	+20/-22	*+23/-19*
March	+0/-31	+0/-8	+7/-7	+8/-0	+40/-0	+73/-0	+16/-13	*+26/-3*
April	+0/-6	+0/-25	+14/-0	+8/-0	+27/-0	+70/-0	+13/-2	*+24/-9*
May	+0/-13	+0/-67	+0/-0	+0/-42	+0/-0	+10/-0	+0/-4	*+3/-38*
June	+0/-13	+0/-0	+0/-6	+0/-0	+7/-0	+0/-0	+2/-6	+0/-0
July	+0/-6	+0/-8	+6/-0	+0/-0	+13/-0	+0/-0	+6/-2	+0/-3
August	+0/-0	+0/-8	+0/-0	+0/-0	+7/-0	+0/-9	+2/-0	+0/-6
September	+0/-31	+0/-33	+0/-0	+0/-8	+13/-0	+9/-0	+4/-11	*+3/-14*
October	+0/-19	+0/-33	+6/-0	+0/-8	+7/-0	+0/-0	+4/-6	+0/-14
November	+0/-21	+0/-30	+0/-8	+0/-10	+27/-0	+9/-9	+10/-10	*+3/-16*
December	+0/-14	+0/-20	+0/-17	+0/-10	+40/-0	+64/-0	+15/-10	*+23/-10*
Annual	+0/-29	+0/-50	+0/-25	+0/-40	+0/-0	+0/-0	+0/-17	+0/-30

Note: Values in *italic* indicate variables that are field significant (at the 5% level).

exhibit field significant increasing trends for the Liard and field significant decreasing trends for both the Athabasca and the Peace. April has field significant increasing trends for the Peace and Liard and field significant decreasing trends for the Athabasca. May has field significant decreasing trends for both the Athabasca and the Peace, though only in the longer analysis period. There are no field significant trends for the months of June through August and only the Athabasca exhibits field significant trends (decreasing) for September and October. In November, there are field significant decreasing trends in the Athabasca and field significant increasing trends in the Liard. In December, there are field significant decreasing trends in the Athabasca and the Peace and field significant increasing trends in the Liard. On an annual basis, there are field significant decreasing trends in the Athabasca and the Peace, but no trends in the Liard.

The Liard River is characterized by increasing flows during the low flow months of November to April with no overall increase in the total (annual) flow. The Athabasca and the Peace are characterized by generally decreasing flows. For the Athabasca, decreasing flows occur from September through to May while for the Peace, the decreasing flows are from December through to May. It is noted from Table 2 that there are differences between the results for the 30 and the 40 year analysis periods. These differences can be explained in part by the greater power of the trend test for the longer analysis period and in part by the differences in the stations that are used for the 30 year versus the 40 year period. Of particular note is the strength of the signal in the May flow for the Athabasca and Peace Rivers. This result is only field significant for the 40 year period.

5 Discussion

The results indicate the presence of a larger number of trends than would be expected to occur by chance. This is illustrated by the number of months for which a field significant number of trends was identified. Differences among the three watersheds imply that the results determined on a combined basis can mask some of the different individual behaviors for the three watersheds. The Liard tends to exhibit increasing trends in the low flow period, but not on an annual basis. Both the Athabasca and the Peace exhibit decreasing flows for the annual period and for several monthly periods as well. On a combined basis, the three watersheds exhibit field significant decreasing trends on an annual basis with generally increasing trends (field significant) in the November to March period and decreasing trends in May (for the longer analysis period only). Combination of the three watersheds yields a greater mixture of increasing and decreasing trends than is obtained for individual watersheds when each is considered separately.

Results of this study indicate that the three watersheds are experiencing more trends than can be explained by chance. Future work will examine possible causes for the trends and will attempt to attribute the trends to the impacts of climate change. Possible causes for the differences in trends results (e.g., annual flow trends for the Liard versus the Peace and Athabasca) will also be examined in future work. The changes noted in the hydrologic regime pertain to both the winter low flow (most notably for the Liard) and the spring freshet (notable for the Peace and the Athabasca, where decreases are observed). Changes of this nature could have important implications for the water resources of the watersheds, especially in terms of water availability of the Athabasca and the Peace. If trends of this nature were to continue, or accelerate with further climate changes, the watersheds could experience localized or widespread water shortages. Limited analysis for other watersheds within the Mackenzie River Basin indicates similar behavior, though in the northerly parts of the Basin, a lack of gauging stations with long records makes determination of trends a challenging task.

6 Conclusion

An analysis of trends in monthly streamflow in the Athabasca, Peace, and Liard River basins reveals generally decreasing streamflows in the Athabasca and Peace Rivers and increasing streamflows in the Liard River. The latter result did not lead to increasing streamflows on an annual basis. Future work will examine attribution of the trends identified and will also investigate trends in modeled data, as opposed to measured data. This will be a first step towards projecting likely streamflow trends based on modeling of climate change impacts.

Acknowledgements

The authors acknowledge the contributions to this research project from Alain Pietroniro, Ric Soulis, Frank Seglenieks, Omar Abdul Aziz, and Juraj Cunderlik.

References

Burn DH, Cunderlik JM, Pietroniro A (2004a) Hydrological trends and variability in the Liard River Basin. Hydrol Sci J 49:53–67

Burn DH, Abdul Aziz OI, Pietroniro A (2004b) A comparison of trends in hydrological variables for two watersheds in the Mackenzie River Basin. Can Water Resour J 29:283–298

Burn DH, Hag Elnur MA (2002) Detection of hydrologic trends and variability. J Hydrol 255:107–122

Spence C (2002) Streamflow variability (1965 to 1998) in five Northwest Territories and Nunavut rivers. Can Water Resour J 27:135–154

Woo MK, Thorne R (2003) Streamflow in the Mackenzie Basin, Canada. Arctic 56:328–340

Yue S, Pilon PJ, Phinney B, Cavadias G (2002) The influence of autocorrelation on the ability to detect trend in hydrological series. Hydrol Process 16:1807–1829

Chapter 20

Re-Scaling River Flow Direction Data from Local to Continental Scales

Lawrence W. Martz, Alain Pietroniro, Dean A. Shaw, Robert N. Armstrong, Boyd Laing and Martin Lacroix

Abstract This paper evaluates methods for aggregating detailed flow pattern data extracted from DEM to larger spatial units suitable for representing the hydraulic and routing characteristics of large basins such as the Mackenzie. Five approaches to up-scaling flow data are discussed: simple averaging of DEM elevation data to coarser resolution, drainage enforcement that minimizes impact on the DEM outside the drainage network area, subdivision of drainage basins at various scales while the DEM resolution is held constant, vector averaging of sub-grid flow directions from a higher resolution DEM and a quasi-expert system approach based on the WATFLOOD parameterization scheme. Simple averaging has a profound impact on basin boundaries and flow patterns and cannot be used to generalize flow pattern data reliably. An objective approach based on vector addition is mathematically appropriate and effective in preserving the essential features of the sub-grid flow patterns. A quasi expert system approach developed by the automation of the WATFLOOD manual method of topographic parameterization is shown to be the superior approach to rescaling flow data for macro or regional scale hydrologic modeling.

1 Introduction

Satisfactory representation of atmosphere land-surface interaction is an important component for successfully modeling climate systems. An inherent problem with incorporating land-surface hydrology into atmospheric systems is one of scale. Hydrologic processes occur at a local scale and are typically modeled at a scale of 100 m to 10 km, while atmospheric models are applied at a large-scale, normally at 25 to 250 km (Pietroniro and Soulis 2003).

There has been an ongoing research effort directed toward finding a satisfactory method of coupling small-scale hydrologic processes and models with large-scale atmospheric processes and general circulation models (GCM) (Armstrong and Martz 2003; Fekete et al. 2001; O'Donnell et al.

1999; Olivera et al. 2002; Shaw et al. 2005a). The Canadian Mackenzie GEWEX (Global Energy and Water Cycle Experiment) Study (MAGS) has focussed research into coupling hydrologic and atmospheric models as a part of research into understanding the flow of energy and water into and through both the hydrologic and atmospheric systems of the Mackenzie River Basin (MRB) in North America.

Distributed or semi-distributed hydrologic models sub-divide the target area into small grid-squares (sub-grids) or sub-units such as sub-basins, so as to preserving the heterogeneity of the basin in the model. The square-grid nature of some hydrologic models allows square-grid data, such as digital elevation model (DEM) data (Fig. 1), to be used as input. The square-grid nature also allows large-scale hydrologic models to be used to parameterize square-grid GCMs.

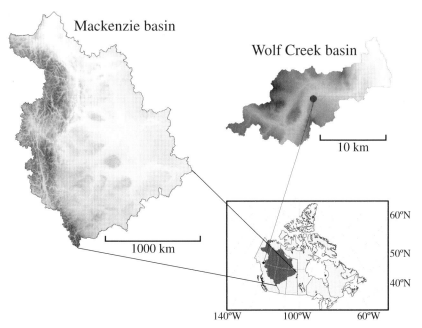

Fig. 1. Examples of study basins: the Mackenzie River Basin (area of 1.8×10^6 km^2) and Wolf Creek Basin (195 km^2)

Issues involved in the aggregation of detailed spatial data into a more generalized representation are known as the Modifiable Area Unit Problem (MAUP). The MAUP attempts to minimize the effects of aggregation of areal data into more simplified regions (Jelinski and Wu 1996). It is also concerned with characterizing areal data into zones which may have different spatial relationships. Aggregation methods usually increase variations in the resulting simplified areal units. This chapter does not explore solutions to the MAUP, rather it evaluates procedures that involve the aggregation of detailed flow patterns into simplified large-scale units.

There are many methods for aggregating data for large-scale hydrologic models, ranging from a simple algorithm that averages elevation data to sophisticated methods that integrate the topographic variability of the aggregated unit. Integrating spatial variability in the aggregated unit will result in flow direction assignment that more closely matches "real world" flow directions. The benchmark against which assigned flow directions are assessed are the 1:50 000 and 1:250 000 National Topographic Series (NTS). Although there are acknowledged limitations to the accuracy of channel placement on topographic maps (Geomatics Canada 1996), the hydrography on NTS maps remain the best available standard against which to evaluate integrated 'stream' networks extracted from DEM. This is particularly true in this study where the scale of analysis results in a focus on major river reaches. Drainage directions for each sub-unit are determined manually using the main channel from the NTS hydrography. These drainage directions are used to evaluate the performance of automated aggregated drainage directions methodologies outlined in the paper.

When large-scale grid is overlaid over a large continental sized basin such as the Mackenzie, there are relatively few grid-squares to assign drainage directions. In this case it may be argued that a manual method is preferable. However, the strength of the automated methods is their scale-independence and their elimination of subjectivity in drainage direction assignment, providing repeatable results for drainage direction assignment over a variety of scales.

This study evaluates five up-scaling methods, all of which use digital elevation model (DEM) data as input. DEM data can be used to describe drainage basin topography and allows physiographic information about a basin to be extracted through automated methods (Martz and Garbrecht 1998). Automatic extraction of drainge basin characteristics has become a valuable tool for topographic parameterization of hydrological models (Lacroix et al. 2002; Shaw et al. 2005b). All the methods outlined in this paper are based upon the derivation of hydrologically significant physiographic parameters from DEM data using the **TO**pographic **PA**rameteri-**Z**ation (TOPAZ) landscape analysis tool (Garbrecht and Martz 1999). TOPAZ consists of modules that identify topographic features; measure topographic parameters; define surface drainage; subdivide watersheds along drainage divides; quantify the drainage network; and parameterize subcatchments (Martz and Garbrecht 1998).

TOPAZ uses the deterministic eight neighbor (D8) method (Fairfield and Leymarie 1991) and downslope flow routing and accumulation concepts (Martz and deJong 1988; O'Callaghan and Mark 1984) to define surface drainage. The channel network is defined as all cells with a contributing area greater than a user specified threshold. In TOPAZ, two parameters, critical source area (CSA) and minimum source channel length (MSCL), are used to control the configuration of the drainage network derived from a DEM (Martz and Garbrecht 1992). The CSA value defines the minimum drainage area (in hectares) below which a source channel is initiated and maintained. The MSCL prunes the channel network of exterior links shorter than a specified threshold value.

Re-scaling methodologies aim to reflect adequately the topographic variability within sub-units of the modeled basin as the data are scaled up. This chapter assesses the extent to which several selected techniques can provide satisfactory results.

2 Study Areas

Two study areas were chosen for this research: (1) the Mackenzie River Basin (MRB), and (2) the Wolf Creek basin, both of which are characteristic of northern Canadian drainage basins. MRB covers 1.8×10^6 km^2 with elevations ranging from sea level to approximately 3400 m. The physical environment of this Basin is provided in Woo et al. (2007). The Wolf Creek basin is located in the Yukon Territory, with a basin area of 195 km^2 and a relief of approximately 1400 m. Vegetation ranges from bare rock at the high elevation, to thick boreal forest. This basin was included because of its accessibility and data records generated by several hydrologic studies as representative of the alpine environment (Pomeroy and Granger 1999).

3 Up-scaling Methodologies

3.1 SLURPAZ

A computerized interface (SLURPAZ) was developed to combine the output of an established digital terrain analysis model (TOPAZ) with digital land cover data required as input by SLURP (Kite 1997), a widely used semi-distributed hydrologic model (Thorne et al. 2007). An interface between a digital terrain analysis model and a hydrologic model is beneficial as it allows prompt analysis at several sub-basin scales to determine the optimal sub-basin resolution that best fits the hydrologic simulation. SLURPAZ makes it possible to derive physiographic parameters for SLURP rapidly and accurately for drainage networks and corresponding sub-basins at varying levels of detail controlled by the user. SLURPAZ differs from the other four approaches put forth in this chapter in that it sub-divides the basin using sub-watersheds rather than a grid (Fig. 1). TOPAZ is used to automatically delineate sub-watersheds at a variety of scales through manipulation of the CSA variable. Through evaluation of model result for a variety of scales or subdivision, the required level of subdivision for optimal model performance is determined (Thorne et al. 2007). This operational method facilitates the examination of the spatial scale or level of detail with which sub-basins must be represented for significant hydrologic processes to be adequately simulated.

3.1.1 Methodology

Wolf Creek was subdivided into a variable number of sub-basins or aggregated simulation areas (ASA) by manipulating the CSA parameter in TOPAZ. TOPAZ also allows the MSCL to be manipulated, though it was left constant for the purpose of this experiment. It was felt that an MSCL of 100 m was short enough to initiate a channel and varying the minimum area (i.e., CSA) would produce sub-basins of varying sizes and numbers. Each resulting set of TOPAZ output data was processed by SLURPAZ to generate SLURP input files. To analyze the effects of varying only the sub-basin scale, the ten hydrologic input parameters listed in Table 1 were held constant. The SLURP outputs from the varying levels of subdivisions were then compared to determine the subdivision at which the best hydrologic simulation was produced.

Table 1. SLURP hydrologic input parameters

1	Initial contents of snow store
2	Initial contents of slow store
3	Maximum infiltration rate
4	Manning roughness surface
5	Retention constant fast store
6	Maximum capacity to fast store
7	Retention constant slow store
8	Maximum capacity to slow store
9	Precipitation factor
10	Snowmelt temperature

3.1.2 Results

Figure 2 displays some TOPAZ generated images of the ASA and their corresponding networks. These figures illustrate how varying the CSA parameter from a high value to a lower value will represent a drainage area

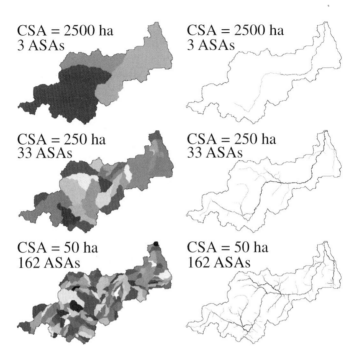

Fig. 2. Subdivision of Wolf Creek basin and channel network produced through manipulation of the critical source area (CSA)

from coarse to fine. Table 2 displays the subdivision statistics created by manipulating the CSA parameter in TOPAZ. By varying the CSA parameter from 18000 ha to 5 ha, ASAs vary in number from 1 to 1588, with the corresponding mean area of the ASA ranging from 183.3 to 0.12 km^2.

Some of the drainage network compositions are listed in Table 2 at all scales. For each level of subdivision, the following variables are listed: highest Strahler order, total length of channels, mean length of channels, overall total drainage density, mean channel sinuosity, bifurcation ratio, length ratio and area ratio. The stream orders range from 1 through 6. For comparison, the 'blue line' stream network from the 1:50000 scale topographic maps results in the basin having an order of 4. Thus, the generated scales with an order of 4 are considered to be the most representative of the 'blue line' network from 1:50 000 scale topographic maps.

Examination of the variation in water balance components, as the number of ASAs increases, reveals one scale effect. At low levels of detail (i.e., approximately 17 ASAs or less), the water balance components of precipitation, evapotranspiration, storage and computed flow fluctuate. This was attributed to the variations in the areal weighting of time series climate input data with variation in number of ASAs. This is an issue because of the limited number of climate stations in an area of such high variability. Another observation is that an optimal level of subdivision seems to be attained at a certain level of detail. At this level, any further subdivision does not enhance model performance significantly.

A minimum number of ASA subdivisions are required to adequately reflect the spatial variability of climatic input data. The Wolf Creek basin has three meteorological stations and the optimal point for satisfactory representation of the water balance components is at around 20 ASAs, with each representing approximately 9.2 km^2 or 5% of the total area (Lacroix et al. 2002) (Fig. 3). The water balance parameters remain constant with further subdivision of the watershed. This water balance optimum is probably related to the relief of the basin. Perhaps for flat terrain such as the prairies, water balance components may be represented at coarser levels of detail than for the moderately steep Wolf Creek basin.

3.2 DEM Aggregation

DEM aggregation is an automated method for extracting topographic information from a DEM to parameterize hydrologic models over a wide range of scales. However, choosing appropriate DEM scales for particular hydrologic modeling applications is limited by a lack of understanding of the effects of scale and grid resolution on land-surface representations. The scale effects of aggregation on square-grid DEMs of a continental basin are examined in this study.

3.2.1 Methodology

The initial DEM of MRB was derived from the HYDRO1k DEM of North America, as a sub-DEM. The DEM has an initial horizontal grid spacing of 1 km and a vertical resolution of 1 m. The initial DEM was aggregated to coarser grid resolutions using simple averaging aggregation to generate additional DEMs of 2, 4, 8, 16, 32 and 64 km for the Basin. TOPAZ was

Table 2. Drainage network composition for all scales

# of ASAs	CSA value	Mean area per ASA [km²]	Highest Strahler Order	Total length of channels [km]	Mean Length of channels [km]	Overall Total drainage density [1/km]	Mean channel sinuosity	Bifurcation ratio	Length ratio	Drainage area ratio
1	1800	183.3	1	2.1	2.1	0.01	1.5	N/A	N/A	N/A
1	2550	183.3	1	32.6	32.6	0.18	1.5	N/A	N/A	N/A
3	2500	61.1	2	33.2	11.1	0.18	1.4	2	2.2	3.2
5	2000	36.6	2	38.5	7.7	0.21	1.4	3	7.6	6.3
7	1000	26.2	2	53	7.6	0.29	1.4	4	4.3	7.2
11	750	16.6	2	59.1	5.4	0.32	1.4	6	5.3	9.3
15	600	12.2	2	66.3	4.4	0.36	1.3	8	5.7	11.2
17	500	10.7	2	71.2	4.2	0.39	1.3	9	6.3	12.9
19	400	9.6	3	76.8	4.5	0.42	1.3	3.5	1.2	3.5
33	250	5.5	3	90.8	2.8	0.5	1.2	4.3	3	5.3
77	100	2.3	4	122.4	1.6	0.67	1.2	3.6	2.3	4
162	50	1.1	4	169.2	1	0.92	1.1	4.4	2.9	5.2
308	25	0.6	4	236.2	0.8	1.29	1.1	5.8	2.8	5.9
413	20	0.4	4	265	0.6	1.45	1.1	6.2	3	6.6
447	18	0.4	5	278	0.6	1.52	1.1	4.1	2.2	4.4
525	15	0.3	5	307.6	0.6	1.68	1.1	4.3	2.2	4.5
571	14	0.3	5	321.4	0.6	1.75	1.1	4.5	2.2	4.5
615	13	0.3	5	337.2	0.5	1.84	1.1	4.6	2.1	4.6
682	12	0.2	5	356.1	0.5	1.94	1.1	4.8	2.1	4.7
744	11	0.2	5	375.1	0.5	2.05	1.1	4.9	2	4.7
824	10	0.2	5	398.7	0.5	2.18	1.1	4.6	2.3	5
1588	5	0.1	6	614.5	0.4	3.35	1.1	4	1.9	4.2

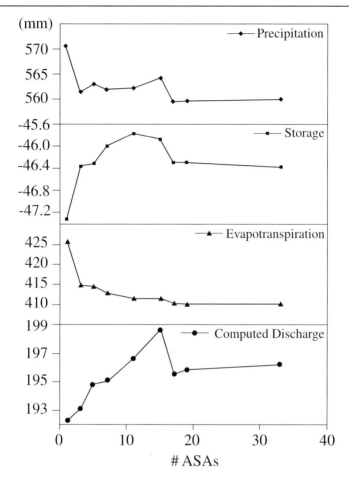

Fig. 3. Variation in water balance components of Wolf Creek basin, for 1 to 33 ASAs

was used to pre-process the DEMs and perform the hydrographic segmentation and parameterization.

To assess the impact of increasing the grid cell size on the delineation of the basin and drainage network, and derived variables, TOPAZ was applied to the base (1 km resolution) and aggregated DEMs using the same CSA (4096 km^2) and MSCL (128 km) parameter values. The CSA and MSCL values used represent the area of one grid cell at 64 km resolution and the length of two grid cells at 64 km resolution, and produce a channel network of maximum drainage density for a grid size of 64 km, the coarsest grid used in this study. Using these parameter values yields drainage networks approximately similar to the blue line on 1:7 500 000 topographic maps.

The effects of varying DEM resolution are examined by considering changes to the spatial distribution and statistical properties of the basin, network and derived topographic variables, including local (e.g. slope) and non-local or global variables (Florisnky 1998; Martz and deJong 1988) such as basin and sub-basin areas. For the purpose of comparison, basin and network properties obtained from the 1 km resolution DEMs are assumed to be the most accurate, and used as a reference for comparing results of the basin and network properties from the aggregated DEMs. Statistical properties reported in this study include basin area, total channel length, highest Strahler order, drainage density, and bifurcation ratio.

3.2.2 Results

Comparisons between basin and channel network statistics for the base reference and aggregated DEMs showed that the values remained reasonably stable (within ±10 %) up to a grid size of 8 km (Fig. 4). Beyond 8 km, values for the aggregated DEMs tended to deviate considerably from those of the base 1 km DEM. Results also showed that values of variables for the basin and channel network tended to be reduced with increasing grid size. With regard to hydraulic slope, the relative instability of topographic information was generally associated with a mean slope of less than 1%. Overall, the behavior of the basin, channel network and topographic variables is unpredictable due to the loss of information from the DEM, and the inherent elevation errors produced through aggregation.

Figure 5 shows variations in the spatial distribution of the basin extent and network at each resolution for the MRB and the Liard, Peace, and Athabasca River sub-basins. Variations in basin extent and the network for the 2 to 8 km resolution grids tend to be relatively small when compared with the basin and network (for the entire MRB) at 1 km resolution. With increased grid size, there is a general decline in basin area, total channel length (calculated as the length of all channel segments) and drainage density. The CSA and MSCL parameters used for this study produce a maximum channel order of 4 for the Mackenzie Basin. At each resolution, the highest Strahler order remains constant and the bifurcation ratio is also fairly consistent (Armstrong and Martz 2003).

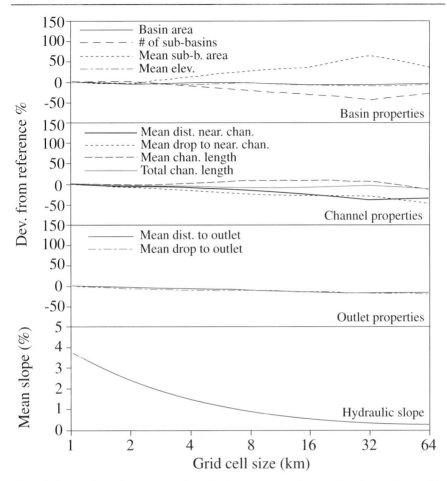

Fig. 4. Comparison between basin and channel network statistics for various grid-cell sizes, Mackenzie River Basin

The substitution of coarse-resolution DEMs for continental-scale hydrology is constrained by the redirection of flow across large flat areas, as observed from major changes in network structure for MRB from 16 to 64 km resolutions. Coarse grid substitution is also constrained by elevation errors (a result of averaging) or obstructions that block flow paths within valley bottoms. This becomes a problem as the valley width approaches the size of the grid cell (e.g., the Peace and Athabasca basins at 2, 4 and 8 km resolutions). The poor definition of flow paths at coarser resolutions is in agreement with the findings of Veregin (1997). Owing to the increased generation of vertical errors with aggregation, network structure tends to

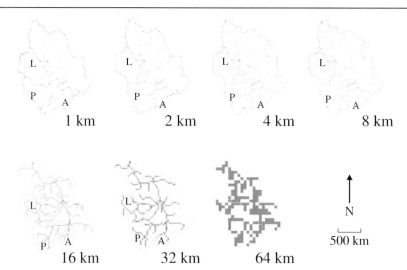

Fig. 5. Extent of sub-basins and drainage network at various aggregation levels for the Mackenzie River Basin. The labeled sub-basins are: Liard (L), Peace (P), and Athabasca (A)

deviate considerably from that derived at finer resolutions, indicating that flow path definition becomes very unreliable.

Compared with the basin and network properties at 1 km resolution, the overall reproducibility of the basin, network and topographic variables is relatively good for grid sizes up to 8 km but considerably less, in general, with grid sizes larger than 8 km. Based on the comparisons of the basins, networks and derived topographic variables for the aggregated DEMs (with those from the base DEMs), it is recommended that a grid size of 10 km or less be used for continental hydrologic applications. The decreasing slope with grid size is expected, and is consistent with previous studies (Wolock and McCabe 2000; Zhang et al. 1999). This study suggests that a mean hydraulic slope of approximately 1% (for the basin) can be used to define a limit to DEM generalization for applications in continental-scale hydrology. This would appear to be a threshold for reproducing basin and network properties from coarser resolution DEMs to within 10% of those derived from finer resolution DEMs.

3.3 Drainage Enforcement

Initial evaluation of the "hydrologically correct" HYDRO1k (2001) data set, found that the drainage network derived from the topographic data using the TOPAZ software was inconsistent with the blue line data in the National Atlas of Canada. Further analysis found minor elevation errors in the HYDRO1k DEM causing drainage network inconsistencies in some flat and low elevation areas. Implementation of the Australian National University Digital Elevation Model (ANUDEM) algorithm (Hutchinson 1989) for drainage enforcement corrected the drainage networks, but it adversely affected the DEM characteristics by substantially lowering high elevation values and created a smoothing effect across the DEM surface.

ANUDEM is a program that enforces the blue line drainage network while generating a raster DEM from a variety of elevation sources (point elevation, contour elevation, existing DEM). ANUDEM calculates DEMs as regular grids from irregularly spaced elevation data points, contour lines and streamline data and automatically removes spurious sinks to produce a hydrologically correct DEM. Streamline data can also be incorporated to obtain drainage enforcement requirements. Input of streamline data is useful where more accurate stream placement is required than those generated by automated methods. All elevation points that conflict with down slope flow along each streamline is removed. The process also uses stream lines as break lines for interpolation so that each stream line is at the bottom of the valley.

3.3.1 Methodology

This study has three objectives: (1) to compare "blue line" or vector stream networks, TOPAZ delineated networks, and ANUDEM drainage enforcement program stream networks in an attempt to isolate differences in stream line characteristics and networks derived from drainage enforcement, (2) to develop a hydrologically sound DEM by smoothing a stream network into the surrounding elevation data using a new algorithm, and (3) to compare the new hydrologically correct DEM to a DEM developed from ANUDEM. MRB was processed using a 2 km DEM that has been aggregated from the HYDRO1k DEM.

Different modules were run using the ArcInfo geographic information system (GIS), ANUDEM and ArcView GIS to develop a hydrologically accurate DEM. Since the input data for ANUDEM can consist of several GIS software formats, ArcInfo GIS input files were chosen as they perform well in terms of computer processing requirements and in speed in generating output results.

Output of the ANUDEM model run provides the initial dataset for the development of an alternative hydrologically accurate DEM with a less intrusive altering of the DEM landscape. The rasterized drainage network provides a dataset in which 5 buffered regions are generated using the ArcInfo GIS EXPAND command. These are at 1, 3, 5, 10 and 20 cell buffer radii. For each of these regions the DEM (generated through the ANUDEM process) is extracted to produce subsets of elevation data that can be incorporated into the original DEM using a distance-weighting procedure. Such an approach preserves the profile and contour curvatures in the area along the stream network. The result is a DEM that has the ANUDEM drainage enforcement applied only to the buffered regions.

The Euclidean distance grid is processed using a distance query based on the buffer width, to calculate distances from all cells in the DEM to the raster blue line network. The Euclidean distance grid and the distance query form the basis for the distance grid algorithm (Eq. 1) that provides distance values for calculating the percentage-weights. Equation (2) uses the distance weighting obtained from Eq. (1) to produce hydrologically correct segments for various buffered regions.

$$W = 1 - (L - D)/L \qquad (1)$$

where W is weight, D is Euclidean distance from channel, and L is the buffer width.

$$Z_n = Z_a[(L - W)/L] + Z_o[1 - (L - W)/L)] \qquad (2)$$

where Z_n is the new elevation value, Z_a is elevation value from original DEM, and Z_o is elevation value from ANUDEM-generated DEM

This procedure gives greater weight to the raster river network in the ANUDEM-values and less for the original inaccurate elevation values; but grading out to the edges, the original DEM-values are weighted more than the ANUDEM imposed values. Figure 6 provides an example. The algorithm derives elevations according to locations in the distance grid which dictates the percentage each DEM (ANUDEM and original DEM) contributes to the new elevation value. For example, if an elevation value is located at a distance that is 20% of the total buffer width distance, then 80% of the elevation value is taken from the ANUDEM output and 20% is used from the original DEM elevation value. Conversely, if an elevation value is located at a distance of 80% of the total buffer width distance, then 20% of the elevation value is taken from ANUDEM and 80% is taken from the original DEM value. This method minimizes the effect of ANUDEM on regions outside the river network, but corrects for inaccurate DEM values within the buffer regions where hydrography does not match the known blue lines.

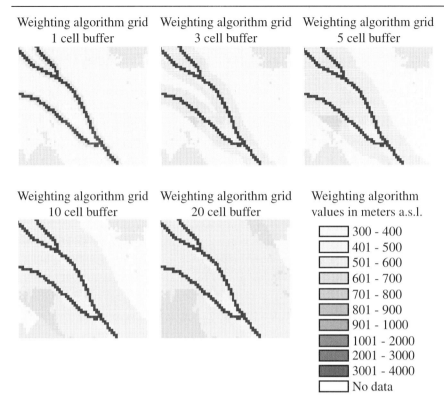

Fig. 6. Examples showing the impact on original DEM by the drainage enforcement weighting algorithm

3.3.2 Results

Following implementation of the new procedure for MRB, TOPAZ provided drainage networks that were consistent with the corresponding blue line networks in the National Atlas of Canada dataset. Table 3 indicates that the new procedure has minimal effects on DEM characteristics. The range of elevation was maintained and the mean elevation of the DEM was statistically unchanged using a 1, 3 and 5 cell buffer width. Using 10 and 20 cell buffers also yielded correct drainage networks, though effects on the DEM characteristics started to increase.

While results on MRB were promising, application of this method to another northern basin, the Snare River basin, produced disappointing results. The processing of the Snare River basin using ANUDEM initially did not cause significant deviations from the original DEM. This is due to the relative flatness of the DEM and the low relief of the area. The drain-

Table 3. Descriptive statistics for the Mackenzie River Basin

Descriptive statistics	MackDEM – 2 km Mackenzie Basin DEM	ANUDEM – Enforced drainage	DEM – Using 1 cell buffer width	DEM – Using 3 cell buffer width	DEM – Using 5 cell buffer width	DEM – Using 10 cell buffer width
Mean elevation [m]	1339	1302	1338	1327	1324	1307
Median elevation [m]	1330	1294	1329	1318	1314	1298
Standard deviation [m]	778	754	770	770	768	760
Kurtosis	-1.06	-1.10	-1.07	-1.08	-1.08	-1.06
Skewness	0.082	0.065	0.079	0.072	0.073	0.083
Range [m^3]	3338	3129	3338	3338	3338	3338
Minimum elevation [m]	11	11	11	11	11	11
Maximum elevation [m]	3349	3140	3349	3349	3349	3349

age network, however, is very complex and exhibits divergent flow as well as chaotic drainage patterns. The blue line data for this area defines many lakes with very short drainage connections. Using buffers of 1, 3, 5, 10 and 20 grid cells did not prove successful. With a high drainage density, even the buffers at 1 grid cell width were very close and in most cases overlapping, rendering the smoothing procedure non-functional. This type of drainage network marks an uppermost limit in terms of hydrography scale and density. TOPAZ also appears to reach its limit as the depression-filling and flat-area algorithms tend to fill the landscape and provide improper drainage networks.

3.4 Vector Addition

Vector addition is an automated method that determines flow routing between sub-grids by providing flow vector information. It involves laying flow vectors for each sub-grid head-to-tail (Thomas and Finney 1979), with the first vector starting on the Cartesian coordinate 0,0. Where the head of this vector ends the tail of the next vector begins. This process is repeated until all vectors have been added head-to-tail. The resultant is drawn from the tail of the first vector to the head of the last vector (Fig. 7a). Vectors can also be weighted to allow certain vectors to influence direction to a greater degree than those not weighted. Figure 7b illustrates how vectors with a greater weight (length) can affect the general direction of the summed vectors.

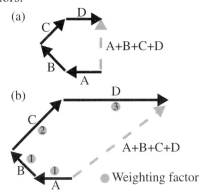

Fig. 7. Examples showing two approaches in vector addition: (a) head-to-tail without weighting, (b) weighted head-to-tail with vector C weighted by a factor of 2 and vector D weighted by a factor of 3

3.4.1 Methodology

In this study, GTOPO30 Global 30 arc second data (cell size of the raster is 30 arc seconds) are used to create the DEM of MRB. The DEM was processed by TOPAZ with a CSA and MSCL value of 409600 ha and 128000 m, respectively, to produce a drainage network in which most sub-grids contained a major channel. A sub-grid size of 230 km was chosen, as it is similar to the grid size of many atmospheric models.

The vector addition flow routing program was written using the Environmental Systems Research Institute (ESRI) software, ArcInfo and its development language Arc Macro Language (AML). The program aggregates detailed flow vector information and related hydrologic variables derived from a DEM using TOPAZ (Fig. 8).

The TOPAZ model outputs a number of raster data sets (Garbrecht and Martz 1999). Three TOPAZ output raster files provide sufficient information for the vector addition flow routing algorithm. The first file contains the drainage direction at each raster cell (local flow vectors) as determined by the D-8 method. The second defines the drainage network above the user-selected basin outlet. It is used in the vector addition algorithm to identify which cells in the sub-grid are channel cells. The final raster file contains the accumulated upstream area (uparea) draining into each cell. The accumulated upstream area is given in number of upstream cells. Cells at the drainage divide have a value of 0. All other cells have a value of

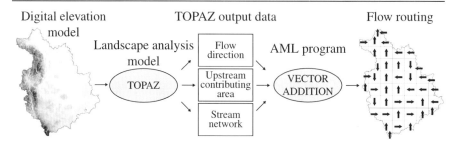

Fig. 8. Procedures in the application of the vector addition flow routing algorithm, using the Mackenzie River Basin as an example

greater than zero that range up to the value of the product of the number of rows and columns of the raster. Uparea values provide the weighting for use in the algorithm.

The vector addition flow routing algorithm determined sub-grid flow directions using four methods: (1) flow vectors in each sub-grid are summed; (2) flow vectors for each sub-grid are summed after each vector has been weighted with upstream area; (3) only flow vectors identified as channel cells are summed; and (4) only flow vectors identified as channel cells are summed after they have been weighted with upstream area.

3.4.2 Results

Ideally, the vector addition flow routing program should be able to assign drainage to all eight adjacent sub-grids. However, if sub-grids are routed using all eight directions, the program produced many instances where a sub-grid assigns flow to a diagonal sub-grid and has that sub-grid return flow back. Rather than manually or arbitrarily reassigning flow directions to resolve this problem, we simply constrained the algorithm to assign flow to the four cardinal directions. This limits the ability to represent channel details but provides hydraulically connected and consistent networks at a coarse resolution.

All Vectors (Non-weighted)

Of the four methods, the summing all flow vectors in the sub-grid yielded the worst representation of the drainage network, producing drainage directions that were in error in 15 of the 49 sub-grids in the MRB (Fig. 9a). Using all flow vectors in the sub-grid determines a drainage direction that merely reflects the general slope of the sub-grid rather than a drainage direction based on the channel network. The circled sub-grid on Fig. 9a illustrates one of the major limitations. This sub-grid has a general eastward decrease in elevation from the Rocky Mountains to the lowlands. Thus, most flow vectors are directed to the east and summing all vectors results in an eastward drainage direction for the sub-grid. However, the main drainage channel along the eastern edge of this sub-grid runs north-south. The correct drainage direction for this sub-grid should be north.

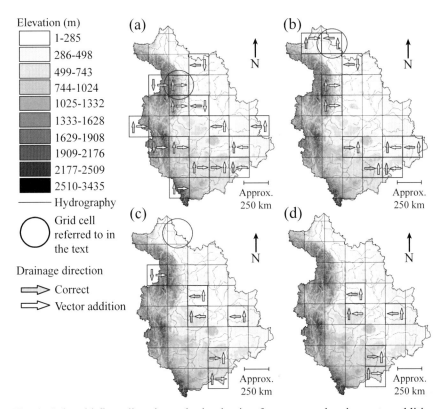

Fig. 9. Sub-grid flow directions obtained using four approaches in vector addition: (a) all vectors (non-weighted), (b) channel vectors (non-weighted), (c) all vectors (weighted), and (d) channel vectors (weighted)

Channel Vectors (Non-weighted)

This method that only examines flow vectors for cells identified as channel cells, improves flow routing but does not produce results that are entirely satisfactory. Flow directions were in error in 10 of 49 sub-grids. The circle in Fig. 9b identifies a sub-grid that is in error because the main channel veers sharply to the west near the outlet of the sub-grid.

All Vectors (Weighted)

Introducing the contributing area as a weighting factor further improves flow routing results, with only 6 of 49 sub-grids being in error (Fig. 9c). The reason for the improvement is that cells near the channels and within the channels have a very high contributing area value so that these cells are weighted more than those in the rest of the sub-grid. Also, cells that are closer to the outlet are assigned a greater weight because they are further downstream. This encourages flow directions being assigned correctly even to sub-grids with channels that veer sharply in a new direction as the channel approaches the sub-grid outlet. The circle in Fig. 9c identifies a sub-grid in error in Fig. 9b being corrected by weighting flow vectors near the sub-grid outlet.

Channel Vectors (Weighted)

Summing only the channel cells in the sub-grid and weighting them with the contributing area produces results that best reflect the channel network in the basin. Only 5 of the 49 sub-grids are in error (Fig. 9d). Erroneous sub-grids produced by this method can only be corrected manually.

In general, the vector addition method generates significantly better flow routing results than the traditional method which uses mean elevations to route flow. Using only those vectors that are identified as channel cells and weighted by the upstream area produce results that are very close to those determined by a quasi-expert system that uses a complex approach.

3.5 WATPAZ

The WATPAZ interface is an expert system that automates the derivation of flow routing for the semi-distributed WATFLOOD hydrologic model which is square-gridded and uses grouped response units (GRU) to subdivide a watershed (Soulis and Segleniek 2007). The interface utilizes digital physiographic data obtained through TOPAZ. It employs detailed physiographic data extracted from a DEM to automate derivation of routing parameters for the hydrologic model at a large scale, while preserving the channel network structure of the basin.

3.5.1 Methodology

WATPAZ up-scales data from DEM cells using the mosaic method, an approach that is much used in atmospheric modeling (Arora et al. 2001). Of particular relevance to this study is the method for determining flow routing parameters in WATFLOOD which takes into consideration the flow pattern variability within each sub-grid.

WATFLOOD models both the vertical and horizontal movement of water within the basin, but the present research is concerned only with the flow routing component. The manual method of obtaining routing parameters for WATFLOOD entails extracting data from a hard copy map. The basin is delineated on the base map and subdivided into segments at the level of detail required in the output, and the size of the meteorological data available for the basin. For each sub-grid the following six parameters are derived (Kouwen 2001), viz., drainage direction, elevation, drainage area, contour density, river classification, and channel density.

The WATPAZ interface consists of eight main programs. The first program obtains information from the user about the spatial nature of the data sets. The second program creates datasets required by the subsequent programs. Six additional programs produce the physiographic parameters used by WATFLOOD, following the algorithms developed for the manual method outlined by Kouwen (2001). Table 4 provides a summary of the output datasets derived from digital data, for use in WATPAZ.

The algorithms employed by TOPAZ for drainage pattern determination, and by WATPAZ for flow routing between segments do not allow ambiguous flow. For example, segments cannot route flow in a loop even within large, flat areas. A final manual continuity check of drainage directions derived by the WATPAZ automated method is recommended.

3.5.2 Results

Manipulating the MSCL and CSA values in TOPAZ completely controls the delineation of drainage in a basin. Flow routing results for MRB were evaluated by comparing the WATPAZ-generated flow routes with those determined using the traditional averaging method. The simple averaging method entails moving flow to an adjacent sub-grid with the lowest average elevation. Several depressions or "pits" were produced using this approach because of a number of sub-grids not having an adjacent cell with lower elevation. These depressions cause the flows to terminate and the hydrologic model will not run under these conditions. The averaging method also incorrectly assigned drainage direction in about half of the segments. The WATPAZ method as applied to MRB was able to resolve those segments identified as pits by the mean method, and assigned flow routing directions that reflect the true hydrography of the segments (Fig. 10). Furthermore, though not presented here, WATPAZ achieves improvements in assigning flow routine directions over a variety of basin scales (Shaw et al. 2005b).

Table 4. TOPAZ output raster data sets used in the WATPAZ program

TOPAZ data	Description	
Bound.out	The data set defines the boundary of the watershed above a user specified outlet point	
Inelev.out	The data set contains the elevation values of the input DEM	
Subwta.out	The data set identifies sub-catchments in the watershed after the channel network has been defined	
Netw.out	The data set defines the drainage network within the watershed boundaries above the user specified watershed outlet	
Uparea.out	The data set contains the accumulated upstream area draining into each cell	

4 Conclusions

Five methods for up-scaling flow information are presented. SLURPAZ combines the output of a terrain analysis model, TOPAZ, with a semi-distributed hydrologic model, SLURP. In the application of SLURPAZ to streamflow simulation, reducing the number of sub-units or ASAs results in a decline in model performance. This can be rectified through optimization but only to some minimum number of ASAs at which level water balance components can no longer be calculated correctly. This lower limit is possibly a function of topographic complexity.

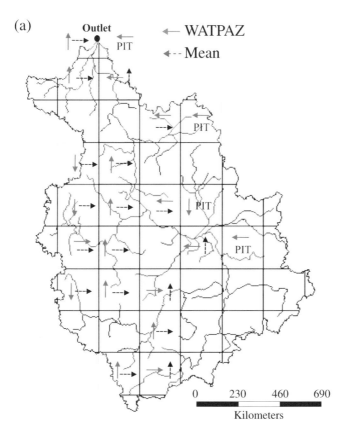

Fig. 10. Sub-grids of the Mackenzie River Basin that (a) possess conflicting drainage directions, and (b) traversed by major rivers and possess conflicting drainage directions

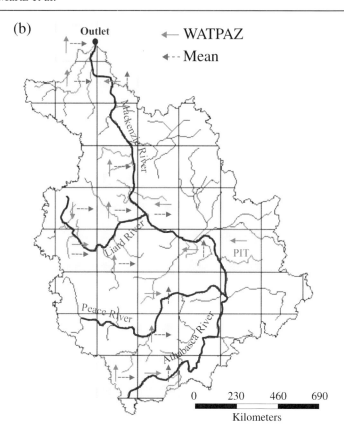

Fig. 10. (cont.)

DEM aggregation results in a rapid reduction in the information content of the digital model and can produce serious error in the delineation of drainage direction. These errors can be overcome to some degree by drainage enforcement using such program as ANUDEM. However, the quality of the physiographic information on the network and sub-basin suffers degradation in the averaging process. Also, some undefined upper limit of drainage network complexity may be related to the capacity of ANUDEM to process very intricate terrain.

Vector addition is a more effective aggregation technique though this method is limited to four direction flow. In preserving sub-grid topographic complexity, it is possible to extract meaningful physiographic properties for each spatial modeling unit. With this method, however, some user intervention is required to ensure that the derived sub-grid drainage fully reflects the "real-world" drainage.

The WATPAZ method, based on the hydrologic parameterization scheme of WATFLOOD, obtains flow patterns that match those that would be determined manually by an expert. It employs detailed physiographic data extracted from a DEM to automate the derivation of routing parameters for hydrologic models at the GCM grid scale, while preserving the hydraulic structure of the basin. This method is superior to the vector addition method in that no user intervention is needed, but the WATPAZ requires a complicated algorithm and high computational overhead.

References

Armstrong RN, Martz LW (2003) Topographic parameterization in continental hydrology: a study in scale. Hydrol Process 17:3763–3781

Arora VK, Chiew FHS, Grayson RB (2001) Effect of sub-grid variability of soil moisture and precipitation intensity on surface runoff and streamflow. J Geophys Res 106(D15):14347–14357

Fairfield J, Leymarie P (1991) Drainage networks from grid digital elevation models. Water Resour Res 27:29–61

Fekete BM, Vörösmarty CJ, Lammers RB (2001) Scaling gridded river networks for macroscale hydrology: development, analysis, and control of error. Water Resour Res 37:1955–1967

Florisnky IV (1998) Combined analysis of digital terrain models and remotely sensed data in landscape investigations. Prog Phys Geog 22:33–60

Garbrecht J, Martz LW (1999) TOPAZ: an automated digital landscape analysis tool for topographic evaluation, drainage identification, watershed segmentation and subcatchment parameterization; TOPAZ overview. US Department of Agriculture, Agriculture and Research Service, Grazinglands Research Laboratory, USDA Agricultural Research Service, El Reno, Oklahoma

Geomatics Canada (1996) Standards and specifications of the national topographic data base. Minister of Supply and Services Canada. Catalogue No. M52-70/1996E

Hutchinson MF (1989) A new procedure for gridding elevation and stream line data with automatic removal of spurious pits. J Hydrol 106:211–232

Jelinski DE, Wu J (1996) The modifiable areal unit problem and implications for landscape ecology. Landscape Ecol 11:129–140

Kite GW (1997) Manual for the SLURP hydrological model. NHRI, Saskatoon, Saskatachewan

Kouwen N (2001) WATFLOOD/SPL8 flood forecasting system. University of Waterloo, Waterloo, Ontario

Lacroix M, Martz LW, Kite GW, Garbrecht J (2002) Using digital terrain analysis modeling techniques for the parameterization of a hydrological model. Environ Modell Softw 17:127–136

Martz LW, deJong E (1988) CATCH: A FORTRAN program for measuring catchment area from digital elevation models. Comput Geosci 14:627–640

Martz LW, Garbrecht J (1998) The treatment of flat areas and depressions in automated drainage analysis of raster digital elevation models. Hydrol Process 12:843–855

Martz LW, Garbrecht J (1992) Numerical definition of drainage network and sub-catchment areas from digital elevation models. Comput Geosci 18:747–761

O'Callaghan JF, Mark DM (1984) The extraction of drainage networks from digital elevation data. Comput Vision Graph 4:375–387

O'Donnell G, Nijssen B, Lettenmaier DP (1999) A simple algorithm for generating streamflow networks for grid-based, macroscale hydrological models. Hydrol Process 13:1269–1275

Olivera F, Lear MS, Famiglietti JS, Asante K (2002) Extracting low-resolution river networks from high-resolution digital elevation models. Water Resour Res 38:1231,doi:10.1029/2001WR000726

Pietroniro A, Soulis ED (2003) A hydrology modelling framework for the Mackenzie GEWEX programme. Hydrol Process 17:673–676

Pomeroy JW, Granger RJ (1999) Wolf Creek Research Basin – hydrology, ecology, environment. National Water Research Institute, Saskatoon, Saskatchewan

Shaw D, Martz LW, Pietroniro A (2005a) Flow routing in large-scale models using vector addition. J Hydrol 307:38–47

Shaw D, Martz LW, Pietroniro A (2005b) A methodology for preserving channel flow networks and connectivity patterns in large-scale distributed hydrological models. Hydrol Process 19:149–168

Soulis ED, Segleniek FR (2007) The MAGS integrated modeling system (Vol. II, this book)

Thomas G, Finney R (1979) Calculus and analytic geometry. Addison-Wesley, Reading, Massachusetts

Thorne R, Armstrong RN, Woo MK, Martz LW (2007) Lessons from macroscale hydrological modeling: experience with the hydrological model SLURP in the Mackenzie Basin. (Vol. II, this book)

Veregin H (1997) The effects of vertical errors in digital elevation models on the determination of flow-path directions. Cartogr Geogr Infom 24:67–79

Wolock DM, McCabe GJ (2000) Differences in topographic characteristics computed from 100-m, 1000-m resolution digital elevation model data. Hydrol Process 14:987–1002

Zhang X, Drake NA, Wainwright J, Mulligan M (1999) Comparison of slope estimates from low resolution DEMs: scaling issues and fractal method for their solution. Earth Surf Proc Land 24:763–769

Chapter 21

Lessons from Macroscale Hydrologic Modeling: Experience with the Hydrologic Model SLURP in the Mackenzie Basin

Robin Thorne, Robert N. Armstrong, Ming-ko Woo and Lawrence W. Martz

Abstract Macroscale models are used increasingly in hydrology to simulate regional responses to external forcing, to evaluate large basin management strategies, and to extend hydrologic data sets. The hydrologic model SLURP (Semi-distributed Land Use-based Runoff Processes) is a semi-distributed model that has been successfully applied to basins of various sizes, notably those in cold regions. The SLURP manual provides explanations of computational algorithms, sets of commonly applicable parameter values, and computational steps required to run the model. Although the manual offers much information, users can benefit from additional information on certain procedures in order to operate the model successfully. In this chapter we share our experiences in operating this model, including the preparation of input data, initialization of variables, optimization of parameters, and validation of model results. We suggest that the lessons learned from the use of SLURP can be applied to other macroscale hydrologic models.

1 Introduction

The performance of macro-scale hydrologic models is influenced by how a basin is divided into sub-units, the scale of investigation, process representation in the model, parameterization, and input data considerations. Although the SLURP (Semi-distributed Land Use-based Runoff Processes) manual offers substantive information on algorithms, parameters, and initialization procedures for the model, experience and knowledge gained from past studies can help users operate the model successfully and effectively. In this paper we share our experiences in operating the SLURP model for a variety of studies conducted as part of the Mackenzie GEWEX (Global Energy and Water Experiment) Study (MAGS).

2 Model Structure

The SLURP model is well tested and has been applied successfully in cold region basins of various sizes in both alpine and lowland environments (Kite et al. 1994). Simulation by SLURP is based on (1) a vertical component consisting of surface water balance and flow generation at daily time intervals, and (2) a horizontal component that includes flow delivery within each sub-unit and channel routing to the basin outlet. Mean elevation, area, and areal percentages occupied by various land cover types are obtained using digital elevation data combined with a land cover map input by the user.

The model divides a large catchment into aggregated simulation areas (ASAs) each encompassing a number of land cover types characterized by a set of parameters on the basis of topography and land cover (Kite 2002). The difference between SLURP and many other hydrologic models which subdivide a basin area into regular grid squares, is that SLURP makes use of topographic divides as the boundaries of individual hydrologic units. This offers a more physical basis than arbitrary grids (Woo 2004) as the individual hydrologic units are based on the stream network derived from the topography. According to Kite (2002), models that reduce a basin to a series of grid squares are attempting to simulate an artificial environment rather than a natural one as the hydraulic routing in grid square models can reduce river basins to a series of plates and waterfalls. Two advantages of using topographically-based sub-units are (1) that the size of the sub-unit can vary more than grid squares, and (2) that lakes and reservoirs can be more easily simulated (Kite 1998).

The basic requirements for an ASA are that the distribution of land cover and elevation within the ASA are known and that the ASA contributes runoff to a definable stream channel (Kite 2002). This information can be obtained by using an automated delineation of the channel network from digital elevation data with the TOPAZ (TOpograhic PArameteriZation) program. TOPAZ processes a raster Digital Elevation Model (DEM) to derive a wide range of topographic and geometric variables that are physically meaningful to watershed runoff processes (Lacroix et al. 2002). A special link between SLURP and TOPAZ has been developed, named SLURPAZ, to enable their seamless integration (Martz et al. 2007).

2.1 Topographic Parameterization and Scale Effects

The automated extraction of topographic information (e.g., basin and drainage network delineation) from a digital elevation model (DEM) for parameterizing hydrologic models has become a common practice. The overall process depends on correctly defining flow directions across the landscape. As a result, the scale of the DEM used to represent the landscape becomes important. Several studies have been conducted regarding the problems and potential solutions to correctly defining flow directions under re-scaling, from local to continental scales (Martz et al. 2007). Common to each of these studies is the use of TOPAZ for the automated extraction of physiographic parameters from a DEM.

Depending on the scale of the DEM used, however, the resulting network and other DEM characteristics can be substantially different at successive grid scales. In applying TOPAZ to the Mackenzie and the Mississippi basins, Armstrong and Martz (2003) found that using a simple averaging approach to upscale a DEM from 1 km to 2, 4, 8, up to 64 km can result in severe flow direction errors (due to elevation smoothing during aggregation), potentially producing dramatic changes in basin and network definition. This renders such an approach to rescaling flow directions unreliable for modeling at larger scales (Fig. 1). An alternative approach used by Laing (2004) to preserve drainage patterns at larger scales employs a method of drainage enforcement to correct for elevation errors that produce incorrect flow directions. In this method, an existing stream network is imposed on the DEM to ensure proper flow patterns, while maintaining the integrity of the elevation data outside a buffered area along the stream channels. This new procedure was successful in producing corrected drainage patterns comparable to that of the 'blueline network', and also resulted in minimal impacts on DEM characteristics for network buffer sizes of 1, 3 and 5 grid cells.

A general problem with the above approaches is that they fail to fully address the sub-unit scale variability needed for flow routing in large-scale models. Recently, an automated method of aggregating flow directions using a quasi-expert system has been introduced by Shaw et al. (2005). The new method is implemented through an interface (WATPAZ) between TOPAZ and the hydrologic model WATFLOOD. The basic principle of the method is to use the hydraulic structure of the basin to preserve the drainage network at larger scales (e.g., GCM grid scales), thus providing more accurately defined flow directions for runoff routing.

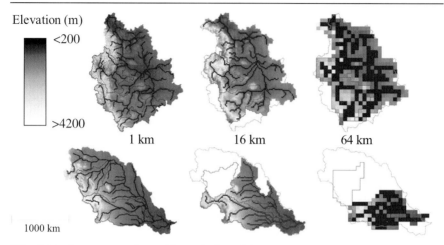

Fig. 1. Drainage networks and boundaries of the (top) Mackenzie and (bottom) the Missouri River Basins derived from DEMs at original resolutions of 1 km and after aggregation to 16 km and 64 km

2.2 Determining the Number of Sub-units

The number of ASAs used in modeling a basin will depend on the basin area and the scales of data available (Kite 2002). The SLURP model can use an unlimited number of ASAs, but to ensure stability the number of ASAs has to equal or exceed the number of land cover classes. Lacroix et al. (2002) examined the impact of changing the number of ASAs used in a SLURP modeling application in Wolf Creek Basin, Yukon. They demonstrated that a minimum number of ASAs is required to correctly model the volume of runoff generated by a basin (Fig. 2). The minimum number is a function of the spatial variability in temperature and precipitation over the basin which is, in turn, largely a function of topographic relief. They also showed that the timing of runoff will improve as the number of ASAs increases up to a maximum beyond which no further improvement is obtained (Fig. 2). This upper limit is ill-defined but appears to be related to runoff travel times. The suggested criteria in determining the appropriate number of ASAs includes: (1) the optimal number should enhance the simulation of streamflow, and (2) it should be the smallest plausible number of ASAs to keep the model runs efficient.

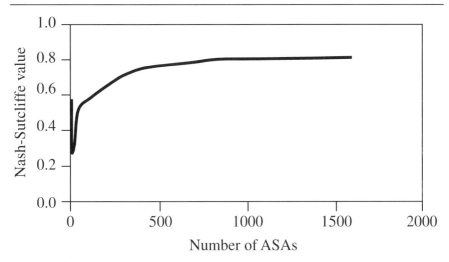

Fig. 2. Variations of the Nash-Sutcliffe statistic with an increase in number of ASAs in Wolf Creek Basin

An experiment was conducted by Thorne (2004) to determine how the Liard Basin, a 250 000 km^2 tributary of the Mackenzie system, can be partitioned into sufficiently distinctive sub-basins. Three criteria were used to select the most suitable number of ASAs, including (1) the best Nash-Sutcliffe (Nash and Sutcliffe 1970) value for the calibrated and (2) validated time periods, and (3) a suitable number of ASAs to reduce the computational time. Results indicated that the Nash-Sutcliffe values increased as the number of ASAs increased. To meet all three criteria, the optimal number of ASAs for the Liard Basin was between 25 and 50.

2.3 Land Cover Representation

Land cover data, commonly derived from satellite data, are used in SLURP as an indicator of vegetation type, soil characteristics, and physiography (Kite 2002). An experiment by Thorne (2004) tried to reduce the number of land classes without significantly jeopardizing the representation of basin land use. The study found that decreasing the number of land classes will increase the computer processing speed as less time is required to gather parameter information for each land cover. Using the criteria of obtaining the best Nash-Sutcliffe statistic for both the calibrated and validated time periods, it was determined that five land cover types (i.e., deciduous, evergreen and mixed forests, water, and tundra) were sufficient for the Liard Basin.

A recent study (Armstrong and Martz, 2007) examined the effects of scale (level of detail) at which land cover is represented on modeling the response of Wolf Creek watershed using SLURP. For this study, the level of detail of a grid-based 30 m land cover map was reduced via several techniques including pixel thinning, smoothing, modal aggregation, and a majority rule method based on hydrologic zone polygons. To provide and maintain a sense of realism throughout the modeling process, the SLURP model was initially calibrated (at Wolf Creek outlet) using literature information, previous research, field data, and remote sensing information, and then manually optimized. This process resulted in satisfactory timing and magnitude of streamflow (peak and mean discharge) compared to observed discharge, but a lower Nash-Sutcliffe statistic (0.66) than is usually preferred.

The initial model calibration run was only used to provide a base reference for comparing further runs using the land cover maps with reduced levels of detail. As shown by the results of the study, in the case of modeling Wolf Creek, reducing the level of detail of land cover data had only a minimal effect on streamflow and model fit statistics at the outlet. This was not the case, however, for the distributed sub-basins (19 in total), which showed local variations in streamflow response to be quite pronounced. This indicates that reducing the level of detail of land cover has negligible effect on the discharge at the outlet for which the model is calibrated, but a more pronounced impact on individual sub-basins.

2.4 Hydrologic Processes

Several weaknesses exist in the representation of hydrologic processes in SLURP. The snowmelt rate, calculated by a degree-day method, can vary within a year, but cannot vary between years. The fast and slow storages within the model simulate the hydrologic processes in the soil layers, yet the model cannot simulate the presence of frozen soil in the winter and spring. Limited infiltration into the soil layers during these two seasons cannot be modeled properly. Glaciers are present in many cold areas but glacier melt is not considered. Finally, a process often observed during the spring and fall seasons in the subarctic is the formation and decay of river ice (Beltaos and Prowse 2001, Hicks and Beltaos 2007) which can seriously affect the timing and magnitude of discharge, but cannot be explicitly simulated by SLURP.

The selection of evapotranspiration method in the model is restricted for the data sparse regions. In these regions the Spittlehouse and Black evapotranspiration method (Spittlehouse 1989), a modified form of the Priestley and Taylor method, is often used due to lack of radiation data that is required for other methods such as the Penman-Monteith (Verhoef and Feddes 1991).

A new function in version 12.2 of the SLURP model is the handling of a "water" land cover type, as either a lake or reservoir, or as a small unregulated lake or even a collection of small water bodies. For the latter, the vertical water balance is carried out as for any other land cover except that some of the parameters are changed to eliminate the functions of the canopy and fast storage, and to ensure that evapotranspiration from the slow storage is assigned to evaporation and not to transpiration (Kite 2002).

A study by van der Linden and Woo (2003a) compared several models of increasing complexity to investigate the role of major hydrologic processes in streamflow simulation. Results indicated that with finer temporal and spatial scale, process representation needs to be more complex. However, it was not always necessary to switch directly from a simple model to a complex one for several reasons: (1) upgrading only a limited number of critical processes in a simple model may be adequate to simulate satisfactory flow features of interest; (2) not all process representations found in a complex model need to be present in a simple model as runoff may be sensitive to only a limited number of processes; (3) a compromise must be considered between the model demand and the availability of reliable input data, particularly for remote areas like the subarctic.

3 Data Ingestion

For cold regions, one of the major limitations is the lack of both long term climate and discharge data with adequate spatial coverage. The use of gridded reanalysis data from climate models or data interpolated from ground observations can offer a solution. For data ingestion into SLURP, the model calculates weighted average climatic inputs for each ASA from any number of climate stations or from gridded data using a weighted Thiessen polygon technique. If climate station data are used, the temperature field is adjusted to account for differences in elevation between the climate stations and the average elevation of the ASA using a specified lapse rate. Precipitation is also adjusted for elevation changes using a specified rate of increase of precipitation with elevation for each ASA (Kite 2002).

4 Variables

To successfully run the SLURP model, several variables are required, such as temperature, precipitation, and runoff. Temperature is needed for the calculation of snowmelt using the degree-day method, and for evapotranspiration computation using the Spittlehouse and Black method, which has been found to yield reasonable results (Barr et al. 1997). The dew point temperature can be approximated by the minimum temperature of the previous day. To estimate net radiation, the ratio of sunshine hours is derived from precipitation data (Kite 2002) using the following assumptions: (1) with zero precipitation, the sky is assumed to be clear; (2) with daily precipitation exceeding 25 mm, the sunshine hours for that day is considered to be zero; (3) for precipitation between 0 and 25 mm, the hours of sunshine are computed proportionately.

Although the SLURP model offers the option of using temperature or a combination of temperature and radiation to calculate snowmelt, Pietroniro et al. (1997) found that the use of combined temperature–radiation data for flow simulation only marginally improves the performance of using temperature-index alone for melt calculation. The use of the robust degree-day method (Hock 2003) instead of an energy balance approach is necessitated by the coarse resolution of the input data and immensity of the catchment that precludes accurate spatial description of radiation and wind fields.

5 Parameters

SLURP requires parameter values for each of the land cover types. Following the classification of the input parameters by van der Linden and Woo (2003b), three types of parameters are generally distinguished. The first group is based on values used by SLURP from past simulations, such as lapse rates, evapotranspiration variables and snowmelt factors. These parameter values are usually set constant since their hydrological sensitivity is medium to low (Kite 2002). The second group has eight parameters that can be obtained by optimization: initial contents of snow and slow storage, maximum infiltration rate, Manning's roughness, retention constant for fast and slow storage, and maximum capacity for the fast and slow storage. All of these parameters are allowed to vary within a specified range to enable some physical credibility to be retained (Kuchment et al. 2000). The third group consists of parameters that can be kept at their default values supplied by the model, including parameters for the canopy properties, leaf area index, porosity, precipitation factor and the rain/snow division temperature.

Van der Linden and Woo (2003b) applied SLURP to the mountainous Liard Basin and its sub-catchment in subarctic Canada, to examine the transferability of parameters for the overall basin, to simulate flows for its sub-basins. Several conclusions were reached: (1) parameters obtained for two sub-basins are similar to each other but different from values generated for the outlet discharge; (2) when parameter values derived for the sub-basins are applied to the Liard catchment, there is improved agreement between the observed and simulated runoff, while an opposite result is observed for the reverse; (3) caution must be exercised when transferring parameters as they largely depend on the climate, topography, land cover type, and compatibility of scale of a basin.

The initial amount of snow and slow storage may be optimized by SLURP, depending on the time of year. Starting the model in the fall season rather than at the beginning of a calendar year has the advantage that there is no snowpack and the soil moisture is low. Then, the initial snow content and the slow storage may be set to close to or at zero and do not require optimization (Kite 2002).

During the snowmelt period, many subarctic catchments have frozen soil conditions, but SLURP does not have a frozen soil routine. To mimic this imperviousness effect of frozen ground in promoting surface runoff, the maximum infiltration capacity can be set to some low value. For example, this value was set to 10 mm/day for the Liard Basin in a study by Woo and Thorne (2006). The presence of river ice in the breakup season influences hydrograph rise along local segments of the channel (Blackburn and Hicks 2002) but macrohydrologic models such as SLURP do not possess the capability to handle the hydraulics of ice breakup.

Thorne (2004) estimated the maximum range for the retention constant and the maximum capacity of the fast and slow stores to address winter flow problems encountered during a preliminary simulation (Fig. 3a). Applying the maximum ranges used by Kite (2002) for the simulation of Kootenay River in British Columbia produced an overestimation of low flow for the Liard Basin, which accumulated over time and attained magnitudes greater than the annual peak flows. This was rectified by re-estimating the maximum range of these storage parameters (Fig. 3b).

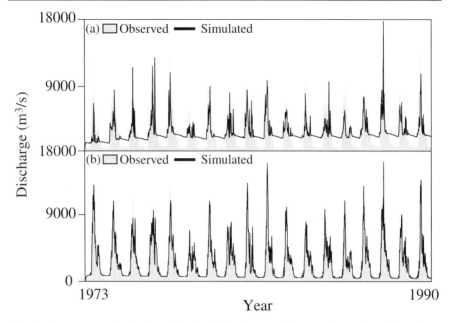

Fig. 3. Long term simulation (1973-1990) of streamflow at the outlet of the Liard Basin. By re-estimating the maximum retention constant and maximum capacity of the slow storage, the accumulation of low flow (a) was reduced to acceptable values (b)

6 Initialization

A spin-up period, an arbitrary number of years prior to the desired simulation time, is often used when simulating discharge to adjust initial parameters to allow the model to reach an equilibrium. Further work is needed to determine the number of years of spin-up needed to achieve satisfactory simulation results.

7 Optimization

In most studies using the SLURP model, manual calibration was performed on the second group of eight parameters mentioned in Sect. 5, to set the limits within which those parameters can vary so as to ensure some physical credibility. The Shuffled Complex Evolution (SCE-UA) Method

(Duan et al. 1994) incorporated into SLURP is used to optimize the parameter values. Thorne (2004) examined the optimization of the parameters under several options. First, the calibration period and the number of iterations were selected. Four calibration runs were executed to insure an adequate calibration of the parameters. The first run included all the parameters except the Manning's roughness which Kite (2002) recommended not be included in the first optimization step. The second calibration run took the parameter values obtained from the first run, calibrated the parameters again, and optimized the Manning's roughness. The third calibration run re-calibrated the parameters along with the routing parameters. The fourth and final run re-calibrated all the parameters. These four calibration runs gave improved results with each optimization. The snowmelt rate and the time of travel for the Muskingum routing are the only factors that are not optimized by the model, but are set manually using the Nash-Sutcliffe statistic as a "goodness of fit" indicator.

For streamflow simulation, results can be evaluated using the Nash-Sutcliffe efficiency measure, the Garrick efficiency measure (Garrick et al. 1978), the Previous-Day criterion (Kite 1991), and the root-mean square error (RMSE). The first three statistics describe how well the daily and seasonal variations in streamflow are simulated (Kite 2002).

8 Validation

When validating simulation results, it is best to use observed discharge record not used in the calibration period. Satellite information, such as Moderate-Resolution Imaging Spectroradiometer (MODIS) of relatively high resolution (5 km) can be used to map a basin snow cover over the snowmelt season. Snow cover extent during the melt season obtained from MODIS can be compared with the areal snow depletion pattern depicted by the hydrologic model. One constraint is model resolution. The SLURP model uses an ASA as its basic simulation unit (where size can greatly vary) and for any particular day, an ASA is considered entirely snow-covered or completely snow-free. Despite this limitation, a comparison of satellite and model results offers a coarse spatial assessment of how well the snow-cover depletion is simulated (Woo and Thorne 2006). These comparisons are useful in indicating the efficacy of snowmelt simulation.

Three sets of climatic input data from weather station observations and reanalysis products were compared for use in the simulation of streamflow for the Liard Basin in subarctic Canada (Thorne and Woo 2006). These data sets are statistically different or show biases for most months. Yet, when they were used in conjunction with specific suites of parameters optimized for the individual data sets, SLURP was able to simulate flows that compare satisfactorily with measured discharge of the Liard catchment. The progressive downstream change in simulated discharge was scrutinized to reveal how and why, despite using inputs that are different, the model can simulate comparable basin outflows. When comparing simulated with measured discharge at several gauging sites within the Liard Basin, it becomes apparent that some ASAs overestimate while others underestimate the measured flow. As one approaches the basin outlet, there are more ASA discharges available to average out these errors and this process usually improves the performance of the simulation for the entire basin. Two interesting conclusions were drawn. (1) The study demonstrated that the pattern of streamflow contribution from within the basin is not reliable even when there is a good overall fit between observed and simulated streamflow at the basin mouth. (2) The optimization capability of a macro-scale hydrologic model can compensate for the differences amongst several climatic input data sets, and this testifies to the flexibility of the model.

9 Conclusions

The SLURP model, like all hydrologic models, performs a vertical water balance over the segmented areas and then links runoff laterally and routes the discharge downstream. The studies discussed here have tested the model vigorously and generated several improvements. We have shown and examined many aspects of the SLURP model such as the scale effects on topographic parameterization; its weaknesses in representing some hydrologic processes (such as routing and snowmelt rates in the model structure); effects of land cover detail on streamflow simulation; dealing with the lack of observed data in cold and mountainous regions; the estimation and approximation of variables; the calibration of parameters with realistic limitations; unique optimization procedures and credible ways of validating model results. Overall, the SLURP model is suitable for modeling the hydrology of cold and/or mountainous regions. By sharing our experiences gained through the use of SLURP we hope other users can benefit. We also suggest that the lessons learned from the use of SLURP can benefit users of other macro-scale hydrologic models as many such models share similar strengths and limitations.

Acknowledgements

We thank Dr. Geoff Kite for the use of the SLURP model.

References

Armstrong RN, Martz LW (2003) Topographic parameterization in continental hydrology: a study in scale. Hydrol Process 17:3763–3781

Armstrong RN, Martz LW (2007) Effects of reduced land cover detail on hydrological model response. Hydrol Process (in press)

Barr AG, Kite GW, Granger R, Smith C (1997) Evaluating three evapotranspiration methods in SLURP macroscale hydrological model. Hydrol Process 11:1685–1705

Beltaos S, Prowse TD (2001) Climate impacts on extreme ice-jam events in Canadian rivers. Hydrol Sci J 46:157–181

Blackburn J, Hicks F (2002) Combined flood routing and flood level forecasting. Can J Civil Eng 29:64–75

Duan Q, Sorooshian SS, Gupta VK (1994) Optimal use of the SCE-UA global optimization method for calibrating watershed models. J Hydrol 158:265–284

Garrick M, Cunnane C, Nash JE (1978) A criterion for efficiency of rainfall-runoff models. J Hydrol 36:375–381

Hicks F, Beltaos S (2007) River ice. (Vol. II, this book)

Hock R (2003) Temperature index melt modelling in mountain areas. J Hydrol 282:104-115

Kite GW (1991) A watershed model using satellite data applied to a mountain basin in Canada. J Hydrol 128:157–169

Kite GW (1998) Land surface parameterizations of GCMs and macroscale hydrological models. J Am Water Resour As 34(6):1247–1254

Kite GW (2002) Manual for the SLURP hydrological model, v12.2. International water management institute, Colombo, Sri Lanka

Kite GW, Dalton A, Dion K (1994) Simulation of streamflow in a macroscale watershed using general circulation model data. Water Resour Res 30:1547–1559

Kuchment LS, Gelfan AN, Demidov VN (2000) A distributed model of runoff generation in the permafrost regions. J Hydrol 240:1–22

Lacroix MP, Martz LW, Kite GW, Garbrecht J (2002) Using digital terrain analysis modeling techniques for the parameterization of a hydrologic model. Environ Modell Softw 17:127–136

Laing BR (2004) Integrating drainage enforcement into existing raster digital elevation models. M.Sc. thesis, University of Saskatchewan, Saskatoon

Martz LW, Pietroniro A, Shaw DA, Armstrong RN, Laing B, Lacroix M (2007) Re-scaling river flow direction data from local to continental scales. (Vol. II, this book)

Nash JE, Sutcliffe JV (1970) River flow forecasting through conceptual models. J Hydrol 10:282–290

Pietroniro A, Hamlin L, Prowse T, Soulis E, Kouwen N (1997) Application of a radiation-temperature index snowmelt model to the lower Liard River valley. Proc 2nd Sci workshop for Mackenzie GEWEX Study (MAGS), March 23–26, 1997, Saskatoon, pp 89–90

Shaw DA, Martz LW, Pietroniro A (2005). Flow routing in large-scale models using vector addition. J Hydrol 307:38–47

Spittlehouse DL (1989) Estimating evapotranspiration from land surfaces in British Columbia. In: Estimation of areal evapotranspiration, IAHS Publ no 177, pp 245–253

Thorne R (2004) Simulating streamflow of a large mountainous catchment using different sets of climatic data. M.Sc. thesis, McMaster University, Hamilton

Thorne R, Woo MK (2006) Efficacy of a hydrologic model in simulating discharge from a large mountainous catchment. J Hydrol 330:301–312

van der Linden S, Woo MK (2003a) Application of hydrological models with increasing complexity to subarctic catchments. J Hydrology 270:145–157

van der Linden S, Woo MK (2003b) Transferability of hydrological model parameters between basins in data-sparse areas, subarctic Canada. J Hydrol 270:182–194

Verhoef A, Feddes RA (1991) Preliminary review of revised FAO radiation and temperature methods. Report 16, Landbouwuniversiteit Wageningen, Wageningen, The Netherlands

Woo MK (2004) Boundaries and border considerations in hydrology. Hydrol Process 18:1185–1194

Woo MK, Thorne R (2006) Snowmelt contribution to discharge from a large mountainous catchment in subarctic Canada. Hydrol Process 20:2129–2139

Chapter 22

Development of a Hydrologic Scheme for Use in Land Surface Models and its Application to Climate Change in the Athabasca River Basin

Ernst Kerkhoven and Thian Yew Gan

Abstract The land surface model ISBA was applied to the Athabasca River Basin. Stand-alone runs were conducted to reconstruct the historic streamflows in the basin. The GEM and ERA-40 archives supplied the meteorological forcings and Ecoclimap provided all the required land surface data. It was found that the way ISBA (hereafter, OISBA) treated the generation of runoff resulted in unrealistic hydrographs. OISBA's hydrological scheme was modified by applying the Xinanjiang distribution to represent variation in soil water retention to formulate new, highly non-linear equations for surface and sub-surface runoff. These modifications significantly improved OISBA's ability to simulate the generation of runoff in the Athabasca River Basin.

This modified version of ISBA (MISBA) was then used to predict changes to the hydrology of the Athabasca River Basin under conditions predicted by a number of SRES climate scenarios for the 21st century. Although most of the scenarios predict increased precipitation in the basin, all the scenarios resulted in significantly decreased streamflows by the end of the century (2070–99), primarily because of a predicted decrease in the size of the winter snowpack due to warmer winters.

1 Introduction

Land surface processes were first included in Global Climate Models (GCMs) in the 1960s, and over the past four decades the development and application of ever more sophisticated Land Surface Models (LSMs) has shown that these water, energy, and carbon exchanges are tightly coupled (Pitman 2003). LSMs are usually based on one-dimensional physics meant for point applications, but are applied to scales on the order of 100 km^2 to 10 000 km^2. Since small-scale variations cannot be averaged out at larger scales, heterogeneity plagues LSM applications as it does almost all numerical modeling. When a LSM is applied to coarse grid cells, therefore, sub-grid heterogeneity should be accounted for. Most studies that had been conducted on the effects of sub-grid parameter variability were based in mid-latitude croplands or grasslands (Noilhan and Lacarrère 1995), usually under summer conditions, and over scales ranging from an 11.7 km^2 watershed (Famiglietti and Wood 1995) to a GCM grid scale of up to 100 000 km^2 (Ghan et al. 1997).

Parameters that tend to have significant heterogeneity are hydraulic conductivity, soil moisture, precipitation, vegetative cover, snow cover, and topography. A simple approach is to assign a single number to each parameter to represent the bulk value in the grid area. Noilhan and Lacarrère (1995) found that averaging the surface parameters produced better results than prescribing surface properties associated with a dominant land use. There are also other ways to account for sub-grid variability. If adequate data is available, a more realistic approach is to divide grids into sub-grids, each with its own set of parameters, or partition a grid cell into tiles, with each tile having distinct land use and physics (Koster and Suarez 1992). Effectively, this means that several parallel simulations are conducted and the resulting fluxes are combined using an areally-weighted average. Alternatively, sub-grid parameter variation can be described statistically (Entekhabi and Eagleson 1989; Sivapalan and Woods 1995).

The earliest LSMs used variations of the simple bucket scheme of Manabe (1969) in which the soil column is treated as a fixed size bucket that produces runoff whenever it fills. More recently, many LSMs use a variable bucket approach based on the Xinanjiang hydrologic model defined by a shape parameter (Yarnal et al. 2001) to represent sub-grid variability of soil moisture and its effect on surface runoff generation. This approach also only considers Dunne runoff mechanisms in which surface runoff occurs when the soil becomes saturated.

Our objective is to simulate the hydrology of the Athabasca River Basin (ARB) below Fort McMurray by coupling an atmospheric model with the land surface model, Interactions between the Soil-Biosphere-Atmosphere (OISBA) developed by Noilhan and Planton (1989) and modified by Kerkhoven and Gan (2006) (MISBA). OISBA/MISBA will be tested in stand-alone mode and then linked with hydrologic routing model to simulate the total streamflow at the basin outlet. The scheme will be applied with climate change scenarios from seven GCMs to simulate flow responses of the Athabasca River.

2 Study Area

The Athabasca River Basin (ARB) is of key interest mainly because of its multi-billion dollar oil sands industry at Fort McMurray. According to Water Survey of Canada the basin area is 133 000 km^2, and its main channel length is about 1154 km (Kellerhals et al. 1972). ARB has a continental climate with daily mean temperature dropping below freezing between mid-October and early April. Typical January temperature is -20°C while July is 17°C. June to October are the wet months, with an average total precipitation of about 300 mm, while winter and spring only receive about 150 mm in an average year. Coniferous, mixed wood and deciduous forests are the dominant vegetation especially in the upland areas (elevation ranging from 350 to 850m) and willow brush, shrubs, black spruce and sphagnum moss dominate the lowland areas which are often poorly drained. Dominant surficial soils are glacial soils (silt, clay and sands), glaciolacustrine soils (clay loam to heavy clay) and glaciofluvial soils (sandy loam to sands).

Natural watersheds in many parts of ARB are characterized by peaty soils with depths that vary from 0.3 m (upland) to over 1 m (lowland). Upland watersheds typically have ground slopes of 0.5% or more, while lowland areas typically have average slope less than 0.5%. Lowland areas normally have thick peat with near-surface groundwater table. As a result, a significant amount of runoff (e.g., could be more than 70%) from lowland watersheds occurs as interflow through deep peat or muskeg (northern wetland), irrespective of the sub-soil types (Golder Associates 2002).

3 Research Methods

LSMs are often run in stand-alone mode with no feedback to the atmosphere, with meteorological data coming from GCM output, global re-analysis datasets, weather model forecasts, or regional re-analysis datasets. The GCM and global data are typically available at spatial scales on the order of 1° to 3° while the regional data are available at spatial scales of 10 to 50 km. Coarse resolution data are easier to acquire and to apply over extended time periods due to modest demand for computer memory, storage, and clock speed. Further, such data usually offer global coverage and cover much longer time periods. However, finer resolution data reproduce superior local variability, such as precipitation.

Two sources of meteorological data were used. The first is the archived forecasts from the Meteorological Service of Canada's atmospheric model, Global Environmental Multiscale Model (GEM), and the second is the ERA-40 historical re-analysis data developed by the European Centre for Mid-range Weather Forecasts (ECMWF). The GEM archive covers western Canada from October 1995 to September 2001 while ERA-40 has global coverage from January 1961 to August 2002. The GEM (ERA-40) data has a spatial resolution of 0.33° (2.5°) latitude and 0.50° (2.5°) longitude and a temporal resolution of 3 (6) hours. The GEM data is typical for weather forecasting and the ERA-40 data is typical for GCM applications.

The LSM used in this study, ISBA of Météo France (referred to as OISBA in this study), explicitly models the energy and water processes at the land surface, which include, but are not limited to, soil water and heat transfer, solid-liquid storage and phase changes, and vegetative interaction with soil water. Recently, the model has been extended to include a sub-grid runoff scheme (Habets et al. 1999), a third soil layer (Boone et al. 1999), and a multi-layer snow scheme (Boone and Etchevers 2001).

To overcome the challenge of estimating many parameters of a complex model, a priori relationships linking parameters with land surface and soil characteristics are now available for many LSMs. The Ecoclimap land use data (Masson et al. 2003), which include all the physical parameters needed to run OISBA, was used to define the requisite surface parameters. Ecoclimap was derived by combining existing land-cover and climate maps, in addition to using the AVHRR satellite data.

Basin characteristics such as areal extent and the drainage network were derived from the 6 arc-second Digital Elevation Model (DEM) of the Peace-Athabasca River Basin. To facilitate cross-referencing the data sets, each DEM square was linked to its nearest land use data square, and each land use data square was linked to its nearest meteorological grid. Each meteorological grid was then divided into a mosaic of land cover tiles and the meteorological data was adjusted based on the average elevation of each tile. Representing the land cover as a mosaic of tiles and adjusting the meteorological data for each tile's mean elevation can account for a large portion of the spatial heterogeneity of land cover and topography. This accounting is primarily limited by the variation in topography within each land cover tile.

3.1 Historic Flows

OISBA was first tested in stand-alone mode in the ARB using the ERA-40 and GEM forecast archives to simulate the interaction between the atmosphere and ARB. Local runoff predicted by OISBA was then input to a hydrological routing model to simulate the total streamflow at the basin outlet. From these simulations, limitations associated with OISBA's treatment of hydrology were identified. A new hydrologic scheme was added to the original OISBA, resulting in the modified version of ISBA (MISBA) (Kerkhoven and Gan 2006).

Downscaling the ERA-40 data could potentially improve the simulation of ARB's streamflow. Even though there are complex, dynamical approaches, only simple statistical downscaling schemes were considered. This approach is computationally modest, parsimonious, but lacking in physical processes. Usually the idea is to develop empirical relationships either between large-scale atmospheric variables and sub-grid elements of local surface environment, or variables of GCM scale to local scale. The complexity of downscaling depends partly on the climate variables considered, e.g., downscaling precipitation is more involved than downscaling temperature data, since precipitation is affected by both local and mesoscale processes, and follows a skewed distribution (Yarnal et al. 2001).

Given that the GEM archive is of much higher resolution than ERA-40 data, instead of using predictors such as geopotential height or sea level pressure, we can directly compare the mean monthly meteorological GEM data for each grid point with the mean monthly data for the nearest ERA-40 grid point during the period that the two datasets overlap (October 1995 to September 2001). Downscaling was achieved by shifting the ERA-40 data to match the monthly mean of each GEM point. For example, if the January precipitation of a GEM point was 10% higher than its closest ERA-40 point during the overlap period, all the January ERA-40 precipitation rates for this point were increased by 10%. Radiation, humidity, air pressure, and wind speed data were handled in the same way while temperature was shifted by the difference in the mean temperature.

This algorithm does not address limitations of ERA-40's temporal scales and spatial variability and therefore should not improve the simulation of summer storms. However, because of the higher resolution, it will better represent the spatial distribution of land cover, topography, and local climate and should therefore improve the simulation of snowmelt and evaporation.

3.2 SRES Climate Scenarios for the 21ˢᵗ Century

ISBA was also run in a stand-alone mode with a number of SRES (Special Report on Emissions Scenarios) climate scenarios for the ARB. The predicted changes to mean monthly temperature and precipitation from seven GCM models (CCSRNIES, CGCM2, CSIROMk2b, ECHAM4, GFDLR30, HadCM3, and NCARPCM) for four SRES climate scenarios (A1FI, A21, B11, B21) over the 1961–90 base period were used to adjust the ERA-40 temperature and precipitation over three 30-year time periods: 2010–39 (early 21st century), 2040–69 (mid 21st century), and 2070–99 (late 21st century). Only two GCMs simulated all four scenarios (HadCM3 and CCRNIES). The other five GCMS only simulated the A2 and B2 scenarios. A total of 18 future climates scenarios were run for each 30-year period (two A1F1 predictions, seven A21 predictions, two B11 predictions, and seven B21 predictions) for a total of 54 simulations.

3.3 A New Hydrological Scheme for LSMs

Habets et al. (1999) developed a sub-grid runoff scheme that statistically considers the sub-grid heterogeneity of the moisture capacity of the soil, x, to follow the Xinanjiang distribution (Zhao 1992),

$$F(x) = 1 - [1 - (x/x_{max})]^{\beta} \qquad 0 \leq x \leq x_{max} \qquad (1)$$

$$\frac{x_{ave}}{x_{max}} = \frac{1}{\beta + 1} \qquad (2)$$

where β is an empirical parameter, and $F(x)$ is the cumulative probability distribution of x. This distribution is completely defined by the maximum (x_{max}) and mean (x_{ave}) values of x. Effectively, this scheme acts like a multi-bucket scheme in which the Xinanjiang distribution defines the distribution of bucket sizes and surface runoff occurs whenever a bucket fills to capacity. When modelers set the parameter β they are effectively defining the maximum bucket size (or soil depth) in the grid. A gravity drainage scheme was also developed to represent subsurface runoff,

$$Q = C_3(w - w_{drain})D \qquad (3)$$

where Q is the subsurface runoff, D is the thickness of the deep soil layer, w is the soil water content, w_{drain} is the minimum soil water content at which drainage will occur, and C_3 is a coefficient that is a function of soil texture. ISBA therefore treats subsurface runoff as a linear reservoir.

The ISBA runoff scheme requires two parameters: the Xinanjiang distribution parameter, and the minimum soil water content for drainage. Both of these parameters need to be calibrated by the user and therefore become problematic when applied to large river basins where they could vary widely. To eliminate this difficulty, we removed these two parameters by making them functions of the soil characteristics. First, runoff was made a function of soil water retention,

$$S = (w - w_r)/(w_{sat} - w_r) \qquad (4)$$

where S is the soil water retention, w_r is the residual water content, and w_{sat} is the saturated water content. The Xinanjiang distribution is used to represent the sub-grid variation in soil water retention. Because the maximum possible retention is 1, and since the model predicts the average water retention at each time step, β can be derived from Eq. (2) for each step as

$$\beta = (1/S_{ave}) - 1 \qquad (5)$$

Any new depth of water added to the soil column is first converted to additional soil water retention, ΔS,

$$\Delta S = \frac{P_{ave}\Delta t}{D_{eff}(w_{sat} - w_r)} \qquad (6)$$

where P_{ave} is the average intensity of rainfall, Δt is the model time step, and D_{eff} is the effective depth of the soil. In the runoff scheme of Habets et al. (1999) the entire root depth was used to calculate runoff, effectively assuming that within the Δt, additional water penetrated the rooting depth. This may not be realistic for vegetation with deep roots and soils with low permeability. To account for this, an effective depth was calculated based on the kinematic wave velocity of the wetting front of Smith (1983),

$$D_{eff} = \frac{K_{sat} - K(w)}{w_{sat} - w}\Delta t \qquad (7)$$

where K_{sat} and $K(w)$ are the saturated and unsaturated hydraulic conductivities respectively.

Adding ΔS uniformly across the distributed initial soil moisture retention and integrating the area where this sum exceeds 1 (i.e., the soil becomes saturated), it can be shown that total surface runoff, in terms of soil moisture retention, S_r, is

$$S_r = \Delta S^{\beta+1}(\beta+1)^{-1} \tag{8}$$

when ΔS is less than or equal to 1. When the additional water exceeds the storage capacity of the soil column to the effective depth, runoff is simply

$$S_r = \overline{S} + \Delta S - 1 = \Delta S - \beta/(\beta+1) \tag{9}$$

In Habets et al. (1999), snowmelt and rainfall were each assumed to be evenly distributed over the grid square. To account for spatial distribution of these important quantities melting snow was assumed to be uniformly distributed over the snow covered area, while rainfall was assumed to follow an exponential distribution after Entekhabi and Eagleson (1989)

$$f(i) = \frac{k}{P_{ave}} \exp\left(-\frac{ki}{P_{ave}}\right) \quad i \geq 0 \tag{10}$$

where k is the fraction of the total area that receives rainfall, and i is the rate of rainfall. For simplicity, k was assumed to be 1. Substituting i for P_{ave} in Eqs. (6), (8), and (9) and integrating over Eq. (10) yields,

$$S_r = \frac{\Delta S^{\beta+1}}{\beta+1} \frac{\gamma(\beta+2, k/\Delta S)}{k^\beta} + \exp\left(-\frac{k}{\Delta S}\right)\left[\Delta S + \frac{k}{\beta+1}\right] \tag{11}$$

where i_{max} is the rate of rainfall when ΔS equals 1,

$$i_{max} = D_{eff}(w_{sat} - w_r)/\Delta t \tag{12}$$

and $\gamma(a,x)$ is Euler's lower incomplete gamma function.

Because i_{max} is equal to the maximum rate that water can infiltrate into the soil, the first term in Eq. (12) represents locations where the rate of precipitation is less than the infiltration capacity and runoff occurs because the soil becomes saturated (Dunne runoff). The second term represents locations where the rate of precipitation exceeds the infiltration capacity of the soil (Horton runoff).

This method eliminates β as a user defined parameter. The other parameter was eliminated by assuming that w_{drain} equals w_r, which can be calculated from the soil texture using one of several empirical functions. Drawing an analogy to the Brooks-Corey equation for hydraulic conductivity of unsaturated soils (Brooks and Corey 1964), Eq. (3) was altered into a function of soil water retention,

$$Q = C_3 D \left(\frac{w - w_r}{w_{sat} - w_r} \right)^{3 + \frac{2}{\lambda}} = C_3 D S^n \qquad (13)$$

where λ is the Brooks-Corey pore-size index which can also be calculated from the soil texture. Assuming that the sub-grid distribution of soil moisture follows the Xinanjiang distribution, the total sub-surface runoff produced is

$$Q = \int_0^1 Q(S) f(S) dS \qquad (14)$$

where, by differentiating Eq. (1),

$$f(S) = \beta (1 - S)^{\beta - 1} \qquad (15)$$

Therefore,

$$Q = \beta C_3 D \int_0^1 S^n (1 - S)^{\beta - 1} dS \qquad (16)$$

The integral in Eq. (16) is similar in form to the Euler's beta function, from which the solution to Eq. (16) can be found to be

$$Q = C_3 D \frac{\Gamma(n+1) \Gamma(\beta + 1)}{\Gamma(n + \beta + 1)} \qquad (17)$$

where $\Gamma(x)$ is Euler's gamma function. This equation is highly non-linear and produces much lower runoff rates under dry conditions than the original ISBA scheme. Under moist conditions, when β approaches 0, the two methods will predict similar runoff rates.

4 Results

4.1 Historic Flows

4.1.1 GEM Simulations

The GEM data was divided into a calibration period (10/1996 to 6/1998) and a verification period (7/1998 to 9/2001). The calibration and verification runs were both initialized on October 1, 1995. One set of calibration and verification runs was made using the Original ISBA (OISBA) scheme and another set of runs was made using the Modified ISBA scheme (MISBA). Simulations were also conducted with the ERA-40 data. For OISBA, the calibrated values for β and w_{drain} from the GEM simulation were used, while for MISBA all parameters were defined by the Ecoclimap data set and so no parameter was calibrated.

Calibration statistics are summarized in Table 1. MISBA outperforms OISBA significantly in terms of all four measures: correlation coefficient, absolute error, root-mean-square error, and log error. While the correlation coefficient and the absolute error are sensitive to all flows, the root-mean-square and log error statistics are particularly sensitive to peak and low flows, respectively. The fact that MISBA is better by all four measures demonstrates that it represents a fundamental improvement over OISBA.

Table 1. Calibration and verification errors (all errors are relative to mean observed flow)

			r^2	Absolute error	RMS error	Log error
GEM	Calibration	OISBA	0.290	0.604	0.808	0.159
		MISBA	0.592	0.402	0.576	0.078
	Verification	OISBA	0.750	0.350	0.501	0.091
		MISBA	0.774	0.263	0.438	0.050
ERA-40	GEM period	OISBA	0.458	0.554	0.880	0.131
		MISBA	0.622	0.355	0.643	0.062
	1961–2002	OISBA	0.323	0.608	0.928	0.128
		MISBA	0.565	0.376	0.628	0.060
ERA/GEM	GEM period	OISBA	0.577	0.441	0.717	0.104
		MISBA	0.771	0.285	0.495	0.050
	1961–2002	OISBA	0.431	0.490	0.754	0.103
		MISBA	0.680	0.313	0.515	0.051

r correlation coefficient, *RMS error* root-mean-square error.

4.1.2 ERA-40 Simulations

Hydrographs from 1/1987 to 12/1995 for the OISBA/ERA-40 and MISBA/ERA-40 simulations are presented in Figs. 1a and 1b. The MISBA hydrograph matches the observed hydrograph much better than OISBA as evidenced by the error statistics (Table 1) which are roughly 40% lower for MISBA than OISBA. The error statistics for the overlap period (10/1995 to 9/2001) reveals similar overall skill, with the ERA-40 simulations generally performing better during the GEM calibration period and worse during the verification period. Unlike the GEM simulations, there is no significant improvement in performance between the GEM calibration and GEM verification periods in the ERA-40 simulations, and therefore Table 1 only shows the error statistics for the full overlap period.

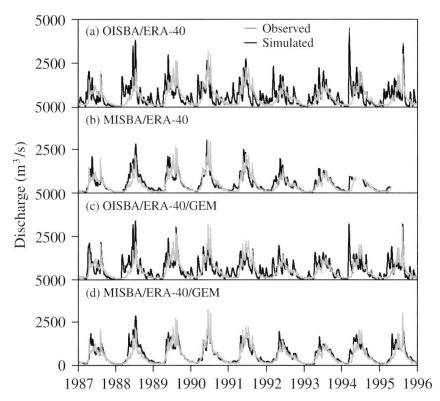

Fig. 1. Observed (grey line) and simulated (black line) hydrographs of 1987–1995 for (a) OISBA/ERA-40, (b) MISBA/ERA-40, (c) OISBA/ERA-40/GEM, and (d) MISBA/ERA-40/GEM cases

Figure 2 shows the daily variation of mean flow and its standard deviation for the OISBA/ERA-40 and MISBA/ERA-40 simulations. Again, MISBA reproduced the mean annual hydrograph and its standard deviation much better than OISBA, particularly the standard deviation which OISBA consistently overestimated. Two-sided t-test and one-tailed F-test were performed for each day for both the OISBA and MISBA simulations. These tests are designed to show statistically whether the means and variances respectively of two samples (simulated and observed streamflow for each day) are drawn from the same population. Test statistics greater than the significant level (e.g., t-test$_{0.025,40}$ = ±2.0211, F-test$_{0.05,40,40}$ = 1.69) cause rejection of the null hypothesis that two samples come from the same population, and vice versa. In Fig. 2, solid circles indicate the days on which the simulations passed the t-test and the F-test. OISBA (MISBA) passed the t-test for 121 (202) days of the year and the F-test 67 (235) days of the year. The results indicate that streamflows simulated using MISBA reflect the statistical properties of the observed flows much better than OISBA.

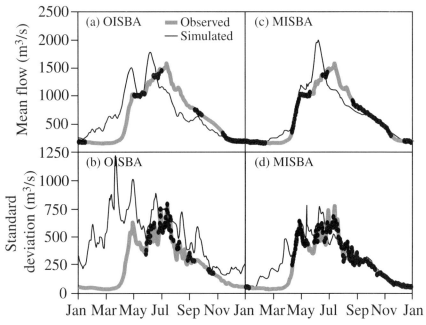

Fig. 2. Observed and OISBA/ERA-40 simulation flows for (a) daily mean flows and (b) standard deviation of flows; and observed and MISBA/ERA-40 simulation flows for (c) daily mean flows and (d) standard deviation of flows. Solid circles indicate days when simulations passed the t-test (mean flow) and F-test (variance)

4.1.3 Downscaled ERA-40/GEM Simulations

As described earlier, the ERA-40 data was downscaled using the GEM forecast archive to produce a new 42-year data set (ERA-40/GEM). For the downscaled OISBA simulation, the calibrated values for β and w_{drain} from the GEM simulation were used. Comparison of the error statistics (Table 1) for the GEM and ERA-40/GEM simulations during the overlap period shows that the ERA-40/GEM simulations are just as accurate as the

GEM simulations. The improvement in the OISBA simulation is not enough to surpass the performance of the MISBA/ERA-40 simulation. Comparing the error statistics for the ERA-40 and ERA-40/GEM simulations shows that the latter are superior by every error measure.

Figures 1c and 1d show the ERA-40/GEM hydrographs from 1/1987 to 12/1995. The OISBA hydrograph is still dominated by a series of peaks and troughs but it is not as extreme as in the OISBA/ERA-40 hydrograph. The most noticeable improvement in the MISBA hydrograph is the reduction of a number of anomalous peaks in the MISBA/ERA-40 hydrograph (Fig. 1b) without compromising the non-anomalous peaks.

Figure 3 is similar to Fig. 2 except it shows the ERA-40/GEM mean daily hydrographs. Although the standard deviations of flow in the OISBA case are greatly improved they are still more severe than the MISBA/ERA-40 simulation (Fig. 2d). Once again, the MISBA/ERA-40/GEM simulation is the best but, as expected, it still cannot consistently account for the observed runoff from convective summer storms partly because of the coarse resolution of ERA-40 data. In Fig. 3, solid circles indicate the days on which the simulations passed the t-test and the F-test. OISBA (MISBA) passed the t-test for 182 (287) days of the year and the F-test 133 (242) days of the year. Again, the MISBA/ERA-40/GEM simulation yields the results while the OISBA/ERA-40/GEM performs much better than the OISBA/ERA-40 simulation but is still not comparable to MISBA/ERA-40.

4.2 SRES Climate Scenarios for the 21st Century

MISBA/ERA-40 was used to simulate a number of SRES climate scenarios for the ARB. In general, the predictions were more sensitive to the model used than the scenario selected. However, most of the models predict continuing decreases in average, maximum, and minimum flows over the next 100 years. Fig. 4a summarizes the GCM predictions for annual temperature and precipitation changes in the ARB. In general, the GCMs predict an increase in both temperature and precipitation. HadCM3 is the wettest, ECHAM4 is the driest, and CCSRNIES is the warmest. CGCM2's predictions fall in the middle. Changes in predicted runoff are weakly correlated with precipitation changes. All 18 GCM scenarios predict decreased streamflow by the end of the 21st century. Two thirds of the scenarios predict stream flows to decline by over 20% (Fig. 4b). On the other hand, the runoff coefficient (ratio of runoff to precipitation) is strongly correlated with changes in temperature. In general, for every degree of temperature increase the runoff coefficient drops by 8% (Fig. 4c).

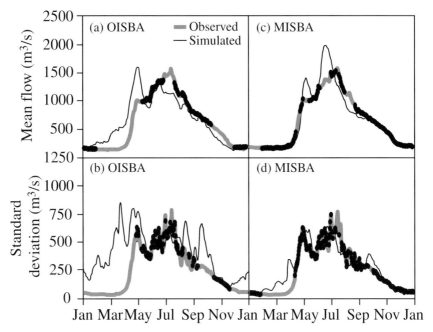

Fig. 3. Observed and OISBA/ERA-40/GEM simulation flows for (a) daily mean flows and (b) standard deviation of flows; and observed and MISBA/ERA-40/GEM simulation flows for (c) daily mean flows and (d) standard deviation of flows. Solid circles indicate days when simulations passed the t-test (mean flow) and F-test (variance)

The mean annual flow is strongly correlated with the mean annual snow accumulation in the basin (Fig. 5). With the exception of the HadCM3 GCM (which has by far the wettest December and January) the scenarios predict a strong decrease in the snow pack over the 21st century resulting in less water available for runoff.

Best-fit curves were developed relating mean annual runoff (Q), mean annual minimum flow (Q_{min}), and mean annual maximum flow (Q_{max}), in m^3 s^{-1}, to mean annual rainfall (P_{rain}), mean annual snowfall (P_{snow}), in mm, and changes to mean annual winter (ΔT_w) and summer temperature (ΔT_s) from the 1960–90 baseline (Figs. 6a, 6b, and 6c). These relationships are

$$Q = 0.276 P_{rain} \exp(-0.00612 \Delta T_s) + 0.604 P_{snow} \exp(-0.165 \Delta T_w) \quad (18)$$

$$Q_{min} = 0.144 P_{rain} \exp(-0.00181 \Delta T_s) + 0.507 P_{snow} \exp(-0.304 \Delta T_w) \quad (19)$$

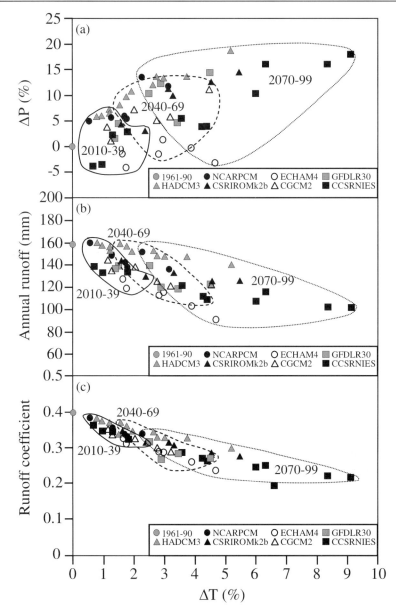

Fig. 4. Temperature changes in the Athabasca River Basin predicted by SRES climate scenarios, plotted against (a) precipitation changes according to the SRES scenarios, (b) mean annual runoff and (c) changes of runoff coefficient

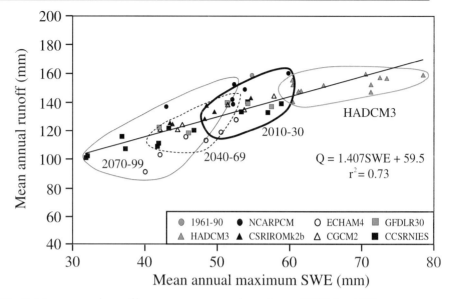

Fig. 5. Mean annual runoff verses mean annual maximum SWE for ARB

$$Q_{max} = 4.25 P_{rain} \exp(0.0396 \Delta T_s) + 12.4 P_{snow} \exp(-0.0719 \Delta T_w) \qquad (20)$$

The coefficients in these equations show that flow in the ARB is more sensitive to changes in winter temperature and snowfall than summer temperature and rainfall.

Figure 7 shows the mean daily stream flow predictions for the A2 scenario for three of the GCMs for the last 30 years of the 21st century. The 1961–90 hydrograph exhibits two distinct peaks. The first is associated with snowmelt in the lowlands and the second is associated with snowmelt in the mountainous southwest. The primary effect of climate change is a shrinking mountain snowmelt peak that occurs approximately two weeks earlier than in the 1961–90 period. In the extreme case of the CCNRIES scenario, the mountain snowmelt peak virtually disappears. All seven GCMs predict significantly lower streamflow from June to November.

Given that among the GCM results, CGCM2's are representative of an average simulation for ARB, we further examined the mean daily streamflow for all the CGCM2 scenarios (Fig. 8). Both scenarios depict similar

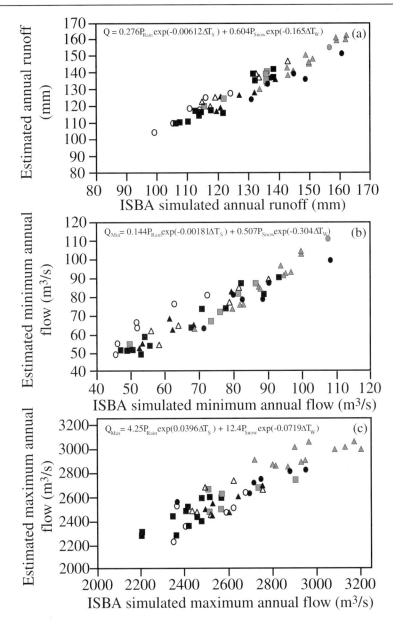

Fig. 6. Generalized relationship between changes in precipitation and temperature, and (a) mean annual flow, (b) mean annual minimum flow, and (c) mean annual maximum flow. P_{rain} and P_{snow} indicate annual rainfall and snowfall in mm, respectively. ΔT_s and ΔT_w indicate changes in summer and winter temperatures in °C, respectively

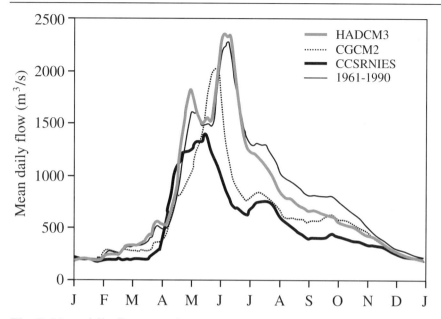

Fig. 7. Mean daily flow rates for 2070–2099 for A2 Climate Scenarios and observed flow rates for 1961–1990

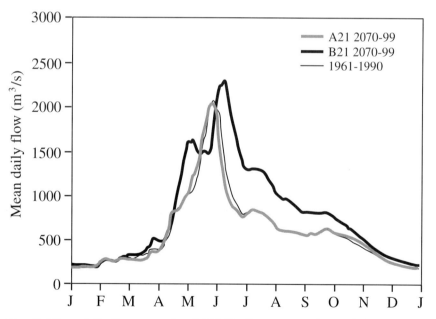

Fig. 8. Mean daily flow rates for CGCM2 climate scenarios and observed flow rates for 1961–1990

patterns. Streamflows become progressively lower as the century progresses. In terms of mean annual flow by the end of the 21st century, the ECHAM4 A21 scenario predicts the largest decrease at -42.8% while HadCM4 B21 predicts the smallest decrease at -4.4% (Table 2). The average change in annual flow by 2070–99 is -22.4%. The HadCM3 and NCARPCM consistently predict the highest flow rates while the ECHAM4 and CCSRNIES predict the lowest. In terms of mean annual maximum flow, CCSRNIES A1FI scenario predicts the largest decrease at -33.6% while HadCM4 B21 predicts the largest increase at +10.0%. The average change in annual maximum flow by 2070–99 is -14.4%. In terms of mean annual minimum flow, the climate scenarios usually predict changes ranging from -10 to +10%, with -6.6% being the average.

5 Discussion of Results

MISBA outperforms OISBA mainly because it more accurately models the subsurface runoff of ARB dominated by interflow. OISBA responds very quickly to precipitation and snowmelt events. It generally produces hydrographs that consist of a series of runoff events with rapid recession but rarely predicts sustained flow over 1000 m^3 s^{-1} in the summer. During the winter it is quite common for OSIBA hydrographs to approach zero interspersed with one or two significant mid-winter runoff events. It is also particularly sensitive to mid-winter snowmelt events, generating unrealistically high runoff responses. MISBA, however, matches the observed streamflow much better, even though it has fewer calibrated parameters. This difference is primarily because MISBA treats subsurface runoff in a nonlinear fashion while OISBA effectively uses a linear reservoir approach. The non-linear approach yields a longer retention time and a more realistic recession of streamflow in ARB, particularly when flows drop below approximately 1000 m^3 s^{-1}. Even with the sub-grid runoff scheme developed by Habets et al. (1999), the soil rarely becomes wet enough to produce noticeable surface runoff. Although dominated by subsurface runoff, MISBA does predict surface runoff during periods of rapid snowmelt and intense rainfall, and this runoff improves the performance of MISBA.

Table 2. Changes in flow statistics for Athabasca River from 2070 to 2099 with respect to 1961–90

Model	Scenario	% change		
		Mean annual flow	Maximum annual flow	Minimum annual flow
NCAR	A21	−13.6	−14.3	8.3
	B21	−4.5	−3.2	12.6
ECHAM4	A21	−42.8	−23.8	−35.2
	B21	−35.3	−18.0	−25.1
GFDCR30	A21	−22.9	−16.2	−14.0
	B21	−25.6	−13.0	−14.9
HADCM4	A1F1	−12.4	1.2	1.0
	A21	−7.9	6.9	−8.1
	B11	−7.1	2.1	−2.6
	B21	−4.4	10.0	0.5
CSIROMk2b	A21	−21.9	−21.1	−7.1
	B21	−21.5	−20.1	−5.5
CGCM2	A21	−24.2	−15.1	0.3
	B21	−24.1	−13.6	−3.3
CCSRNIES	A1F1	−36.6	−33.6	−4.9
	A21	−36.4	−32.3	−5.4
	B21	−27.7	−27.0	−6.1
	B11	−32.7	−28.4	−9.9
Average		−22.3	−14.4	−6.6

From the observed record it can be seen that peak flows occur at three different times. The first is in late April and is due to the melting of snow in the lowlands, the second occurs in June and is due to snowmelt in the mountainous southwest, and the third is in July and is due to convective summer storms. It is this last period that usually results in the annual maximum flow rate. Both OISBA and MISBA reproduce the first two peaks but miss the third one almost entirely. This is attributed to the coarse resolution of the ERA-40 data which tends to diffuse the convective storms both spatially and temporally and therefore significantly underestimates their intensity. The OISBA simulation adds a new peak in March that does not correlate to any observed flows in the basin.

The similarity of the error statistics for the GEM simulations and the downscaled ERA-40/GEM simulations suggests that the simple algorithm used here accounts for the majority of the heterogeneity between the ERA-40 and GEM scales. Although the anomalous March peak is greatly reduced from the OISBA/ERA-40 simulation it still represents a significant mode for producing a maximum annual flow and thus compromises the quality of the OISBA/ERA-40/GEM predictions for these flows.

A reduction in snow pack in the SRES climate scenarios for the 21st century is primarily due to increases in the winter temperature that result in less snow accumulation and increased sublimation. The correlation between winter (December–January) precipitation and maximum snow pack (r =+0.35) is much lower than the correlation between winter (December–January) temperature and maximum snow pack (r=-0.80). As the lowland snowmelt event becomes weaker and the mountain snowmelt comes earlier, the double-peak behavior of the mean annual hydrograph disappears. Mean annual flows are predicted to decrease by almost 25% by the last 3rd of the century. The high flow season also becomes much shorter. Historically, in an average year streamflow is expected to stay over 1000 m^3 s^{-1} for nearly 4 months from late April until mid-August. For both climate scenarios, the CGCM2 predict a high flow season that lasts less than 2 months from early May to mid-June. Between increasing temperature (ΔT) and mean maximum annual flow, with the exception of the HadCM3 GCM, all the models predict significant decreases in mean annual flow peaks with rising temperatures (Fig. 4b). This is mainly due to the decrease in the volume of spring runoff caused by a reduced snow pack and enhanced evaporation under warmer climates.

6 Summary and Conclusions

The original (OSIBA) and modified (MISBA) versions of a land surface model were applied to a large northern basin, the Athabasca River Basin (ARB) to simulate streamflow using gridded meteorological data from a numerical weather prediction model, the Canadian GEM model, and a GCM-scale re-analysis data called ERA-40. Heterogeneity of land cover was accounted for by using a variation of the mosaic approach while the effects of heterogeneity of rainfall, and soil moisture on surface and subsurface runoff were treated statistically. New highly non-linear formulations for surface and subsurface runoff were derived using the assumption that sub-grid variation of soil water retention follows the Xinanjiang distribution, with two user-defined parameters now treated as internal functions. Despite this reduction in the number of calibrated parameters, MISBA performed significantly better than OISBA. Although the method proposed was applied to ISBA, it can easily be applied to any LSM as long as soil texture data are available.

Simulations using the GCM-scale ERA-40 data showed that it is possible to reconstruct the observed streamflow in the large basin without downscaling the data. The simulations were particularly effective in reproducing the onset of snowmelt runoff, autumn and winter baseflow recession, and the annual variation of mean and minimum flows. The simulations could not account for runoff produced by convective storms due to the low temporal and spatial resolutions of ERA-40 data, but the dominant annual spring flood due to snowmelt is well reproduced.

A simple algorithm was employed to downscale the ERA-40 data to the scale of the GEM data, thus improving the accounting for local variation of land cover, topography, and climate. Again, MISBA/ERA-40/GEM significantly outperformed OISBA/ERA-40/GEM. However, the spatial variability of convective storms still cannot be accounted for, and improvement of summer flow predictions would require downscaling ERA-40 data to a scale appropriate for convection.

The predicted changes to mean monthly temperature and precipitation from seven GCMs for four SRES climate scenarios were used to simulate streamflow responses over three 30-year time periods (2010–39, 2040–69, 2070–2100). A total of 54 simulations were performed for ARB. Most of the simulated results indicate continuing decreases in average, maximum, and minimum flows of the Athabasca River over the next 100 years.

Acknowledgements

Florence Habets of Météo France provided the ISBA source code. The Ecoclimap data set was provided by Aaron Boone of Météo France. Kit Szeto of Environment Canada provided the GEM forecast archive. The ERA-40 re-analysis data set was obtained from the European Centre for Mid-range Weather Forecasts' data server, and the 6 arc-second Digital Elevation Model of the Peace-Athabasca Basin was provided by NWRI, Saskatoon, Canada.

References

Boone A, Calvet J-C, Noilhan J (1999) The inclusion of a third soil layer in a land surface scheme using the Force-Restore method. J Appl Meteorol 38:1611–1630

Boone A, Etchevers P (2001) An inter-comparison of three snow schemes of varying complexity coupled to the same land-surface model: local scale evaluation at an Alpine site. J Hydrometeorol 3:374–394

Brooks RH, Corey AT (1964) Hydraulic properties of porous media, Hydrology papers, no. 3, Colorado State University, USA

Entekhabi D, Eagleson PS (1989) Land surface hydrology parameterization for atmospheric models including subgrid scale spatial variability. J Climate 2:816–831

Famiglietti JS, Wood EF (1995) Effects of spatial variability and scale on areally averaged evapotranspiration. Water Resour Res 31:699–712

Ghan SJ, Liljegren JC, Shaw WJ, Hubbe JH, Doran JC (1997) Influence of subgrid variability on surface hydrology. J Climate 10:3157–3166

Golder Associates (2002) Regional surface water hydrology study by re-calibration of HSPF model. Golder Associates, Calgary

Habets F, Noilhan J, Golaz C, Goutorbe JP, Lacarrère P, Leblois E, Ledoux E, Martin E, Ottlé C, Vidal-Madjar D (1999) The ISBA surface scheme in a macroscale hydrological model applied to the Hapex-Mobilhy area Part 1: model and database. J Hydrol 217:75–96

Kellerhals R, Neill CR, Bray DI (1972) Hydraulic and geomorphic characteristics of rivers in Alberta, Edmonton. Research Council of Alberta

Kerkhoven E, Gan TY (2006) A modified ISBA surface scheme for modeling the hydrology of Athabasca River Basin with GCM-scale data. Adv Water Resour 29:808–826

Koster R, Suarez MJ (1992) Modeling the land surface boundary in climate models as a composite of independent vegetation stands. J Geophys Res 97(D3):2697–2715

Manabe S (1969) Climate and the ocean circulation: 1. The atmospheric circulation and the hydrology of the Earth's surface. Mon Weather Rev 97:739–805

Masson V, Champeaux J-L, Chauvin F, Meriguet C, Lacaze R (2003) A global database of land surface parameters at 1 km resolution in meteorological and climate models. J Climate 16:1261–1282

Noilhan J, Lacarrère P (1995) GCM grid-scale evaporation from mesoscale modeling. J Climate 8:206–223

Noilhan J, Planton S (1989) A simple parameterization of land surface processes for meteorological models. Mon Weather Rev 117:536–549

Pitman AJ (2003) The evolution of, and revolution in, Land Surface Schemes designed for climate models. Int J Climatol 23:479–510

Sivapalan M, Woods RA (1995) Evaluation of the effects of general circulation models' subgrid variability and patchiness of rainfall and soil moisture on land surface water balance fluxes. Hydrol Process 9:697–717

Smith RE (1983) Approximate soil water movement by kinematic characteristics. Soil Sci Soc Am Proc 47:3–8

Yarnal B, Comrie AC, Frakes B, Brown DP (2001) Developments and prospects in synoptic climatology. Int J Climatol 21:1923–1950

Zhao RJ (1992) The Xinanjiang model applied in China. J Hydrol 135:371–381

Chapter 23

Validating Surface Heat Fluxes and Soil Moisture Simulated by the Land Surface Scheme CLASS under Subarctic Tundra Conditions

Lei Wen, David Rodgers, Charles A. Lin, Nigel Roulet and Linying Tong

Abstract This study tests the ability of CLASS (Canadian Land Surface Scheme) to simulate sensible and latent heat fluxes, and soil moisture at two tundra sites in the Trail Valley Creek basin, Northwest Territories, Canada. These sites are underlain by continuous permafrost and feature mineral earth hummocks with organic soil in the inter-hummock zones. Two versions of CLASS were used, one with and the other without an organic soil parameterization developed for peatland conditions. CLASS was driven in a stand-alone mode and the results were compared with measurements obtained at each site during three summer months. Results from the peatland version of CLASS showed significant improvement over the standard version though both cases underestimated latent heat and overestimated sensible heat fluxes. CLASS used in this study is a one-dimensional column model and cannot explicitly represent lateral flow. The observed soil moisture content remained almost constant at both sites over the study period. As one site is in a local depression and the other at the bottom of a valley, it is reasonable to assume that the constant moisture content was maintained by lateral flow from adjacent hillslopes.

1 Introduction

The atmospheric component of weather or climate models is coupled to a land surface scheme (LSS) to provide the lower boundary condition for the atmosphere (Ritchie and Delage 2007). Sensible and latent heat fluxes are two major feedbacks from a LSS to the atmosphere, and most LSSs also simulate soil moisture. CLASS (Canadian Land Surface Scheme) is an advanced LSS (Verseghy 1991; Verseghy et al. 1993). It is coupled to the Canadian Regional Climate Model (CRCM) (Caya and Laprise 1999), and the coupled model CRCM/CLASS is used in the Mackenzie GEWEX Study (MacKay et al. 2007). CLASS can also be run in a stand-alone mode. The soil moisture parameterization in the standard version of

CLASS (v2.7) used is for mineral soils and does not have a complete organic soil parameterization. Letts et al. (2000) and Comer et al. (2000) developed a new parameterization for organic soil (peatland) and tested it for ten organic soil sites located at Schefferville in northern Québec, Canada, and at Chippewa in north central Minnesota, USA. The results show significant improvements over the standard version of CLASS; but the tests have not been extended to the arctic landscape.

Rich organic materials and high soil moisture contents are often found in soils of the tundra and subarctic regions of the Mackenzie valley (Aylsworth and Kettles 2000). The presence of a peat cover strongly influences the energy balance and ground heat flux which governs ground freeze-thaw (Woo et al. 2007). The objective of this study is to test the ability of CLASS to simulate surface heat fluxes and soil moisture in a tundra environment with an organic layer on the mineral soil substrate.

2 Study Area

The study sites are located in the Trail Valley Creek (TVC) basin near the Mackenzie River delta, approximately 50 km north of Inuvik and 80 km south of Tuktoyaktuk, NWT, Canada. This basin has a subarctic climate with short summers and long cold winters, maintaining continuous permafrost which can be found at a depth of about 0.4 m at the two study sites (Marsh, personal communication). Hummocky terrain is prevalent in the basin (Fig. 6 in Woo and Rouse 2007), which also includes ponds, tundra vegetation, and some open woodlands (Marsh and Pomeroy 1996).

The first site (upland site, 68°44'27"N; 133°29'11"W) is located in a local depression surrounded by hillslopes, and is about 80 m above the Trail Valley Creek. The inter-hummock zones are narrow and are composed of a layer of sphagnum peat overlying mineral soil (Quinton and Marsh 1998). Dwarf birch (*Betula glandulosa*) is the dominant vegetation that covers approximately 60% of the site, and ranges in height from 0.20 to 0.65 m. The other 40% of the site is equally divided between a cover of low vascular plants (e.g., sedges and berry species) and bare soil. The second site (lowland site, 68°44'17"N; 133°30'9"W) lies at the bottom of a valley on the south bank of the TVC. This site has microtopography and soil similar to the upland site. The dominant vegetation (80%) comprises sedges (*Carex* species) and sphagnum moss, while dwarf birch occupies 20% of the terrain and reach heights of 0.3 m. Data collected at these two sites during CAGES (Canadian GEWEX Enhanced Study) period of July to September 1999 (Petrone et al. 2000) were used to validate sensible and latent heat fluxes simulated by CLASS. Note that the observed latent heat flux is the residual in the energy balance, which contains all of the error of the measurements of net radiation, the ground heat flux, and the sensible heat flux.

3 Methods

Both the standard and the peatland versions of CLASS were used in this study. Soil and vegetation parameters are defined at the beginning of the model integration. CLASS is driven by seven measured meteorological variables, viz., total long-wave and incoming short-wave radiation, precipitation rate, air temperature, wind speed, atmospheric specific humidity, and surface pressure.

CLASS needs specific information on soil texture and vegetation at the test sites, and initial states of soil temperature and moisture content. The two versions of CLASS have identical vegetation parameters but different organic soil treatments and hence different soil parameter values. Based upon site information, two vegetation canopy types (grass and deciduous) are identified for each site, and vegetation parameter values (except for maximum roughness length) are determined from the look-up table for CLASS (Verseghy et al. 1993). The maximum roughness length for grass is 0.012 m, as measured by Tilley et al. (1997). For the shrubs, the roughness lengths are 0.065 m and 0.03 m at the upland and lowland sites respectively (McCrae and Petrone 1996). At the two sites, the soil textures are set to be the 'equivalent' to organic soil in the standard version of CLASS for the first two layers, while fibric and hemic organic soils are used in the peatland version. Mineral soil texture is assigned to the third layer in both versions and the percentages of sand and clay content for both sites are set at 32% and 48%, based upon the site surveys. Soil temperature and moisture content are initialized with observation values.

Simulations were performed with CLASS run in a stand-alone mode from July 2 to September 30, 1999. The integration time step is 30 minutes. The simulated mean diurnal cycle of the surface heat fluxes and soil moisture were evaluated and compared against observations. Soil moisture was measured using Time Domain Reflectometry (TDR) probes, carefully calibrated and maintained by the operator (Petrone, personal communication).

4 Results

Figure 1 shows the mean diurnal cycle of the latent and sensible heat fluxes simulated by the two versions of CLASS. At each site, both versions underestimated the latent heat flux and overestimated the sensible heat flux, but the peatland version gave results that were closer to the observed values. There is almost no evening dew formation simulated by CLASS. In contrast, significant dew formation was observed, as indicated by a 20 W m^{-2} negative latent heat flux at midnight over the upland site. The underestimation of latent heat flux by the peatland version of CLASS was also reported by Bellisario et al. (2000) for two organic soil sites in the northern Hudson Bay Lowland, Canada.

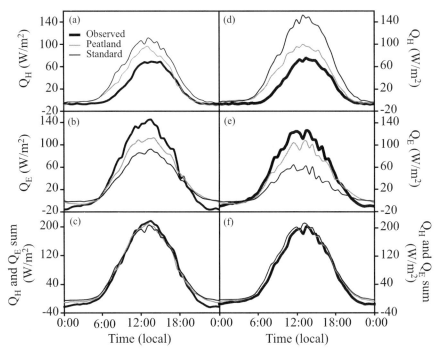

Fig. 1. Comparison of observed and simulated average diurnal cycles of sensible (Q_H) and latent (Q_E) heat fluxes for the period July 2 to September 30, 1999, at the upland and lowland sites using the two versions of CLASS: (a) upland Q_H, (b) upland Q_E, (c) upland sum of Q_H and Q_E, (d) lowland Q_H, (e) lowland Q_E, and (f) lowland sum of Q_H and Q_E

The two versions of CLASS give similar sums of simulated sensible and latent heat fluxes, which compare well with observations. The major difference between the two versions is thus the partitioning of the sensible and latent heat fluxes. Similarity of the flux total is a result of the surface heat balance: at steady state, the sum of the latent and sensible heat flux is balanced by the sum of the net radiation and the ground heat flux. The simulated ground heat flux is close to the observations (figures not shown). Identical net long-wave and incoming short-wave radiation were used to drive both versions with the vegetation canopy albedo being also identical. Thus the accuracy of the simulated sensible heat flux will increase with an improved latent heat flux.

Quantitative comparison of CLASS results was made with the observations using the diagnostic statistics of Willmott (1982, 1984). For each run, the mean observed value ($\overline{O}$) and predicted value ($\overline{P}$) were calculated together with the mean bias error (MBE) and root mean square error (RMSE). The index of agreement (d) was also used

$$d = 1 - \left[\sum_{i=1}^{N}(P_i - O_i)^2 / \sum_{i=1}^{N}(|P_i - \overline{O}| + |O_i - \overline{O}|)^2 \right] \quad (1)$$

where N is the number of observations, P_i and O_i are the predicted and the observed values. The statistic d is superior to the correlation coefficient as the latter is insensitive to a variety of potential additive and proportional differences between observed and predicted values (Willmott 1984).

Results of this quantitative comparison (Table 1) confirm that the peatland version of CLASS performed better than the standard version. The overall MBE values are improved by 50%. For the upland site, the peatland version gives RMSE and index of agreement values of 47.14 W m^{-2} and 0.88, respectively, compared to values of 54.39 W m^{-2} and 0.82 for the standard version. A similar improvement was obtained at the lowland site. Thus, the effectiveness of the modifications made by Letts et al. (2000) is clearly demonstrated, and the soil parameters used in this version are appropriate for modeling organic soil. These parameters represent high porosity soil conditions, and contribute to the better modeling of soil moisture content at the site.

Figure 2 shows the simulated and measured soil moisture contents for the two versions of CLASS. In contrast to the standard version, the peatland version dries much more slowly at the beginning of the integration, and retains higher soil moisture for the second half of the simulation. It also responds more strongly to precipitation. The simulated soil moisture contents are closer to observations after August 13 when more frequent

Table 1. Comparison of observed and modeled turbulent fluxes simulated using standard CLASS and peatland CLASS.

	Standard CLASS			Peatland CLASS		
	MBE [W m^{-2}]	RMSE [W m^{-2}]	d	MBE [W m^{-2}]	RMSE [W m^{-2}]	d
Upland						
Q_H	19.20	44.82	0.81	9.77	36.45	0.84
Q_E	-12.38	54.39	0.82	-6.16	47.14	0.88
Lowland						
Q_H	34.68	66.62	0.70	14.64	47.15	0.77
Q_E	-23.37	65.86	0.66	-6.79	54.68	0.81

Q_H sensible heat flux, Q_E latent heat flux, *MBE* mean bias error, *RMSE* root mean square error, *d* index of agreement.

and heavier precipitation occurred. However, neither version of CLASS can reproduce the measured soil moisture content, especially for the first half of the simulation period. Optimization of CLASS soil parameters yielded only marginal improvement in the simulated soil moisture content.

The observed moisture contents in the top 15 cm soil layer at both sites were relatively unchanged throughout the study period. This is especially true for the upland site where the measured moisture content (~33%) remained remarkably constant, showing only a modest fluctuation during significant precipitation events. At the lowland site, the nearly constant soil moisture increased sharply from 47% to 60% after a major precipitation event on August 13. As the upland site is in a local depression and lowland site is at the bottom of a valley, it is reasonable to assume that the constant moisture content was maintained by lateral flow from adjacent slopes. The sharp rise in moisture content at the lowland site after the heavy rain event may be indicative of flow convergence at the bottom of the valley. Field evidence lends further support to the postulation of persistent lateral flow towards the depressions and low-lying areas. Quinton and Marsh (1998) noted that "hummocks attenuate hillslope drainage as a result of the hydraulic gradient between the hummocks and the inter-hummock area. Specifically, water flows into the hummocks during periods of high flow when the direction of the hydraulic gradient is toward the hummocks, and released back into the inter-hummock area during periods of low flow, when the direction of the hydraulic gradient is reversed".

CLASS is not capable of receiving or releasing water through lateral flow, as it is a one-dimensional column model and lateral flow is a two-dimensional process. The two versions of CLASS used in this study were

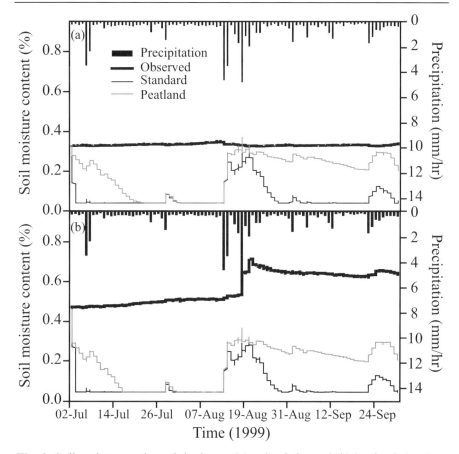

Fig. 2. Soil moisture and precipitation at (a) upland site and (b) lowland site. Precipitation is shown at the top of the figure. Observations are averages over the top 15 cm while simulated soil moisture is from the top layer (10 cm) of CLASS

the only versions available at the time though a more advanced version of CLASS will include the treatment of lateral flow (Soulis and Seglenieks 2007). Nevertheless, a nudging module in which the moisture content of the top soil layer is set to field capacity at the end of each time step (Rodgers 2002) may be useful in parameterizing the effects of lateral flow in the coupled model CRCM/CLASS for a specific site.

To further explore the role of soil moisture content in the underestimation of latent heat flux, we examined the fluxes for a dry period (July 2 to August 13) and a wet period (August 14 to September 30). Using results from the peatland version of CLASS (since this version gave better overall results), statistical analyses were performed separately for each period. At both sites, the latent heat flux underestimation is greater during the dry period than during the wet period. During the dry and wet periods, the MBE values for the upland site are -9.33 W m^{-2} and -2.95 W m^{-2}, respectively, while for the lowland site they are -13.26 and -0.37 W m^{-2}, respectively. The low soil moisture content could thus be the major cause of the underestimation. Furthermore, the simulated evaporation would be at its potential rate when soil moisture content exceeds its field capacity (the amount of water held in soil after excess or gravitational water has drained away) which is 29.5% in this study. As the CLASS simulated soil moisture is much below the field capacity (Fig. 2), an insufficient soil moisture supply is the likely cause of the latent heat flux underestimation.

5 Conclusions

Observations made between July and September 1999 provided an opportunity to evaluate the simulation of sensible and latent heat fluxes and soil moisture by two versions of CLASS, for a continuous permafrost environment with earth hummocks and organic soil in the inter-hummock zones. The simulations are run in a stand-alone mode, with and without the new organic soil parameterization, and the results are compared with measurements at an upland and a lowland site. Results from the peatland version of CLASS showed significant improvement over the standard version as revealed by the surface heat fluxes and the improved diagnostic statistics (mean bias error, root mean square error, index of agreement). There is an underestimation of latent heat fluxes and overestimation of sensible heat fluxes by both versions. With regard to soil moisture, the observed moisture contents at both sites were almost constant and at field capacity throughout the study period, most likely due to persistent lateral inflows from the slopes adjacent to the sites. In contrast, simulated results are too dry in both the standard and peatland versions. Since CLASS is a one-dimensional column model, it cannot explicitly resolve horizontal water flow. Our results nevertheless illustrate the importance of organic soil parameterization and soil moisture for the modeling of a subarctic tundra environment. Although the standard version of CLASS performs reasonably well at both sites, the incorporation of an organic soil parameterization improves the ability of CLASS to model the energy and moisture regimes.

Acknowledgements

The authors thank Phil Marsh and Wayne Rouse for generously supplying data for this study.

References

Aylsworth JM, Kettles IM (2000) Distribution of peatlands. In: Dyke LD, Brooks GR (eds) The physical environment of the Mackenzie Valley, Northwest Territories: a base line for the assessment of environmental change. Geol Surv Can B 547:49–55

Bellisario LM, Boudreau LD, Verseghy DL, Rouse WR, Blanken PD (2000) Comparing the performance of the Canadian land surface scheme (CLASS) for two subarctic terrain types. Atmos Ocean 38:181–204

Caya D, Laprise R (1999) A semi-implicit, semi-Lagrangian regional climate model: the Canadian RCM. Mon Weather Rev 127:341–362

Comer NT, LaFleur PM, Roulet NT, Letts MG, Skarupa M, Verseghy D (2000) A test of the Canadian land surface scheme (CLASS) for a variety of wetland types. Atmos Ocean 38:161–179

Letts MG, Roulet NT, Comer NT (2000) Parameterization of peatland hydraulic properties for the Canadian land surface scheme. Atmos Ocean: 38:141–160

MacKay MD, Bartlett PA, Chan E, Verseghy D, Soulis ED, Seglenieks FR (2007) The MAGS regional climate modeling system: CRCM-MAGS. (Vol. I, this book)

McCrae M, Petrone RM (1996) Description of the vegetation at Trail Valley Creek. Unpublished manuscript

Marsh P, Pomeroy JW (1996) Meltwater fluxes at an arctic forest-tundra site. Hydrol Process 10:1383–1400

Petrone RM, Rouse WR, Marsh P (2000) Comparative surface energy budgets in western and central subarctic regions of Canada. Int J Climatol 20:1131–1148

Quinton WL, Marsh P (1998) The influence of mineral earth hummocks on subsurface drainage in the continuous permafrost zone. Permafrost Periglac Process 9:213–228

Ritchie H, Delage Y (2007) The impact of CLASS in MAGS monthly ensemble predictions. (Vol. I, this book)

Rodgers D (2002) Validating Canadian land surface scheme heat fluxes under subarctic tundra conditions. M. Sc. thesis, Department of atmospheric and oceanic sciences, McGill University, Canada

Soulis ED, Seglenieks FR (2007) The MAGS integrated modeling system. (Vol. II, this book)

Tilley JS, Chapman WL, Wu W (1997) Sensitivity tests of the Canadian land surface scheme (CLASS) for arctic tundra. Ann Glaciol 25:46–50

Verseghy DL (1991) CLASS – A Canadian land surface scheme for GCMs, I. Soil model. Int J Climatol 11:111–133

Verseghy DL, McFarlane NA, Lazare M (1993) CLASS – A Canadian land surface scheme for GCMs, II. Vegetation model and coupled runs. Int J Climatol 13:347–370

Willmott CJ (1982) Some comments on the evaluation of model performance. B Am Meteorol Soc 63:1309–1313

Willmott CJ (1984) On the evaluation of model performance in physical geography. In: Gaile GL, Willmott CJ (eds) Spatial statistics and models. D. Reidel Publishing, pp 443–460

Woo MK, Mollinga M, Smith SL (2007) Modeling maximum active layer thaw in boreal and tundra environments using limited data. (Vol. II, this book)

Woo MK, Rouse WR (2007) MAGS contribution to hydrologic and surface process research. (Vol. II, this book)

Chapter 24

The MAGS Integrated Modeling System

E.D. (Ric) Soulis and Frank R. Seglenieks

Abstract The Mackenzie GEWEX Study (MAGS) integrated modeling system was developed to couple, with full feedback, selected atmospheric and hydrologic models, with the expectation that the imposed consistency will enhance the performance of both models and so mitigate the lack of data for northern basins. As each modeling community moved towards using a common land surface scheme based on the Canadian land surface scheme CLASS, a new mesoscale distributed hydrologic model (WATCLASS) was created, using CLASS for vertical processes and the routing algorithms from WATFLOOD. The version of CLASS used in the atmospheric models was modified to reflect the experience with WATCLASS. Changes were made primarily to the soil water budget and included improvements in the between-layer transfer procedures, the addition of lateral flow, and the enhancement of the treatment of cold soil. The drainage database for the Mackenzie River Basin (MRB) was built from GTOPO-30 digital elevation model and the CCRS-II AVHRR-based landcover classification. Streamflow simulations using the WATCLASS model are compared to measured values for both the MAGS research basins and the major tributaries of the Mackenzie. As well as streamflow, simulated internal state variables from WATCLASS are compared to detailed measurements taken in the research basins. Finally, the water balance of the MRB is examined and the change in storage within the basin is compared to satellite data.

1 Introduction

This chapter presents the integrated modeling system developed for the Mackenzie GEWEX Study (MAGS) to provide a framework that meets the diverse needs of the project, which range from long-term water and energy balances to distributed hour-by-hour feedback for atmospheric models. This system is not the first attempt to model the Mackenzie River Basin (MRB). Special purpose models in support of navigation, hydropower, and pipeline construction have long existed (Fassnacht 1997; Solomon et al. 1977; Soulis and Vincent 1977). It is, however, the first attempt at a general model that covers the watershed in sufficient detail to draw upon local studies for model development, to produce consistent understanding of the entire Basin, and to enhance the ability to estimate future conditions.

The fundamental goal of the modeling system is to capture the physics of the watershed adequately, especially with respect to cold processes, in order to close, with full feedback, the water and energy budgets for all such applications throughout the MRB, as well as to provide cold-soil algorithms suitable for use in atmospheric models. The modeling system is based on coupling hydrologic and atmospheric models. The key step is using a common land surface scheme to act as an interface between them. The expectation is that by constraining both models to be consistent, each will improve, mitigating the shortage of input data.

The purpose of this chapter is to describe the model development and to demonstrate its application to the water balance of the Mackenzie system. Performance of this coupled model is compared to that of a conventional hydrologic model with respect to hydrographs at locations throughout the watershed. There are no similarly comprehensive observations of the atmospheric fluxes, but the effect of the coupling of the models on local estimates of evapotranspiration is presented.

2 Model Development

The particular challenge in model development was to merge the different atmospheric and hydrologic modeling traditions in such a way that the strengths of both were not compromised while keeping the model agile and robust. The target was to generate a model that met Environment Canada's standards and that was appropriate for use in its operational forecasting and climate simulation systems.

It was clear from the outset that the approach had to involve a combination of both the land surface schemes of the day and the hydrologic models of the day. The land surface schemes, such as SSIB (Goward et al. 2002), BATS (Yang and Dickinson 1996), MOSES (Essery et al. 2001), and CLASS (Verseghy 1991), were developing evaporation schemes that only paid cursory attention to the soil water budget. This was consistent with their emphasis on atmospheric fluxes. Runoff was only of interest in monthly or annual scales as a validation (Avissar and Verstraete 1990; Huang and Liang 2006). On the other hand the hydrology models, VIC (Wood et al. 1992), TOPMODEL (Beven and Kirkby 1979), and WATFLOOD (Kouwen et al. 1993), were all developing sub-grid representations for runoff generation generally involving addressing variations of landcover and trying to capture the dynamics of variable saturation (Wood 1991). As a result, a merging of the two development paths involved producing a sub-grid representation for the land surface schemes that achieved the hydrology objectives but preserved the interfaces to the atmospheric models.

2.1 Modeling Strategy

Since MAGS was a Canadian initiative, all the models considered were Canadian. The atmospheric models used were the CMC Regional Finite Element (RFE) which was the operational weather prediction model of the day, later replaced by the Global Environmental Model (GEM) (Cote et al. 1998); the CMC Meso-scale Compressible Community (MC2), a research forecast model (Benoit et al. 1997); the CCC-GCM, a general circulation model (Flato et al. 2000); and the CRCM, a regional climate model developed by the University of Quebec for Environment Canada (Caya and Laprise 1999, MacKay et al. 2007). The hydrologic models used were WATFLOOD (Kouwen et al. 1993), SLURP (Kite 1995), and CHRM (Pomeroy et al. 2007). These represented the most physically-based semi-distributed models available during the course of the project.

Both sets of models meet at the land surface where they have a similar need for parameterizations of the vertical water and energy exchanges. In an atmospheric model, a set of such routines is referred to as a land surface scheme. The most developed Canadian scheme was the Canadian Land Surface Scheme (CLASS) (Verseghy 1991) which uses a landcover-based approach. WATFLOOD has a simple vertical water and energy budget and a very well developed routing scheme. There is enough similarity in architecture that makes it practical to use CLASS as a link between the atmospheric and hydrologic models.

The modeling strategy aimed at establishing CLASS as a link between the two sets of models (Fig. 1). Each column in the figure represents a stage of model development that was given its own level number. The first step was to run the atmospheric models separately and use their output to force WATFLOOD (Level 0). The next step was to add a common version of CLASS to each model independently. This involved simply revising the Environment Canada physics libraries for the atmospheric models (Level 1) but for hydrologic modeling it required the creation of an entirely new model called WATCLASS (Level 2). The final step involved using CLASS as a bridge between models, providing complete coupling (Level 3). The assumption was that the two-way feedbacks would provide high quality input and that the requirement for matching boundary conditions would increase the predictive power of the coupled model. During the course of MAGS, Levels 0, 1 and 2 were achieved and the framework for progressing with Level 3 was established.

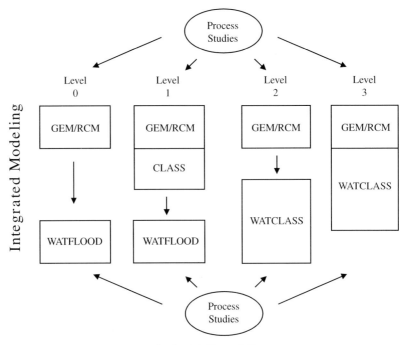

Fig. 1. Mackenzie GEWEX Study (MAGS) modeling strategy

2.2 Development of WATCLASS

Several objectives determined the approach to the design of WATCLASS. The first was to minimize the changes to the basic CLASS architecture and the second was to introduce as much physics as practical, especially in the soil water budget, but with as few calibration coefficients as possible because of the limited data available. As well, CLASS is most interested in atmospheric fluxes and thus has a detailed energy balance and all state variables are intrinsic including soil moisture. WATFLOOD focuses on streamflow and thus has a rudimentary but robust treatment of the energy balance and all state variables are bulk values.

2.2.1 Sub-grid Modeling

While both models make use of the distribution of landcover for sub-grid process modeling, CLASS uses a properties-summed-by-area single landscape unit approach and WATFLOOD uses a fluxes-summed-by-area multiple hydrologic unit approach called the Group Response Unit (GRU) approach (Kouwen et al. 1993). CLASS originally allowed one landscape category per grid cell. This is similar to the hydrologic response unit (HRU) approach (Wood 1991) except that, although a single land cover element is used to represent a grid, its properties are a blend of a generic set of landscape types. For example, a unique albedo was determined for each grid using an area-weighted sum of the albedos for the component tree types. Furthermore, there was limited provision for runoff in CLASS resulting in excess surface ponding and drainage from the soil column being discarded.

WATCLASS uses a set of generic grouped response units that are more sophisticated than in the original GRU approach. They can represent any landscape unit for which the hydrologic response can be defined, such as peat plateau or glacier. The most common, however, is the classical hillslope representation. All landscape units regardless of type are assigned a local slope to provide a gradient for lateral flow. Thus the modeling could be said to be using Modeling by Aggregating Sloped tiles or the MAGS tile approach.

There is a potential for improvement to the within-grid runoff collecting algorithms. While CLASS successfully produces long term water balances partly because of its careful treatment of the canopy processes, it is not as successful in short term runoff generation, resulting in poor soil water balances and poor hydrographs. The MAGS tile approach improves the local water balances and the WATFLOOD grid-to-grid routing algorithm converts runoff to streamflow at points of interest. However, WATCLASS within-grid routing still uses the WATFLOOD assumption that the subgrid runoff is immediately transferred to the point on the main channel where it enters the grid element. The extra travel time spent in the main channel is to account for the subgrid collection time. This treatment is adequate in certain situations such as mixed landcover in temperate zones where a channel network is the major collection mechanism. This is not true in the North. For example, in wetland areas flow occurs in wide and slow moving fens (Quinton and Hayashi 2007) and in Shield areas flow is significantly delayed because of lake storage (Woo and Mielko 2007).

2.2.2 Soil Water Movement

The attention paid to the water movement within-elements was rudimentary in the original version of CLASS, in contrast to the well developed treatment of vegetation. Changes were made to bring the soil water budget to the same level. Briefly, slope was added to the CLASS soil element to provide the gradient for lateral flow (Fig. 2); surface flow was calculated using Manning's equation, and interflow using a control-volume parameterization of Richard's equation (Soulis et al. 2000). Drainage (recharge) was calculated as before using Darcy's law but was transferred to a grid scale linear reservoir to simulate local ground water. Cold-soil parameterizations were enhanced as follows. Effective saturation and saturation conductivity were adjusted to reflect ice content. Conductivity was adjusted for temperature effects and a partial snowcover algorithm was introduced. Other parameterizations included adjustment for the effect of temperature on viscosity and a provision for a separate depth at which 100% snowcover is assumed (D_{100}) for old snow and fresh snow. The most significant changes are presented briefly in the following subsections.

2.2.3 Lateral Flow

Interflow is a runoff mechanism of particular importance in northern regions mainly because of the presence of ice. Ground ice in seasonal frost and in permafrost blocks the vertical pathways in favor of near-surface lateral flow. No such provision was provided for in the original CLASS.

There are no well developed parameterizations for lateral flow. However, there is considerable modeling experience that typically uses an exponential relationship between near-surface storage and outflow. These parameters are typically determined by calibration, which is generally not practical for regional modeling of systems like the Mackenzie. The objective was, therefore, to develop a set of equations that describe these mechanisms rigorously, relying only on the material properties and the bulk state variables of soil moisture and temperature. WATCLASS uses essentially the same physics for vertical flow as the original CLASS but treats the soil column as a sloping aquifer. This follows a kinematic wave approach with modifications to reflect the extreme variation of soil properties with depth of northern soils. An analytic solution to Richard's equation is used to approximate the expected value of the soil moisture distribution for a given average soil moisture. The boundary values of soil moisture are used to define the interflow. The parameterization, shown in Fig. 3, is approximated by

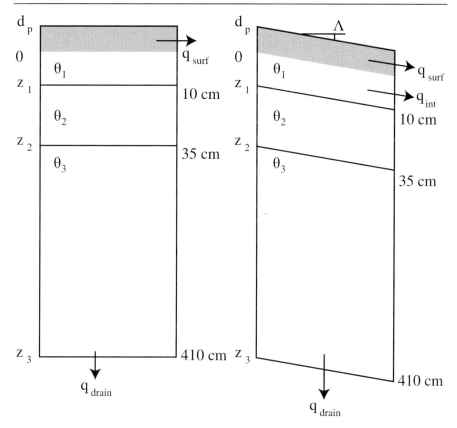

Fig. 2. Modified CLASS sloped soil algorithm

$$q_{INT} = 2D_D \cdot K_0(\theta_S) \cdot S^f \cdot \varepsilon H \cdot \Lambda_I \qquad (1)$$

where S is effective saturation, H is aquifer thickness, Λ_I is aquifer slope, D_D is drainage density, f is a coefficient, and ε is an aquifer efficiency that ranges from 0 to 1, given by

$$\varepsilon = (1 - e^{-\lambda H})/\lambda H \qquad (2)$$

where λ is a conductivity decay coefficient (Soulis et al. 2000). The values for f depend upon λH and other soil properties and typically range from 1 to 4, which is consistent with modeling experience. Prior to the introduction of the interflow algorithm, the model predicts a period of saturated flow. This is because the seepage face remains saturated for some time following the start of the recession.

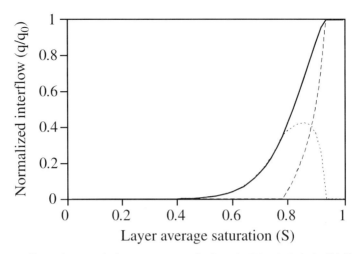

Fig. 3. Interflow characteristic curve example (c = 8, K' = 0.67,) (solid line). The short dashed curve is for unsaturated flow from the seepage face. The long dashed curve is for saturated flow from the seepage face

2.2.4 Drainage

The same approach that was used to model interflow was applied to drainage. The original CLASS used Campbell's equation (Verseghy 1991) and assumed Darcian flow for flow from the bottom of the soil column. Consequently, due to the sensitivity of drainage to soil moisture, the drainage showed little variation. As soon as the third soil layer dropped below saturation the drainage would decline rapidly; therefore the only option was to set the drainage parameters such that the drainage was virtually a constant value. The WATCLASS approach is flexible enough that saturated flow continues until the soil moisture drops below the field capacity, after which flow declines quickly. Where this transition occurs depends upon K', which is the ratio of horizontal conductivity at the bottom of the layer divided by horizontal conductivity at the top of the layer (Fig. 4).

2.2.5 Between-layer Conductivity

The between-layer-conductivity calculation is the most important change related to estimating soil properties. The original implicit finite difference scheme used an arithmetic mean of layer properties at the boundaries that resulted in chronic supersaturation of the receiving layers. Furthermore, it made the model extremely difficult to calibrate because the layer with the the reverse. In the current scheme, flow at the boundaries is controlled by

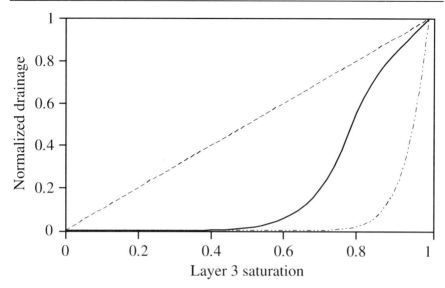

Fig. 4. Drainage characteristic curve example (c = 8, K' = 0.67) (solid line). The dashed straight line is for linear recession curve. The dashed curve line is for matrix flow

the harmonic mean of the conductivity of the adjacent layers (Fig. 5). This is a common practice in ground water models (McDonald and Harbaugh 1988) as it ensures equal fluxes at the boundary. Furthermore, the boundary conductivity is dominated by the lesser of the two values and thus limits the flow in and out of the final layers to physically plausible values.

2.2.6 Cold-soil Processes

The original CLASS had limited sensitivity to ice conditions in the soil. Ice content was assumed to reduce pore space but other soil properties did not change. In WATCLASS, conductivity and porosity were both adjusted to reflect the presence of ice. Porosity was allowed to change when ice was formed to simulate frost heave. Effective porosity is calculated as follows:

$$\phi_{effective} = \max(\theta_{liq} - \theta_{min}, \phi - \theta_{ice} - \theta_{min}, 0) \qquad (3)$$

where θ_{liq}, θ_{min}, and θ_{ice} are volumetric content of liquid water, hydroscopic water, and ice, respectively. Similarly, the degree of saturation is adjusted for ice content (Fig. 6). When ice is present it is considered to be part of the soil matrix that reduces the pore space, thereby increasing effective

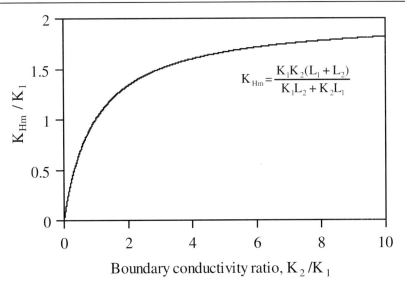

Fig. 5. Normalized harmonic mean. For uniform soil and an evenly spaced grid $K_{Hm} = K$, for $K_2 \gg K_1$, $K_{Hm} = 2*K_1$. Dominated by the smaller of K_1 and K_2, K_{Hm} ensures consistent fluxes across the layer boundaries

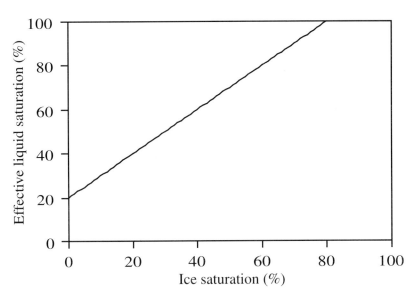

Fig. 6. Effective saturation vs. ice content (ice free liquid content of 20%). When ice is present it is considered to be part of the soil matrix. Super-saturation is permitted to allow for the expansion of ice

saturation, as well as reducing pore size and connectivity which decreases the saturated conductivity.

Conductivity is adjusted by a modified form of the impedance factor (Zhao and Gray 1997):

$$k_{ice} = k_0 \cdot (1 - f_{ice})^d \tag{4}$$

where k_0 is the saturated hydraulic conductivity with no ice present, f_{ice} is the ice fraction, and d is an exponent (Fig. 7). Horizontal flow is restricted by reducing the thickness of the transmitting layer, which is directly related to the ice fraction. Thus d has the value of 1. Vertical movement is through the connected pores in the frozen soil. Assuming that these pathways lose cross-section more than length and that laminar flow dominates, d has a value of 2.

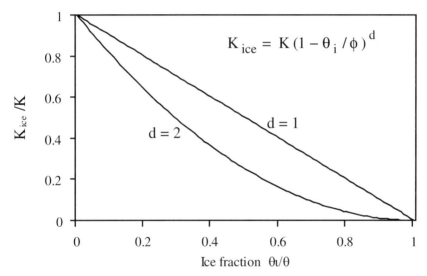

Fig. 7. Ice impedance factor vs. ice fraction. For horizontal flow, most reduction in flow is due to the reduction in the cross section of the layer, thus d =1. For vertical movement, the flow must pass through the frozen layer. Assuming that ice preferentially reduces the pore cross-section and that the flow conditions are laminar, d = 2

CLASS accounts for the aging of the snow pack through changing the values for snow albedo and snow density (Verseghy 1991). WATCLASS extended this to adjust for the effect of partial snow cover on a snow covered area as shown in Fig. 8. Separate values for the depth at which 100%

snowcover is assumed (D_{100}) were applied depending on the age of the snow pack.

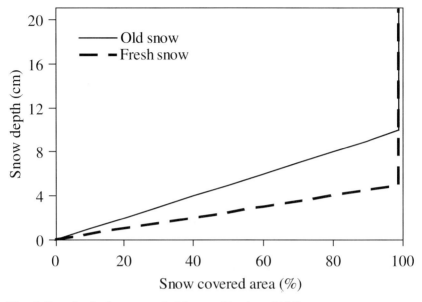

Fig. 8. D_{100} for fresh snow and old snow (Davison 2004)

3 The MAGS Modeling System

Modeling the MRB system is essentially keeping track of the components of the energy and water budgets throughout space and time. Two important preliminary steps are selecting a space and time resolution, and populating the corresponding databases.

3.1 Space and Time Resolution

An integral component of the WATFLOOD/WATCLASS modeling framework is the segmentation of a basin into gridded computational units. Unlike the atmosphere, when the land-surface is subdivided the gridding process requires locations to be explicitly considered in order to maintain the proper drainage network and sub-grid properties. This segmentation process condenses the data to a format that preserves as much of the input information as possible, while greatly reducing the program memory requirements (Kouwen et al. 1993). The choice of resolutions in this study reflected the requirements of both atmospheric and hydrologic modeling.

A one-hour time step was selected for the temporal resolution. Time steps for atmospheric modeling are typically 30 minutes to one hour, as dictated by the need to resolve diurnal cycles. Hydrologic phenomena evolve more slowly than atmospheric phenomena and longer time steps are acceptable, but one hour is often used for convenience especially in distributed models (Kouwen et al. 2005). In terms of spatial resolution, MAGS was committed by GEWEX to establish monthly balances at a scale of 100 km (Szeto et al. 2007). It is good modeling practice to use resolutions that are approximately one order of magnitude less than the target resolution, which suggested a grid spacing of 3 to 30 km.

Because of the heterogeneity of the Earth's surface, hydrologic processes occur over small domains, often much less than 1 km. Bookkeeping becomes impractical at this resolution and a grid must be imposed that, in effect, determines a separation between explicit modeling and sub-grid distribution-based modeling. Sub-grid travel time considerations dictate that the area of one grid should be no more than approximately 4% of the area of the smallest tributary of interest. Channel routing considerations limit grid spacing to no more than 50 km. Ten kilometers is about the low end of the scale at which atmospheric phenomena can be resolved and it is about the size at which landscape units are detectable by the boundary layer. Grid size for global atmospheric models at the outset of MAGS was about 250 km but decreased rapidly as computer technology improved. A 20 km grid was selected after consideration of the size of the domain, the target resolutions, model considerations, and the available data sources.

3.2 Input Databases

The Global 30 Arc-Second Elevation Data Set (GTOPO-30), available from the US Geological Survey (USGS 1997) and distributed by the EROS Data Center (EDC) Land Processes DAAC, provides 30 arc-second elevation data globally. This is equivalent to a spatial resolution of approximately 900 m in the Mackenzie region. For this study, digital elevation data for the Mackenzie Basin were extracted from GTOPO-30 (Fig. 1 in Trischenko et al. 2007) and verified using detailed local DEMs at Fort Simpson and Wolf Creek.

During the life span of MAGS, several AVHRR-based 1 km landcover classifications became available (Hansen et al. 2000; Steyaert and Knapp 1999). Six AVHRR-based datasets, four global and two Canadian, were evaluated using LANDSAT images at the three MAGS research basins (Pietroniro and Soulis 2000). The Canadian CCRS-II classification (Cihlar and Beaubien 1998) performed the best but still only achieved 30% accuracy on a pixel-by-pixel basis and minor land classes were often missed altogether. However, when averaged over the 20 km grid used in this study, the error in landcover fractions was typically within 5% absolute error.

The CCRS-II product contained 31 land cover classes. However, a number of these classes can be combined for atmospheric and hydrologic investigations (Fig. 1 in Trischenko et al. 2007). For the current generation of hydrologic models, differences in the hydrologic response between similar land classes (e.g., mixed intermediate uniform forest and mixed intermediate heterogeneous forest) cannot be determined without detailed field investigations, and parameter sets that quantify hydrologic response are only available for broad land cover categories. Thus, the 31 land cover classes of the original classification were aggregated to seven classes, viz., water, wetland, agricultural, tundra, coniferous forest, mixed forest, and glacier.

3.3 Drainage Database

WATFLOOD and WATCLASS require a basin to be properly divided into segments or grids, with each grid square containing information on the following characteristics: river elevation, drainage area, drainage direction, river classification, contour density, routing reach number, and land cover class (see Kouwen et al. 1993 for detailed description of the characteristics). These characteristics are referred to as the drainage database. The drainage database was created using the Topographic Parameterization (TOPAZ) (Martz et al. 2007) and the Waterloo Mapping Program (WATMAP) (Seglenieks et al. 2004). These programs use the topographic and land cover databases to derive the drainage database. An example of one of the characteristics from the drainage database is the drainage network (Fig. 2 in MacKay et al. 2007).

4 Model Evaluation

Streamflow simulated by the WATCLASS model for several research basins and for the Mackenzie River and its major tributaries were compared to measure values. In addition, certain state variables were examined to evaluate model performance. For streamflow comparison, we used the

Nash coefficient (Nash and Sutcliffe 1970) which is a widely accepted measure of goodness of fit. This coefficient has a range from negative infinity to 1.0, with values less than 0.0 indicating that the observed mean is a better predictor than the model while a value of 1.0 shows a perfect fit for the model (Legates and McCabe 1999).

4.1 Streamflow – Research Basins

The Smoky River, a sub-watershed of the Peace system was used for preliminary testing of WATCLASS. This river drains an area of 3840 km^2 at the foothills of the Rocky Mountain northwest of Edmonton. The basin is largely covered by alpine forest. A comparison of the streamflow simulation results using the original version of CLASS 2.6 to those using WATCLASS with its enhanced hydrology shows the marked improvement when using WATCLASS (Fig. 9). Typically values of the Nash coefficient were less than zero for the original version of CLASS and 0.86 for WATCLASS. Annual runoff errors were in the range of 10 mm per year, about 5% of the annual amount.

Two MAGS research watersheds were also used for testing and validation of WATCLASS. The Wolf Creek research basin is located in the headwaters of the Yukon River approximately 15 km south of Whitehorse, Yukon Territory. It has a drainage area of 195 km^2 and consists mainly of small open spruce forest and low-lying shrubland with some areas of boreal forest and alpine tundra (Martz et al. 2007). Sample results for Wolf Creek streamflow of 1996–2001 show that the Nash coefficient was 0.68 (Fig. 10a). The missed peak flows are consistent with known beaver dam breaks or the clearing of ice from lake outlet (cf., Hicks and Beltaos 2007), both of which are not currently modeled in WATCLASS. Another research basin (area 57 km^2), that of theTrail Valley Creek, is located approximately 50 km northwest of Inuvik, NWT (Fig. 1 in Marsh et al. 2007), in the zone of continuous permafrost at the fringe of forest–tundra transition. The stream occupies an abandoned glacial meltwater channel carved into a plateau. Sample result for Trail Valley Creek (Fig. 10b) indicates a good fit between the simulated and measured streamflow of 1996, with a Nash coefficient of 0.89.

4.2 Streamflow – Mackenzie River Basin (MRB)

The next step in the evaluation of WATCLASS was its comparison to WATFLOOD over the entire MRB for a 10 year period (1994–2003). It

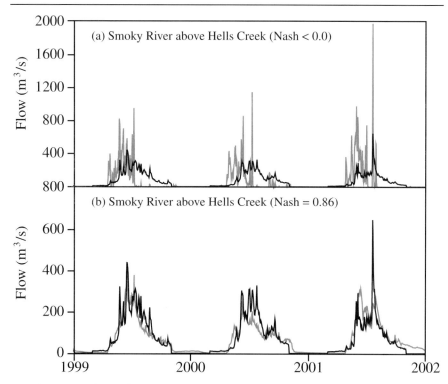

Fig. 9. Comparison of WATCLASS using (a) original and (b) enhanced hydrology, and applied to Smoky River above Hells Creek. (black/thin line is measured, grey/thick line is simulated)

should be noted that WATFLOOD has been used for over three decades and has been calibrated for basins around the world. In contrast, WATCLASS has only been in development during the 10 years of this study and has only been applied to a small number of basins. Thus the land cover parameters are much more developed for WATFLOOD and should be expected to make better simulations of streamflow.

During the course of MAGS, six sets of forcing data became available. The first set consisted of output from the Environment Canada operational weather forecast model. At the beginning of the project this was the only gridded data source available over the entire basin that contained all the necessary fields to run both WATFLOOD and WATCLASS. From January 1994 to June 1998 the Environment Canada operational weather forecast model was the Regional Finite Model (RFE) (Benoit et al. 1989) and since July 1998 it has been the Global Environmental Multiscale (GEM) model (Cote et al. 1998). The RFE and GEM models have problems correctly simulating the precipitation. To compensate for this, the simulated precipitation from the RFE/GEM models was adjusted using a set of measured monthly precipitation data (Louie et al. 2002) that was available between 1994 and 2000. Only the precipitation data were corrected and all other fields were left unchanged. This data set is referred to as the adjusted RFE/GEM data.

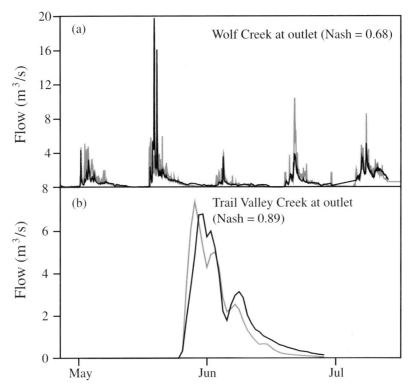

Fig. 10. WATCLASS simulation for (a) Wolf Creek basin outlet, and (b) Trail Valley Creek in 1996 (black/thin line is measured, grey/thick line is simulated)

Both WATFLOOD and WATCLASS were calibrated on the streamflow gauge located on the Liard River at Fort Liard for the years 1994–96 using the RFE/GEM data. This location was chosen because it is the largest unregulated watershed within the MRB. The hydrologic parameters for both models were adjusted to minimize the differences between the measured and simulated flows for the chosen streamflow gauge. These parameters were then used for simulations on other areas of the watershed.

During the latter part of the project other data sets became available: observed data from Environment Canada, NCEP/NCAR reanalysis 1 data (Kalnay et al. 1996), ECMWF reanalysis ERA-40 data (Uppala et al. 2005), and CRCM4.0c data (MacKay et al. 2006). These datasets were used to simulate hydrographs using the calibration parameters sets derived using the RFE/GEM data.

Figure 11 presents the hydrographs of the measured and simulated streamflow at the four major gauging sites in the MRB using the adjusted RFE/GEM data. The Nash values associated with the simulations using the other forcing datasets are also presented (Table 1). Note that the observed data from Environment Canada does not contain enough information to run WATCLASS and thus only WATFLOOD results are available using this dataset.

The simulations using the raw RFE/GEM data and the adjusted RFE/GEM data show that greater precipitation of the adjusted data led to better simulated flows (higher Nash values). The observed forcing dataset creates a streamflow simulation that matches well with the measured values, producing the highest Nash values for the WATFLOOD model. The ERA-40 forcing dataset results in a simulation with good overall fit and Nash values approaching 0.80. Both the NCEP/NCAR forcing data and the CRCM 4.0c forcing data contain an excess of precipitation, causing simulated streamflows that are consistently higher than the measured values and thus low Nash values. Overall, the Nash values for WATFLOOD and WATCLASS simulation were comparable.

4.3 State Variables – Research Basins

It is important to evaluate hydrologic models based not only on streamflow results, but also on several major internal variables. While hydrograph comparisons are very important, streamflow is the aggregation of many different hydrologic processes, and it is possible to incorrectly simulate internal variables that would cancel out and yet produce correct streamflow simulations. For example, an overestimation of snowmelt runoff could be compensated by an overestimation of evaportranspiration, resulting in correct simulated streamflows but for the wrong reasons. If, however, internal variables can be shown to compare well to measured values, then it will increase confidence in the streamflow simulations. It is also important to compare the model results to independent sources, as model to model comparisons are inherently risky.

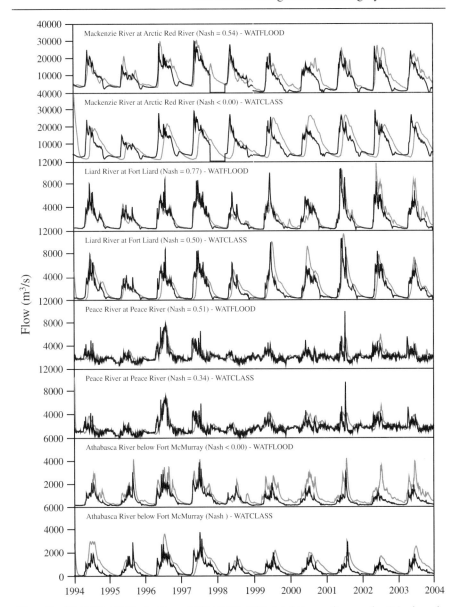

Fig. 11. WATFLOOD (left) and WATCLASS (right) results on the Mackenzie Rive Basin using RFE/GEM adjusted forcing data (black/thin line is measured, grey/thick line is simulated)

Table 1. Nash coefficient for simulations at major streamflow gauges of the Mackenzie River Basin using different forcing data sets for (a) WATFLOOD and (b) WATCLASS

Streamflow gauge	Forcing data set					
	Obs.	GEM	GEM adj.	ERA-40	NCEP	CRCM
(a) WATFLOOD						
Mackenzie at Arctic Red R.	0.67	0.16	0.54	0.76	<0.0	0.11
Liard R. at Fort Liard	0.88	0.3	0.77	0.79	<0.0	<0.0
Peace R. at Peace River	0.58	0.21	0.51	0.62	<0.0	<0.0
Athabasca R. at F. McMurray	<0.0	<0.0	<0.0	<0.0	<0.0	<0.0
(b) WATCLASS						
Mackenzie at Arctic Red R.		<0.0	<0.0	0.63	<0.0	<0.0
Liard R. at Fort Liard		0.51	0.5	0.8	<0.0	<0.0
Peace R. at Peace River		0.11	0.37	0.77	<0.0	<0.0
Athabasca R. at F. McMurray		<0.0	<0.0	0.31	<0.0	<0.0

The Wolf Creek research basin, intensely monitored during the MAGS project, offers high quality dataset for direct comparison of many internal variables. An example is provided in Fig. 12 that compares the simulated with the measured values of snow water equivalent, soil moisture and temperature in the upper soil layer. In general, the simulated and the measured values are in good agreement. This is a sampling of the internal state variables that can be examined when using advanced distributed hydrologic models such as WATFLOOD and WATCLASS. Further model development will require the comparison of an expanded set of internal variables to detailed field measurements.

4.4 Comparison with GRACE Results

During the last few years of MAGS, a remote sensing technique became available that allowed for a direct calculation of storage within the Basin. Previously, remote sensing tools that estimated basin wetness used surrogates such as snow covered area or extent of surface water. The Gravity Recovery and Climate Experiment (GRACE) (Tapley et al. 2004) produces integrated geopotential anomalies that relate directly to stored water. Figure 13 compares the simulated storage in the MRB using the WATFLOOD and WATCLASS models to the storage calculated using the GRACE satellites. Although the two datasets only intersect for two years of the study, it is encouraging to see the general agreement between two very different methods of calculating basin storage.

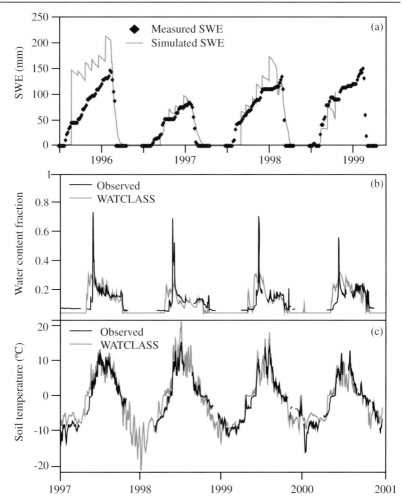

Fig. 12. Comparison of selected internal variables simulated and measured at the Wolf Creek research basin: (a) snow water equivalent, (b) water content fraction in the upper soil layer, and (c) soil temperature in the upper soil layer

5 Application

5.1 Water Budget – Mackenzie River Basin

The ability of WATFLOOD and WATCLASS in closing the water budget for the MRB was examined by calculating the evaporation and runoff for each grid square of the Basin. Combining this information with the pre-

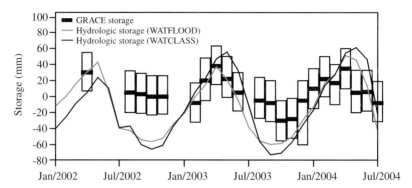

Fig. 13. Terrestrial water storage anomalies in the Mackenzie River Basin using GRACE and WATFLOOD (adapted after Yirdaw)

cipitation from the forcing dataset allowed the storage to be derived for each grid square during each time step. In many studies, it is assumed that this storage term will sum to zero over the course of an annual cycle. However, this does not allow for the basin to have a wetting or drying cycle that lasts over several years.

Water budgets for the MRB using the adjusted RFE/GEM forcing data (Table 2) suggest that the Basin was gaining moisture during 1994–2004, though the average change of storage over the 10 years was less than 1% of the average precipitation and this magnitude is within the error in the measurements. Interestingly, the annual variation in basin storage is much larger in the WATCLASS model, and the amount of annual variation in storage is comparable to that found in a study of the Bityug River in Russia (Chebotarev 1966).

In general, the WATCLASS model calculates more evaporation and therefore less runoff than the WATFLOOD model (Table 2). The term precipitation minus evaportranspiration (P–E) is also shown so that it can be compared to the atmospheric P–E commonly provided by atmospheric studies. In particular, atmospheric P–E was calculated for the MRB based both on GEM output (Strong et al. 2002) and ERA-40 (Schuster 2007) output. These were compared to the hydrological P–E simulated with WATCLASS (Fig. 14) and the results show good agreement.

Table 2. Mackenzie River Basin water budget results from (a) WATFLOOD, and (b) WATCLASS, using RFE/GEM adjusted forcing data (all values in mm)

Water Year	Water budget results				
	Precipitation	Evaporation	Hydrologic P-E	Local runoff	Δ Storage
(a) WATFLOOD					
1994-95	370.3	242.9	127.4	130.7	-3.3
1995-96	468.6	259.4	209.2	186.5	22.7
1996-97	479.0	260.3	218.7	217.1	1.6
1997-98	372.1	261.4	110.8	131.5	-20.7
1998-99	432.6	220.8	211.8	195.3	16.4
1999-2000	446.8	231.4	215.5	209.5	5.9
2000-01	422.8	212.9	209.9	219.5	-9.6
2001-02	463.5	209.0	254.4	245.3	9.2
2002-03	457.7	222.3	235.4	237.5	-2.2
2003-04	461.0	216.7	244.3	240.9	3.5
Average	437.4	233.7	203.7	201.4	2.4
(b) WATCLASS					
1994-95	369.8	253.4	116.4	100.7	15.7
1995-96	467.5	255.6	211.9	107.3	104.6
1996-97	476.9	258.0	218.8	166.6	52.3
1997-98	372.3	274.2	98.0	141.0	-42.9
1998-99	431.1	238.2	193.4	153.6	39.8
1999-2000	443.4	234.6	208.8	182.6	26.1
2000-01	423.1	229.3	193.8	267.9	-74.2
2001-02	464.1	234.3	229.7	266.9	-37.2
2002-03	457.6	240.1	217.6	275.3	-57.7
2003-04	496.6	241.3	255.3	244.5	10.8
Average	440.3	245.9	194.4	190.6	3.7

5.2 Estimation of Changes in Evapotranspiration

One of the project goals was to assess the impact of integrated modeling on the atmospheric fluxes, particularly latent heat. BOREAS, a large scale international interdisciplinary experiment in the northern boreal forests of Canada between 1994 and 1996 (Sellers et al. 1997), offered a BOREAS Northern Study Area (NSA) dataset suitable for testing the sensitivity of evapotranspiration using WATCLASS. Table 3 summarizes the results for the BOREAS. For these, WATCLASS was run in three modes on selected points in the NSA: the first simulations case uses CLASS with vertical processes only, the second case integrated sub-grid lateral flow was intro-

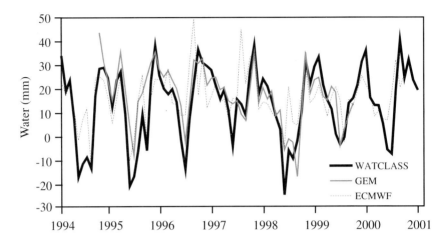

Fig. 14. Comparison of hydrological P–E and atmospheric P–E

Table 3. Evapotranspiration for the BOREAS study area using different WATCLASS modes. Entries are seasonal average for 1994–96 for selected points in the Northern Study Area

	No sub-grid lateral flow, no enhanced cold-soil physics [mm]	Sub-grid lateral flow, no enhanced cold-soil physics [mm]	[% difference]	Sub-grid lateral flow, enhanced cold-soil physics [mm]	[% difference]
(a) Dry forest					
Spring	77.1	91.9	19.1	168.9	119.0
Summer	155.6	133.5	-14.3	202.9	30.4
Fall	10.3	2.6	-74.9	24.4	137.0
Annual	243.2	229.1	-5.8	400.4	64.7
(b) Wet forest					
Spring	114.5	101.7	-11.2	148.9	30.1
Summer	148.4	123.5	-16.8	155.6	4.8
Fall	9.1	-2.7	-130.0	13.2	44.9
Annual	274.3	223.6	-18.5	320.6	16.9
(c) Wetland					
Spring	117.6	105.1	-10.7	159.2	35.3
Summer	154.8	126.1	-18.5	165.5	6.9
Fall	10.2	-0.8	-108.1	13.3	30.7
Annual	284.8	231.5	-18.7	340.8	19.7

duced, and the third case used cold-soil physics. The introduction of lateral flow reduced the water available for evapotranspiration, which was decreased by as much as 20%. On the other hand, cold-soil processes more than offset this in wet environments. Treatment of ice appears to restrict the flow of water with the result that evapotranspiration increased by 20%.

6 Conclusion

An integrated modeling system, WATCLASS, was successfully implemented by merging a land surface scheme, CLASS, with the hydrologic model WATFLOOD. This corresponds to Level 2 modeling in the MAGS modeling strategy that would couple atmospheric and hydrologic models through a common land surface scheme. The coupled model includes the rigorous treatment of the land surfaces of CLASS, the new treatment of the soil water budget, and WATFLOOD's established routing capabilities into a powerful model that produces reasonable hydrographs over a wide range of conditions and scales. Nash coefficients increased from <0 to an acceptable range of 0.6 to 0.8 over the life of the study. Water balance closure over ten year runs were typically within 5% of the precipitation and largely independent of drainage area.

One of the by-products of the comparison is the simulation of the water-balance using six sets of forcing data from a variety of reanalysis and observation datasets. Both WATFLOOD and WATCLASS produced simulations showing an increase in water storage in the MRB over the simulation period of 1994–2004. The increase in storage was radically different between the first and second half of the simulation period. In the first half, the increase in storage was 50 mm per year for WATCLASS and 10 mm per year for WATFLOOD. For the remaining years, both models showed an average storage increase of about 2 mm per year.

Basin runoff estimates match those made by WATFLOOD but the latent heat fluxes were significantly different from those produced by CLASS. The introduction of lateral flow mechanisms reduced annual evapotranspiration for all land cover types by as much as 20%. Changes to the cold-soil physics had little impact on dry environments, but for wet environments this reduction changed to a net increase of 20%.

The work is continuing on WATFLOW. Tests include the use of significant soil horizons (i.e., the wetting front, the freezing and thawing fronts, and the water table), all of which have significant effect on water movement. Testing of a provision for transfers between separate elements to simulate the storage processes is underway, and progress is being made to streamline the calibration procedure.

Acknowledgements

The system reflects 10 years of model and database development by many researchers. The model is based on a combination of CLASS developed by Diana Verseghy and WATFLOOD developed by Nick Kouwen. Ken Snelgrove and Ted Whidden coded early versions of the model. Al Pietroniro and Murray MacKay provided institutional support. Many students' theses contributed to the algorithms. S.Z. Yirdaw provided information for Fig. 13. The construction of the drainage database relied on data from MSC, NWRI, USGS, CCRS, as well as contributions by members of MAGS.

References

Avissar R, Verstraete MM (1990) The representation of continental surface processes in atmospheric models. Rev Geophys 28:35–52

Benoit R, Chartier Y, Desgagné M, Desjardins S, Pellerin P, Pellerin S (1997) The Canadian MC2: a semi-Lagrangian, semi-implicit wideband atmospheric model suited for finescale process studies and simulation. Mon Weather Rev 125:2382–2415

Benoit R, Côté J, Mailhot J (1989) Inclusion of a TKE boundary layer parameterization in the Canadian Regional Finite-Element Model. Mon Weather Rev 117:1726–1750

Beven KJ, Kirkby MJ (1979). A physically based variable contributing area model of basin hydrology. Hydrol Sci B 24:43–69

Caya D, Laprise R (1999) A semi-implicit, semi-Lagrangian regional climate model: the Canadian RCM. Mon Weather Rev 127:341–362

Chebotarev NP (1966) Theory of stream runoff. Israel Program for Scientific Translation, Jerusalem.

Cihlar J, Beaubien J (1998) Land cover of Canada 1995, version 1.1. Digital data set documentation, Natural Resources Canada, Ottawa, Ontario

Côté J, Gravel S, Méthot A, Patoine A, Roch M, Staniforth A (1998) The operational CMC–MRB Global Environmental Multiscale (GEM) Model, part I: design considerations and formulation. Mon Weather Rev 126:1373–1395

Davison B (2004) Snow accumulation in a distributed hydrological model. Master's thesis, University of Waterloo

Essery R, Best M, Cox P (2001) Moses 2.2 technical documentation. Hadley Centre, Met Office, Bracknell, UK

Fassnacht SR (1997) A multi-channel suspended sediment transport model for the Mackenzie Delta, Northwest Territories. J Hydrol 197:128–145

Flato GM, Boer GJ, Lee WG, McFarlane NA, Ramsden D, Reader MC, Weaver AJ (2000) The Canadian Centre for Climate Modeling and Analysis Global Coupled Model and its climate. Clim Dynam 16:451–467

Goward SN, Xue Y, Czajkowski KP (2002) Evaluating land surface moisture conditions from the remotely sensed temperature/vegetation index measurements, an exploration with the simplified simple biosphere model. Remote Sens Environ 79:225–242

Hansen M, DeFries R, Townshend JRG, Sohlberg R (2000) Global land cover classification at 1 km resolution using a decision tree classifier. Int J Remote Sens 21:1331–1365

Hicks F, Beltaos S (2007) River ice. (Vol. II, this book)

Huang M, Liang X (2006) On the assessment of the impact of reducing parameters and identification of parameter uncertainties for a hydrologic model with applications to ungauged basins. J Hydrol 320:37–61

Kalnay E, Kanamitsu M, Kistler R, Collins W, Deaven D, Gandin L, Iredell M, Saha S, White G, Woollen J, Zhu Y, Leetmaa A, Reynolds B, Chelliah M, Ebisuzaki W, Higgins W, Janowiak J, Mo KC, Ropelewski C, Wang J, Jenne R, Joseph D (1996) The NCEP/NCAR 40-year reanalysis project. B Am Meteorol Soc 77:437–471

Kite GW (1995) The SLURP model. In: Singh VP (ed) Computer models of watershed hydrology, Chap. 15. Water Resources Publications, Colorado, USA, pp 521–562

Kouwen N, Danard M, Bingeman A, Luo W, Seglenieks F, Soulis ED (2005) Case study: watershed modeling with distributed weather model data. J Hydrol Eng–ASCE 10:23–38

Kouwen N, Soulis ED, Pietroniro A, Donald J, Harrington RA (1993) Grouped response units for distributed hydrologic modeling. J Water Res Pl–ASCE 119:289–305

Legates DR, McCabe GJ Jr (1999) Evaluating the use of "goodness-of-fit" measures in hydrologic and hydroclimatic model validation. Water Resour Res 35:233–241

Louie PYT, Hogg WD, MacKay MD, Zhang X, Hopkinson RF (2002) The water balance climatology of the Mackenzie Basin with reference to the 1994/95 water year. Atmos Ocean 40:159–180

MacKay MD, Bartlett PA, Chan E, Derksen C, Guo S, Leighton H (2006) On the simulation of regional scale sublimation over boreal and agricultural landscapes in a climate model. Atmos Ocean 44:289–304

MacKay MD, Bartlett PA, Chan E, Verseghy D, Soulis ED, Seglenieks FR (2007) The MAGS regional climate modeling system: CRCM-MAGS. (Vol. I, this book)

Martz LW, Pietroniro A, Shaw DA, Armstrong RN, Laing B, Lacroix M (2007) Re-scaling river flow direction data from local to continental scales. (Vol. II, this book)

McDonald MG, Harbaugh AW (1988) A modular three-dimensional finite difference ground-water flow model. In: Techniques of water-resources investigations of the United States Geological Survey, Book 6, Chap. A1

Nash JE, Sutcliffe JV (1970) River flow forecasting through conceptual models, I. A discussion of principles. J Hydrol 10:282–290

Pietroniro A, Soulis ED (2000) Comparison of global land-cover databases in the Mackenzie Basin. Proc IAHS Remote Sensing and Hydrology, April, 2000, Santa Fe, New Mexico, pp 46–49

Pomeroy JW, Gray DM, Brown T, Hedstrom NR, Quinton WL, Granger RJ, Carey SK (2007) The Cold Regions Hydrological Model: a platform for basing process representation and model structure on physical evidence. Hydrol Process (in press)

Quinton WL, Hayashi M (2007) Recent advances toward physically-based runoff modeling of the wetland-dominated central Mackenzie River Basin. (Vol. II, this book)

Seglenieks F, Soulis ED, MacKay M (2004) Producing the drainage layer database for North America (WATGRID). AGU-CGU Joint Assembly, May 17–21, Montreal, Canada. Eos Trans. AGU, 85 (17), Jt. Assem. Suppl. Abstract H43C-06

Sellers PJ, Hall FG, Kelly RD, Black A, Baldocchi D, Berry J, Ryan M, Ranson MK, Crill PM, Lettenmaier DP, Margolis H, Cihlar J, Newcomer J, Fitzjarrald D, Jarvis PG, Gower ST, Halliwell D, Williams D, Goodison B, Wickland DE, Guertin FE (1997) BOREAS in 1997: experiment overview, scientific results, and future directions. J Geophys Res 102(D24):28,731–28770.

Schuster M (2007) Characteristics of the moisture flux convergence over the Mackenzie River Basin for the water-years 1990–2000. (Vol. I, this book)

Solomon SI, Cadou CFX, Jolly JP, Soulis ED (1977) Regional flood analysis for proposed Arctic Gas Pipeline Route. Proc Canadian Hydrology Symposium: 77, Edmonton, Alberta, pp 210–220

Soulis ED, Snelgrove K, Kouwen N, Seglenieks F, Verseghy D (2000) Towards closing the vertical water balance in Canadian atmospheric models: coupling of the land surface scheme CLASS with the distributed hydrological model WATFLOOD. Atmos Ocean 38:249–255

Soulis ED, Vincent DG (1977) Statistics of individual storm events from daily rainfall records. Proc 2nd Hydrometeorology Conference, October 1977, Toronto, Ontario, pp 202–207

Steyaert L, Knapp D (1999) BOREAS AFM-12 1-km AVHRR Seasonal Land Cover Classification, Data set. Available on-line from Oak Ridge National Laboratory Distributed Active Archive Center, Oak Ridge, Tennessee, U.S.A (http://www.daac.ornl.gov)

Strong GS, Proctor B, Wang M, Soulis ED, Smith CD, Seglenieks F, Snelgrove K (2002) Closing the Mackenzie Basin water budget, water years 1994/95 to 1996/97. Atmos Ocean 40:113–124

Szeto KK, Tran H, MacKay M, Crawford R, Stewart RE (2007) Assessing water and energy budgets for the Mackenzie Basin. (Vol. I, this book)

Tapley BD, Bettadpur S, Ries JC, Thompson PF, Watkins M (2004) GRACE measurements of mass variability in the earth system. Science 305(5683):503–505

Trishchenko AP, Khlopenkov KV, Ungureanu C, Latifovic R, Luo Y, Park WB (2007) Mapping of surface albeldo over Mackenzie River Basin from satellite observations. (Vol. I, this book)

Uppala SM, Kållberg PW, Simmons AJ, Andrae U, da Costa Bechtold V, Fiorino M, Gibson JK, Haseler J, Hernandez A, Kelly GA, Li X, Onogi K, Saarinen S, Sokka N, Allan RP, Andersson E, Arpe K, Balmaseda MA, Beljaars ACM, van de Berg L, Bidlot J, Bormann N, Caires S, Chevallier F, Dethof A, Dragosavac M, Fisher M, Fuentes M, Hagemann S, Hólm E, Hoskins BJ, Isaksen L, Janssen PAEM, Jenne R, McNally AP, Mahfouf J-F, Morcrette J-J, Rayner NA, Saunders RW, Simon P, Sterl A, Trenberth KE, Untch A, Vasiljevic D, Viterbo P, Woollen J (2005) The ERA-40 re-analysis. Q J Roy Meteor Soc 131:2961–3012

U.S. Geological Survey (1997) GTOPO30 Global 30 Arc Second Elevation Data Set Available at http://edc.usgs.gov/products/elevation/gtopo30/gtopo30.html.

Verseghy DL (1991) CLASS – A Canadian land surface scheme for GCMS. I. Soil model. Int J Climatol 11:111–133

Woo MK, Mielko C (2007) Flow connectivity of a lake–stream system in a semi-arid Precambrian Shield environment. (Vol. II, this book)

Wood EF (1991) Global scale hydrology: advances in land surface modeling. Rev Geophys 29:193–201

Wood EF, Lettenmaier DP, Zartarian VG (1992) A land-surface hydrology parameterization with subgrid variability for General Circulation Models. J Geophys Res 97(D3):2717–2728

Yang Z-L, Dickinson RE (1996) Description of the Biosphere-Atmosphere Transfer Scheme (BATS) for the Soil Moisture Workshop and evaluation of its performance. Global Planet Change 13:117–134

Zhao L, Gray DM (1997) A parametric expression for estimating infiltration into frozen soils. Hydrol Process 11:1761–1775

Chapter 25

Synopsis of Mackenzie GEWEX Studies on the Atmospheric–Hydrologic System of a Cold Region

Ming-ko Woo

Abstract The atmospheric–hydrologic system of the Mackenzie River Basin (MRB) shares many traits special to the world cold regions. MAGS investigators used a variety of research methods (field investigations, remote sensing, data analyses and modeling) to characterize, understand and predict the cold climate phenomena. Research emphasized the atmospheric and hydrologic processes in the MRB which exhibit pronounced seasonality. In general, the atmosphere is most dynamic in the cold season and hydrologic activities are particularly vigorous in the snowmelt period. The Basin experiences large climate variability that may be linked to shifts in the direction and intensity of airflows. Climate warming signals are also being recognized and a number of studies examined the future warming effects on wildfires, snow, frost, lakes, river ice and streamflow. This chapter synthesizes the contributions of MAGS to the knowledge on the atmosphere, hydrosphere and cryosphere of the cold region.

1 Introduction

In the decade of Mackenzie GEWEX Study (MAGS), we have unraveled many perplexities of the cold environment identified in the atmospheric and hydrologic fields. Investigations were multi-scaled, by dint of research obligation and reality of resource constraints. At one end are the studies of water and energy budgets to fulfill the requirement of the GEWEX continental-scale experiments to characterize the hydroclimate of a cold region (Woo et al. 2007a). At the other extreme are site-specific process studies such as frozen soil infiltration and active layer thaw. Many other studies range between small basin investigations and mapping or flux studies at a regional level. The complex Mackenzie River Basin (MRB) is an excellent field laboratory for process and modeling studies as it has most of the attributes common to the circumpolar areas. By concentrating its research endeavor in the MRB, MAGS was able to address crucial research problems pertaining to the atmosphere, hydrosphere and cryosphere of the cold region.

Atmospheric conditions of the Mackenzie region influence the climate of other areas, as perturbation of the jet stream in this region reverberates around the globe, and the anticyclones intensified over the region can affect the whole of North America. The Mackenzie River, as the largest in the North American continent to discharge directly into the Arctic Ocean, brings in large quantities of freshwater that influence the polar sea ice regime and the thermohaline circulation of the Ocean. Changes in the climate and the hydrologic behavior of the MRB are of a global concern.

This chapter provides a summary of MAGS results included in this two-volume book and those published elsewhere. In light of their contributions to cold region atmospheric and hydrologic sciences, research of the Mackenzie region is entirely relevant to other circumpolar areas. Not only is the knowledge of most of the physical processes transferable to all cold regions, but the models developed can be applied broadly also.

2 Research Methods

MAGS employed a wide assemblage of research methods that is required by a large cross-disciplinary program. Novel and conventional approaches were used to address data acquisition, analysis and modeling issues. Each domain of study had its variant set of conditions, and each research question had its specific methodological requirements. Generally considered, MAGS research methodology encompassed four broad categories of field investigation, remote sensing, data analysis and modeling.

2.1 Field Investigations

With strong support from Environment Canada, MAGS researchers had excellent access to the climatic and hydrometric data archives of Canada. Additional weather and streamflow data were collected in connection with the field datasets needed by particular projects. Several types of field observations were made.
- multi-year monitoring programs (e.g., meteorological towers and rain gauges, stream-gauging stations, instrumented buoys in lakes, ground temperature arrays)

- single-season measurements (e.g., McMaster University portable radar for cloud studies at Fort Simpson in 1998–99, rawindondes in Great Bear Lake)
- special surveys (e.g., snow surveys to ground-truth satellite-derived data; ice survey along the Athabasca River)

Most of the data collected by MAGS are available on website http://www.usask.ca/geography/MAGS/lo_Data_e.htm

2.2 Remote Sensing

Satellite data were acquired from several agencies, including the Canada Centre for Remote Sensing, the EOSDIS National Snow and Ice Data Center Distributed Active Archive Center (NSIDC DAAC), NASA Langley Research Center Atmospheric Sciences Data Center and the Goddard Space Flight Center Data Center, and Environment Canada. A range of products was used by MAGS investigators for regional-scale mapping (e.g., soil wetness, surface temperature and albedo, snow), surveillance (e.g., of river ice), cloud and radiation studies, trend detection (e.g., snow cover change) and validation of model results (e.g., of the Canadian Regional Climate Model). A special flight over northern MRB was commissioned in 1999, with sensors on board a National Research Council of Canada Twin Otter aircraft to measure energy fluxes along the flight transects.

2.3 Data Analyses

Three groups of data were available for process studies, statistical analyses, development of empirical relationships and as inputs for simulation experiments.
- Measured data: The Environment Canada archives were an important data source, providing climatic and hydrometric data (e.g., HYDAT), and also datasets such as the national Large Fire Database of the Canadian Forest Service. Other data sources include ground temperature data from the Geological Survey of Canada. Spatially distributed data were also derived from point measurements (e.g., CANGRID data produced by the former Atmospheric Environment Service of Environment Canada)
- Reanalysis products and numerical weather predictions: The principal data sources were the National Centers for Environmental Prediction (for NCEP and NARR data), the European Centre for Medium-Range

Weather Forecasts (for ERA-15 and ERA-40 data), and the Canadian Meteorological Centre (for SEF before 1998, and GEM afterwards)
- Scenario outputs: Climate change scenarios were used for a number of impact studies, including streamflow, river ice and forest fire. These scenario outputs were obtained from General Circulation Models of several centers such as the Canadian Centre for Climate Modelling and Analysis

Recorded data were used for the analyses of climate variability, climate and streamflow trends and their impacts. These data, together with field measurements, enabled the development of empirical relationships. Reanalysis data permitted regional analyses of water and energy budgets and the associated moisture and energy fluxes, and the study of anticyclone developments and storms. Recorded and reanalysis data also supplied inputs to macro-scale hydrologic models.

2.4 Modeling

A suite of models was developed or improved under MAGS. The models range from simple algorithms to highly involved computational schemes. The main purpose of modeling is for prediction, but there is a preference that the algorithms have a physical or theoretical support. Many of the empirical algorithms tend to robust and have low demands on data (e.g., rainfall vs elevation, frozen soil infiltration, ground freeze-thaw) necessitated by the usual paucity of data in the largely remote cold regions.

There is a diversity of modeling approaches at various scales and for different subjects.
- models based on or made use of results from process studies (e.g., forest fire prediction, blowing snow, river ice hydraulics and ice jam flood forecasts, lake dynamics, ground freeze-thaw)
- techniques and algorithms that improve model performance, particularly in a cold region context (e.g., TOPAZ, MISBA)
- coupled models that incorporate feedback mechanisms (e.g., CLASS linked with operational forecast model or with the Canadian RCM, and with hydrologic model WATFLOOD)
- application of models developed or modified, to investigate cold climate phenomena (e.g. snow distribution and melt in small catchments, river flow from large subarctic basins)

2.5 The CAGES Coordinated Effort

Coordinated effort was made to study the atmospheric–hydrologic system of MRB for the water year that spanned August 1998 to September 1999. During this period, known as the Canadian GEWEX Enhanced Study (CAGES) water year, special sets of data were collected in support of investigations on aspects of the energy and water cycle. Some results arising from studies of the CAGES period appeared as a Special Issue in the Journal of Hydrometeorology (Marsh and Gyakum 2003).

3 The Mackenzie Seasons

Physical setting of the MRB and its atmospheric linkages with neighboring areas govern the climatic and hydrologic conditions, as summarized in Szeto et al. (2007a) and Woo and Rouse (2007). High latitude location gives rise to extreme seasonal imbalance in the radiation regime, with large radiative losses in the dark winter and long hours of radiation receipt in the summer. Coldness is intense and persistent in the long winters, enabling the development of frozen ground and retention of a snow and ice cover on the land and water surfaces for protracted periods each year. Continentality is another consideration that makes the MRB a preferred area for anticyclonic development in the cold seasons and renders many areas semi-arid. The MRB receives airflows from the Pacific and also from the Gulf of Mexico during the summer. These airflows transport moisture and heat into and out of the Basin. Both moisture influx and internal recycling of moisture are major sources of precipitation. Much of the precipitation is deposited in the mountainous areas in southwestern MRB, and a secondary maximum is in the northwestern part of the Basin. Atmospheric forcing interacts with a diversity of terrain and land surfaces to produce a broad range of hydrologic processes. The cold climate processes operating in the MRB have a strong seasonality flavor, with different atmospheric and hydrologic activities being more pronounced or subdued at different times of the year.

3.1 The Cold Season

Atmospheric processes are most active in the winter season while hydrologic activities are more subdued. Long winter nights with little or no solar radiation input and large radiative losses produce net radiation deficit at the surface. Extreme coldness prevails in all parts of the MRB, with daily temperatures that can plummet below -30°C (e.g., Fig. 1 in Woo et al. 2007a). Surface cooling usually gives rise to a persistent and deep temperature inversion layer that characterizes much of the Basin (Cao et al. 2007).

Lying to the lee of the Cordillera, MRB is subject to several significant orography related effects. The high mountains impinge the westerly air flow to form a semi-permanent pressure ridge during the winter. Together with the influence of the Alaskan region and the western Arctic Ocean (two areas of pronounced anticyclonic activity), the MRB is itself a major area of anticyclogenesis. Ioannidou and Yau (2007) noted that these anticyclones have a short life span of about four days and tend to have a southeastward mobility.

The western Cordillera forces uplift of the onshore airflow from the Pacific, leading to heavy winter precipitation in the coastal areas. On the leeward side, the descending air warms adiabatically and often rides on the temperature inversion (Szeto 2007). Even within the rain shadow area to the lee of the Rocky Mountains, there is precipitation enhancement by synoptic forcing with upslope flows from the east and northeast. Smith (2007) found that monthly precipitation–elevation relationships during the cold season were statistically significant when precipitation was not much below normal. Between autumn and spring, extratropical cyclones can transport mid-level moisture from subtropical and mid-latitudes in the Pacific Ocean to the west coast of temperate North America. These events, popularly known as the 'pineapple express', give rise to extreme precipitation along the coast and sometimes into the MRB.

Snowfall is accumulated for 5 to 8 months with little intervening melt during the long winter. However, snow on the ground is subject to redistribution, particularly in the tundra areas where strong northerly winds are prevalent. There, terrain controls the erosion and deposition of snow drifts, and Marsh et al. (2007) showed that such snow relocation can be modeled for a small basin. Snow interception by boreal forests is an important winter process (see Fig. 2 in Woo and Rouse 2007). For some forests, sublimation of the intercepted snow may remove up to 30–45% of the annual snowfall (Pomeroy et al. 1999).

Blowing snow sublimation is especially important in the open tundra. Déry and Yau (2007) noted that surface sublimation is higher in early fall and in the spring than in mid-winter. Using the PIEKTUK model, they calculated an annual loss of 29 mm snow water equivalent to blowing snow and surface sublimation. The rates of blowing snow sublimation provided by Déry and Yau (2007) and by Gordon and Taylor (2007) are lower than that given by Pomeroy et al. (1993), partly due to the inclusion of negative thermodynamic feedbacks in PIEKTUK model, and a neglect of blowing snow interception by vegetation.

3.2 Spring

Owing to the large latitudinal and altitudinal extents of the large cold Basin, spring can arrive early in April in the southern plains, but is delayed for about two months in the far north and at high elevations. Several days of rise of daily maximum air temperature above the freezing point often heralds the coming of spring. Spring is associated with snow melt but under high solar radiation and particularly on south-facing slopes, melting and sublimation can commence before air temperature reaches $0°C$. Brown-Mitic et al. (2001), analyzing data collected from flights over northern MRB, found that the tundra landscape converts more than 80% of the net radiation to non-turbulent fluxes, primarily for snow melt and warming of the soil. On a local scale, the terrain in the tundra strongly affects the spatial variation of energy fluxes and when the snow cover is fragmented, advection of sensible heat from bare grounds accelerates melting of snow patches (Marsh et al. 2007). In forested areas, melt energy is much reduced as the stand density increases, largely due to the interception of solar radiation by the canopies. At high latitudes, tree shadows are much more elongated than in the temperate areas so that large parts of the forest are often deprived of direct beam radiation input. Melt energy within a stand is large for open woodlands (Giesbrecht and Woo 2000) and for denser forests, the energy decreases due to reductions in short-wave radiation and turbulent fluxes, but becomes spatially less variable within the forests (Pomeroy et al. 2007).

Ground thaw usually begins soon after the snow disappears. Freeze-thaw rates are greatly influenced by soil moisture (both ice and water) content which affects the soil thermal conductivity. Thus, ground thaw is retarded in the presence of an organic cover with hydrologic and thermal properties different from mineral soils, or where ground ice is abundant to consume much ground heat flux for ice melt (Woo et al. 2007b).

Recharge of soil moisture is inhibited by frozen soils which generally are impervious to infiltration (Pomeroy et al. 2007). On terrestrial surfaces, the abundance of surface water and saturated soils following snowmelt allows strong and sustained evaporation to occur. Small and medium-size lakes begin to lose their ice cover after the snow is depleted but large lakes remain ice covered until early summer. Among the three great lakes, ice melt begins first on Lake Athabasca, followed by Great Slave Lake about two weeks later and Great Bear Lake about two months later (Bussières 2007). Nagarajan et al. (2004) showed that the presence of myriad lakes reduces the domain-averaged surface sensible heat flux by 7–9% but raises the surface latent heat flux by 18–80%.

By releasing most or all of the snow accumulated in the winter, usually within several weeks, snowmelt triggers surface runoff. Not all facets of the landscape produce snowmelt runoff, however. Little surface flow occurs on hillslopes that lose much snow to sublimation or in areas with dry and porous soils that facilitate infiltration (Carey and Woo 1999). Other than these exceptions, spring is the period of extensive surface runoff. Water is delivered rapidly through various modes of flow. Overland flow is common on bedrock uplands of the Canadian Shield (Spence and Woo 2002). Gullies, rills, cracks between hummocks, and soil pipes concentrate the flow to bring water quickly downslope. When the organic soils thaws, subsurface flow is effective in the matrix of the porous soil. In the wetlands, bogs drain internally or have ephemeral surface flow routes to the channel fens which are the main conduits to the stream channels (Quinton and Hayashi 2007; Quinton et al. 2003). Small lakes overflow as they are filled by runoff from hillslopes and by meltwater from the snow on the lake ice (Mielko and Woo 2005).

Runoff reaching the rivers often encounters a river ice cover that does not break up until weeks after the initiation of snowmelt. Thermal breakup is largely through ice melt. Rapid runoff induced by intense melt and heavy rain can produce sufficient hydrodynamic forces to lift and fragment the ice. A combination of thermal weakening and mechanical fracturing of the ice leads to mechanical breakup (Hicks and Beltaos 2007). Ice jams are created as the fractured ice drifts and accumulates downstream. Severe ice runs associated with ice jam formation and release usually have the potential of causing large floods, the prediction of which is crucially important to the river-side communities (Mahabir et al. 2007).

Rivers in the MRB generally produce their high flows in the spring. For large basins with considerable altitudinal range, an early onset of spring followed by gradual warming would extend the snowmelt season, whereas a late spring followed by intense melt that occurs synchronously at different elevations will cause sharp hydrograph rise and pronounced peak flows (Woo and Thorne 2006). After the spring freshet, streamflow declines as the meltwater is depleted and as evaporation increases. Such a seasonal flow pattern, referred to as the nival regime because of the prominent influence of snowmelt on discharge, is exhibited by most rivers in the Basin.

3.3 Summer

In the summer, large scale atmospheric forcing is weak. In addition to the airflow from the Pacific, MRB receives airflows from the Arctic Ocean and from the Gulf of Mexico (Liu et al. 2007), the latter sometimes is a source of exceptionally large precipitation for the southern Basin. Brimelow and Reuter (2007) found that when the high frequency Great Plains low-level jet events are caught up with the infrequent lee cyclogenesis, rapid transport (<10 days) of subtropical air from the Gulf brings considerable moisture to produce extreme rainfall events.

Summer wildfires are mostly started by lightning (Kochtubajda et al. 2007). Extreme lightning events occur with the dominance of an anomalously strong pressure ridge in the upper troposphere before the onset of the events, usually triggered by lee-cyclogenesis. The probability of severe wildfire increases with low moisture content and instability in the lower troposphere. Wildfires produce much aerosols into the atmosphere. Guo et al. (2007a) found that these aerosols and summer cloud radiative forcing can affect the shortwave and longwave radiation budgets by 30–50% both at the top of atmosphere and at the surface for the MRB region.

Large solar radiation input (high intensity and long days) warms the air, the lake water and the ground. Strong surface radiative forcing is conducive to convective processes, especially with the many lakes and wetlands in central and eastern Basin that provide ample moisture to the atmosphere through evaporation. On the east slopes of the mountains, exposure to strong solar radiation leads to local heating and upward flux of sensible heat. This warms the air above the slopes to induce the mountain-plains circulation (Smith and Yau 1993). Air and moisture are drawn in from the Interior Plains and convective precipitation that results would add moisture to the surface to enhance evaporation. Thus, there is a strong recycling of moisture. Szeto et al. (2007b) noted that about half of the summer precipitation in the downwind areas is derived from local evaporation and even on annual basis, the recycling ratio (fraction of total precipitation derived from local evaporation) is almost 23%.

Evaporation is a dominant summer process. The largest evaporation rates are in the southern Basin with higher temperatures and denser vegetation cover than the north. Facilitated by the loss of an ice cover, evaporation is a major process through which water is lost from lakes of all sizes (Oswald et al. 2007). Of note is evaporation from large lakes which do not show diurnal patterns but experience episodic moisture exchange with the atmosphere at a periodicity of about 3 days (Blanken et al. 2007). Large lakes are effective as a heat storage which induces seasonal lags in regional energy balance (Schertzer et al. 2007). Water storage in large lakes also regulates streamflow by modulating and delaying peak flows and extending the low flow into late summer and even the cold season. For the smaller lakes in this semi-arid environment, however, evaporation may draw down the lake storage to the extent that lake levels fall below the outflow thresholds. This leads to cessation of lake outflow and interruption of discharge from the lake–stream system (Woo and Mielko 2007). Wetlands also suffer much water loss through evaporation and flow connectivity is severed as the water level drops (Quinton and Hayashi 2007).

There is general drying of the landscape and deepening of the water table. In many cases, this leads to the loss of flow connection between segments of the hillslopes (Leenders and Woo 2002) or shrinkage in the flow contributing area (Quinton and Marsh 1999). Headwater streams may also cease to flow. A depletion of water storage in the soil and lakes requires ample recharge in order to re-generate runoff. The fill-and-spill concept expresses the need to fill the storage to some threshold level (e.g., outlet elevation of a lake) before water can spill down the basin (Spence and Woo 2003, 2007). In this case, large rain events or a prolonged duration of moderate rainfall may revive streamflow.

3.4 Autumn

In autumn, MRB is a major source area for the development of continental cyclones because moisture is still abundant, the polar front is located nearby and the climatological high pressure ridge is not yet strong (Szeto et al. 2007a). The Basin experiences increasing cloudiness and precipitation. Hudak et al.'s (2007) radar study of clouds and precipitation showed that clouds occurring in autumn and spring were generally deeper and more multi-layered than the mid-winter clouds. Maximum cloud cover is found over the western mountains, while the Interior Plains are less cloudy possibly due to the effect of air subsidence.

As air temperature drop below $0°C$, ground freezing begins. Freeze-back of the soil is mainly through heat conduction process and frost penetration is highly dependent on the soil moisture content that controls ground ice formation. Lateral subsurface flow in slopes can advect heat to retard frost penetration (Carey and Woo 2005), and the seepage of subsurface water above ground produces icing as the water freezes on the slopes.

The arrival of autumnal rainstorms and a decrease in evaporation may lead to a rise of streamflow after its summer decline. Later in the fall, snowfall replaces rain as the main form of precipitation, though there may be intermittent melt until the onset of winter. Then, river discharge recedes to its winter low level. River ice freeze-up follows (Hicks and Beltaos 2007) with the formation of plate (or border) ice in tranquil waters and frazil ice in the turbulent flows, until an ice cover is established over many reaches of the rivers. Large accumulation of frazil ice may block the flow to cause freeze-up floods. Ice is also formed as the lakes get colder. Lake ice first appears on small lakes and ponds. Large lakes such as Great Slave and Great Bear lakes, being dimictic (Schertzer et al. 2007), undergo lake overturn again in September or October (the first overturn is usually in July), depending on the climatic condition. Large lakes do not acquire a complete ice cover until late autumn so that evaporation losses continue into early winter. Walker et al. (2000) reported that Great Slave Lake freezes about 10 days later than Great Bear Lake due to the more southerly location and an influx of river water mainly from Slave River.

4 Atmospheric–hydrologic System under Present and Future Climates

4.1 Variability and Trends

Zhang et al. (2000) analyzed the temperature records for selected climate stations across Canada and found significant recent (1950–98) warming in the west and cooling in the east. The precipitation trend is less conspicuous. Through feedback mechanisms involving local topography and other factors, the climate response of MRB to large scale circulation is amplified and warming is particularly evident in the Basin.

While long term trends may not be confirmed definitively by short data series (say, <100 years), there are obvious fluctuations that give rise to interannual climate variability; and the climate variability of the MRB is one of the largest in the world (Szeto 2007). Such variability is linked to the large-scale circulation over the North Pacific, with the teleconnection signals being strongest in the cold season. Szeto et al. (2007a) argued that recent warming in MRB largely reflects a 'jump' in the climatic state as the Basin responded to a shift in the Pacific Decadal Oscillation in the mid-1970s. After the shift, there was an increase in southwesterly onshore airflow from the Pacific Ocean to continental North America. There, the coastal mountains modify the temperature response in the MRB by forcing moisture and latent heat to be released over the west-facing slopes, thereby effectively enhancing the transport of sensible heat but reducing the influx of moisture into the Basin. Thus, after the shift, winters were generally warmer and winter precipitation was lower. These effects correspond with the noted warming trend in winter temperature (Serreze et al. 2000).

Changes in climate forcing are reflected in the Basin snow cover. Derksen et al. (2007a) analyzed snow cover data for 1945–2004 and noted little change in the onset date of snow cover, but snowmelt has become earlier in the mountainous areas. These changes are consistent with the April warming trend and with the minimal change in air temperature in the fall. Derksen et al. (2007a) further recognized an increase in the interannual variability of the April snow cover after about 1975, coinciding with a shift in the Aleutian Low.

Changes in pressure pattern accompanied by shifts in airflow can lead to severe drought events in the Canadian prairies, including southern MRB. Liu et al. (2007) noted that a reduction in zonal moisture transport from the Pacific Ocean during the cold season diminishes the winter precipitation, and a weakening of the meridional moisture transport from the Gulf of Mexico in the summer leads to a decrease in rainfall. A combination of these events increases the likelihood of severe droughts, as happened to the western and central prairies in 2001–02.

An analysis of streamflow in continental northern Canada led Déry & Wood (2005) to conclude that there has been a decline in the annual flow of many rivers west of the Hudson Bay. Abdul Aziz & Burn (2006) examined the monthly flows of the MRB and found a diversity of trends for its sub-basins, notably an increase in the winter flows and weakly decreasing flows in summer and late fall. Burn and Hesch (2007) confirmed that the Athabasca, Liard and Peace river basins are experiencing more trends than can be explained by chance. Woo et al. (2006) noted that streamflow regime is controlled not only by the climate, but also by such factors as basin location, topography and storage characteristics. Thus, there may not be coherent spatial patterns to show areas where rivers manifest a trend. On the other hand, being unable to detect a streamflow trend does not necessarily imply an absence of climate change signal which may only be masked by local confounding factors.

4.2 Implications for the Future

All Global Climate Models (GCM) point to a temperature rise in the 21st Century due to the projected higher concentrations of atmospheric greenhouse gases resulting from increased human activities. Mean annual temperature of the boreal zone may increase by 3–4°C, though there are discrepancies among the predictions from various GCM or from different assumed greenhouse gas emission scenarios. ACIA (2005) has examined comprehensively the impacts of climate change on the cold environment. Here, we select examples studied under MAGS.

Global warming is likely accompanied by changes in storm activities. Intensification of summer thunderstorms together with lengthening of the thaw season, for example, can affect the intensity and frequency of wildfire (Flannigan et al. 2007) which will impact the productivity and diversity of biologic communities.

Warming is expected to shorten the snow covered duration and promote deeper ground thaw in permafrost terrain. Lawrence and Slater (2005) simulated permafrost degradation under climate warming and their results suggest that only 1 million km^2 of the current 10.5 million km^2 of the shallow permafrost (down to 3.43 m depth) will remain by the end of this century. These simulated results were questioned by Burn and Nelson (2006) based on historical and field evidence. Furthermore, the presence of a top organic layer, even as thin as 10 cm, can effectively reduce ground thaw (Woo et al. 2007b), hence preventing the permafrost from widespread degradation. Thawing of organic materials, however, may mobilize the soil organic carbon sequestered in the upper permafrost layer as microbial respiration increases (Kling et al. 1991) and this has significant feedback on the climate through gas release. For the MRB where there are extensive wetlands with organic soils, deepening of the active layer can promote vertical drainage. This has considerable effects on wetland storage, causing changes to the hydrologic behavior of the extensive wetland and lake system in the Interior Plains.

A warmer climate will extend the ice-free season of lakes which, together with higher summer air temperatures, will raise the lake temperature. For very large lakes, an extended duration of open water conditions in late fall and early winter will produce lake-effect snowfall (Rouse et al. 2007a) and this can have hydrologic and ecological implications to the terrestrial areas downwind of the lakes. For shallow lakes in permafrost areas, higher water temperature will be transmitted to the lake bottoms to promote deeper active layer thaw or talik (unfrozen zone in permafrost) formation below the lakes. Tundra lakes are particularly sensitive to changes in the permafrost. A recent study by Smith et al. (2005) found that in the continuous permafrost zone of western Siberia, there has been an increase in both the number and the area of tundra lakes, attributable to climate warming and thermokarsting. In discontinuous permafrost zone where the permafrost is thin, deepening of the active layer may lead to lake drainage. These results from Siberia are equally applicable to the tundra and northern Interior Plains of MRB.

Lake evaporation will increase in the future. As the eastern and central MRB has numerous lakes of different sizes, a regional increase in lake evaporation yields more moisture to the atmosphere (Rouse et al. 2007b). This would intensify precipitation recycling in the warm season (Szeto et al. 2007b), thus possibly accelerating the water cycle. A large evaporation loss not compensated by snowmelt freshet and summer rain will lead to large storage deficits of the lake–stream drainage systems in the Interior Plains and in the Canadian Shield. If not compensated by large freshets in the snowmelt period and by summer rain, there will be more frequent disruption of flows along these drainage networks.

Winter and spring warming can seriously affect the river ice regime. Mid-winter thaw will remove portions of the snow accumulation, yielding lower peak flow in the spring and reducing the probability of severe ice-jam flooding during breakup (Beltaos et al. 2007). Maximum ice thickness will decrease, and the ice-covered duration will be shortened (Andrishak and Hicks 2007).

For large drainage basins, Kerkhoven and Gan (2007) simulated the flow of Athabasca River under climate warming scenario. Their result shows that streamflow will decrease progressively in this century. Woo et al (2007c) simulated the flow of the Liard River, and found that warmer winters will increase the frequency of rain and intermittent snowmelt, while spring warming can advance the timing of breakup to extend the snowmelt season, leading to less intense freshets. Summer flow will decrease due to enhanced evaporation. However, they suggest that the impacts of human activities on future streamflow regime should not be ignored, as is already well demonstrated by the effects of reservoir regulation on the flow regimes of Peace River in the MRB (Woo and Thorne 2003) and of the many rivers in Siberia (e.g., Yang et al. 2004).

5 Lessons Learned from MAGS

Recall that the primary purposes of MAGS were to understand and model the high-latitude energy and water cycles, and to improve our ability to assess the water resources under the influence of climate variability and change. These objects were accomplished with considerable success. Moreover, MAGS provided an invaluable opportunity for Canadian researchers to work together towards better understanding the cold region processes, to develop appropriate techniques for data collection and analysis, and to train highly qualified personnel. The lessons learned from the MAGS experience should be documented and the research gaps acknowledged, for the benefit of future studies of the cold region.

5.1 Data and Analysis

Field data collected for various MAGS projects have been indispensable to the process investigations and they are a valuable legacy dataset for future research, including comparative studies with other cold regions. An enhanced observation period (CAGES of 1998–99) was highly successful. During this period, specially commissioned data collection projects such as flights over northern MRB or IPIX radar deployment at Fort Simpson have extended point observations through horizontal or vertical sweeps. The gathering of field data should remain an important research approach and for large programs, our CAGES experience is worthy to be repeated. What is regrettable, notably for cold regions worldwide, is the attrition of monitoring stations. For these areas sensitive to climatic forcing and human development, the monitoring network must be maintained and even strengthened, and the data quality must be controlled.

Modeled data proved to be very useful to large scale studies. A comprehensive climatology of the water and energy budgets was obtained for MRB (Schuster 2007; Szeto et al. 2007c) and streamflow simulation for large catchments used such data as inputs. However, there are large differences among reanalysis datasets which when used in MAGS studies, caused considerable uncertainties in the results. Satellite data offer spatial information not otherwise available for large areas. They have been used in MAGS for mapping, analysis of radiation fields, studies of snow cover trends, and validation of model outputs (e.g., Bussières 2007; Derksen et al. 2007a; Guo et al. 2007b; Leconte et al. 2007; Trischenko et al. 2007). Continued upgrading of satellite capabilities and resolution will allow improved thematic mapping and analyses of atmospheric and hydrologic phenomena. Further use of satellite data should be explored for real-time applications, such as river ice breakup and flooding (see Fig. 8 in Woo and Rouse 2007).

5.2 Processes

Through MAGS, some cold region processes were discovered and the interpretation of many atmospheric and hydrologic phenomena was enhanced. Findings from process investigations constitute a major part of this book. We recognize that much remains unknown about the many physical processes operating in the MRB and in other cold regions, and there are strong demands for future works by individuals and through concerted efforts.

With deeper understanding of the processes, we are better equipped to appraise climate change and variability, and can be more confident in projecting their impacts on the cold region. Furthermore, economic development of the cold regions is inevitable as human settlement, resource exploitation and transportation networks expand in the circumpolar areas. All these subjects are of current and future concern, not only to the scientific and environmental communities, but also to the society as a whole.

5.3 Models

Modeling effort of MAGS has benefited from the knowledge gained through process studies. The opportunities for collaboration and the availability of field and model-derived data also contribute to the development or the improvement of a number of models. There has been a tendency to replace empirical equations by more physically based or theoretically grounded algorithms. Models, new or adapted, were tested for use in the cold environments (e.g., Kerkhoven and Gan 2007; Ritchie and Delage 2007; Thorne et al. 2007; Wen et al. 2007). Unfortunately, data availability remains a constraint and model robustness in many cases takes precedence over detailed representation of processes in many operational models.

One modeling goal of MAGS was to provide an integrated system that would couple atmospheric and hydrologic models, with bridging provided through a land surface scheme. The coupling of CRCM or with CLASS and of CLASS with WATFLOOD has been successful (MacKay et al. 2007; Soulis and Seglenieks 2007), but a complete linkage with the requisite feedbacks among the atmosphere, land and hydrologic components needs more work. Several models have been applied for operational forecasts of such hazards as ice-jam floods (Mahabir et al. 2007), wildfire (Kochtubajda et al. 2007) and heavy snowfall (Dupilka and Reuter 2007), as well as for long-term prediction of climate change impacts. The practical aspect of research can be further emphasized.

5.4 Scaling

While MAGS made limited conceptual contributions to the issue of scaling, some techniques have been developed to extend areally-limited data (e.g., Martz et al. 2007). The intent of International GEWEX is to obtain information on a large (or global) scale, but processes occurring on a local level are important to the understanding of large-scale phenomena. Furthermore, the interpolation and extrapolation of point, transect or small-area data are subject to considerable but unknown degree of uncertainty due to the lack of validation datasets. Ground truthing satellite information (e.g., Derksen et al. 2007b) and comparing airborne with meteorological tower data are examples of scaling up the ground observations.

5.5 Collaboration and Training

Through many meetings and collaboration, atmospheric and hydrologic researchers were made aware of the performance and limitations of data and models in each other's disciplines. These activities led to fruitful interfacing of projects and productive cross-disciplinary research ventures. Interactions with stakeholders provided management and operational benefits to our users, and augmented community well being (e.g., information exchange with residents of the MRB, see Woo et al. 2007d).

Throughout the duration of MAGS, scientists and engineers remained open minded to possible disciplinary biases and were appreciative of the contribution of other investigators. It was in this spirit of receptiveness to cross-disciplinary collaborative approach that we trained the new generations of students. It is our enriched pool of experts and expertise that best represents the legacy of MAGS.

References

Abdul Aziz OI, Burn DH (2006) Trends and variability in the hydrological regime of the Mackenzie River Basin. J Hydrol 319:282–294

ACIA (2005) Arctic Climate Impact Assessment. Cambridge University Press, Cambridge, UK

Andrishak R, Hicks F (2007) Impact of climate change on the Peace River thermal ice regime. (Vol. II, this book)

Beltaos S, Prowse T, Bonsal B, Carter T, MacKay R, Romolo L, Pietroniro A, Toth B (2007) Climate impacts on ice-jam floods in a regulated northern river. (Vol. II, this book)

Blanken PD, Rouse WR, Schertzer WM (2007) The time scales of evaporation from Great Slave Lake. (Vol. II, this book)

Brimelow JC, Reuter GW (2007) Moisture sources for extreme rainfall events over the Mackenzie River Basin. (Vol. I, this book)

Burn CR, Nelson FE (2006) Comment on "A projection of severe near-surface permafrost degradation during the 21st century" by David M. Lawrence and Andrew G. Slater. Geophys Res Lett 33:L21503, doi:10.1029/2006GL027077

Burn DH, Hesch N (2007) Trends in Mackenzie River Basin streamflows. (Vol. II, this book)

Bussières N (2007) Analysis and application of 1- km resolution visible and infrared satellite data over the Mackenzie River Basin. (Vol. II, this book)

Brown-Mitic G, MacPherson IJ, Schuepp PH, Nagarajan B, Yau MK, Bales R (2001) Aircraft observations of surface-atmospheric exchange during and after snowmelt of different arctic environments: MAGS 1999. Hydrol Process 15:3585–3602

Cao Z, Stewart RE, Hogg WD (2007) Extreme winter warming over the Mackenzie Basin: observations and causes. (Vol. I, this book)

Carey SK, Woo MK (1999) Hydrology of two slopes in subarctic Yukon, Canada. Hydrol Process 13: 2549–2562.

Carey SK, Woo MK (2005) Freezing of subarctic hillslopes, Wolf Creek basin, Yukon, Canada. Arct Antarct Alp Res 37:1-10

Derksen C, Brown R, MacKay M (2007a) Mackenzie Basin snow cover: variability and trends from conventional data, satellite remote sensing, and Canadian Regional Climate Model simulations. (Vol. I, this book)

Derksen C, Walker A, Toose P (2007b) Estimating snow water equivalent in northern regions from satellite passive microwave data. (Vol. I, this book)

Déry SJ, Wood EF (2005) Decreasing river discharge in northern Canada. Geophys Res Lett 31, L10401, doi:10.1029/2005GL022845

Déry SJ, Yau MK (2007) Recent studies on the climatology and modeling of blowing snow in the Mackenzie River Basin. (Vol. I, this book)

Dupilka ML, Reuter GW (2007) On predicting maximum snowfall. (Vol. I, this book)

Flannigan MD, Kochtubadjda B, Logan KA (2007) Forest fires and climate change in the Northwest Territories. (Vol. I, this book)

Giesbrecht MA, Woo MK (2000) Simulation of snowmelt in a subarctic spruce woodland: 2. Open woodland model. Water Resour Res 36:2287–2295

Gordon M, Taylor PA (2007) On blowing snow and sublimation in the Mackenzie River Basin. (Vol. I, this book)

Guo Q, Leighton HG, Feng J, Trischenko A (2007a) Wildfire aerosol and cloud radiative forcing in the Mackenzie River Basin from satellite observations. (Vol. I, this book)

Guo Q, Leighton HG, Feng J, MacKay M (2007b) Comparison of solar radiation budgets in the Mackenzie River Basin from satellite observations and a regional climate model. (Vol. I, this book)

Hicks F, Beltaos S (2007) River ice. (Vol. II, this book)

Hudak D, Stewart R, Rodriguez P, Kochtubajda B (2007) On the cloud and precipitating systems over the Mackenzie Basin. (Vol. I, this book)

Ioannidou L, Yau MK (2007) Climatological analysis of the Mackenzie River Basin anticyclones: structure, evolution and interannual variability. (Vol. I, this book)

Kerkhoven E, Gan TY (2007) Development of a hydrologic scheme for use in land surface models and its application to climate change in the Athabasca River Basin. (Vol. II, this book)

Kling GW, Kipphut GW, Miller MC (1991) Arctic lakes and streams as gas conduits to the atmosphere: implications for tundra carbon budgets. Science 251:298–301

Kochtubajda B, Flannigan MD, Gyakum JR, Stewart RE, Burrows WR, Way A, Richardson E, Stirling I (2007) The nature and impacts of thunderstorms in a northern climate. (Vol. I, this book)

Lawrence DM, Slater AG (2005) A projection of severe near-surface permafrost degradation during the 21st century. Geophys Res Lett 32:L24401, doi:1029/2005GL025080

Leconte R, Temimi M, Chaouch N, Brissette R, Toussaint T (2007) On the use of satellite passive microwave data for estimating surface soil wetness in the Mackenzie River Basin. (Vol. II, this book)

Leenders EE, Woo MK (2002) Modeling a two-layer flow system at the subarctic, subalpine treeline during snowmelt. Water Resour Res 38(10):1202, doi:10.1029/2001WR000375

Liu J, Stewart RE, Szeto KK (2007) Water vapor fluxes over the Canadian Prairies and the Mackenzie River Basin. (Vol. I, this book)

MacKay MD, Bartlett P, Chan E, Verseghy D, Soulis ED, Seglenieks FR (2007) The MAGS regional climate modeling system: CRCM-MAGS. (Vol. I, this book)

Mahabir C, Robichaud C, Hicks F, Robinson Fayek A (2007) Regression and logic based ice jam flood forecasting. (Vol II, this book)

Marsh P, Gyakum JR (2003) The hydrometeorology of the Mackenzie River Basin during the 1998–99 Water Year. J Hydrometeorol 4:645–648

Marsh P, Pomeroy J, Pohl S, Quinton W, Onclin C, Russell M, Neumann N, Pietroniro A, Davison B, McCartney S (2007) Snowmelt processes and runoff at the arctic treeline: ten years of MAGS research. (Vol. II, this book)

Martz LW, Pietroniro A, Shaw DA, Armstrong RN, Laing B, Lacroix M (2007) Re-scaling river flow direction data from local to continental scales. (Vol. II, this book)

Mielko C, Woo MK (2005) Snowmelt runoff processes in a headwater lake and its catchment, subarctic Canadian Shield. Hydrol Process 20:987–1000

Nagarajan B, Yau MK, Schuepp PH (2004) The effects of small water bodies on the atmospheric heat and water budgets over the Mackenzie River Basin. Hydrol Process 28:913–938

Oswald CJ, Rouse WR, Binyamin J (2007) Modeling lake energy fluxes in the Mackenzie River Basin using bulk aerodynamic mass transfer theory. (Vol. II, this book)

Pomeroy JW, Essery R, Gray DM, Shook KR, Toth B, Marsh P (1999) Modelling snow-atmosphere interactions in cold continental environments. In: Tranter et al. (eds) Interactions between the cryosphere, climate and greenhouse gases. IAHS Publ. No. 256, IAHS Press, Wallingford, UK, pp 91–102

Pomeroy JW, Gray DM, Landine PG (1993) The Prairie Blowing Snow Model: characteristics, validation, operation. J Hydrol 144:165–192

Pomeroy JW, Gray DM, Marsh P (2007) Studies on snow redistribution by wind and forest, snow-covered area depletion and frozen soil infiltration in northern and western Canada. (Vol. II, this book)

Quinton WL, Marsh P (1999) A conceptual framework for runoff generation in a permafrost environment. Hydrol Process 13:2563–2581

Quinton WL, Hayashi M (2007) Recent advances toward physically-based runoff modeling of the wetland-dominated central Mackenzie River Basin. (Vol. II, this book)

Quinton W, Hayashi M, Pietroniro A (2003) Connectivity and storage functions of channel fens and flat bogs in northern basins. Hydrol Process 17:3665–3684

Ritchie H, Delage Y (2007) The impact of CLASS in MAGS monthly ensemble predictions (Vol. I, this book)

Rouse WR, Binyamin J, Blanken PD, Bussières N, Duguay CR, Oswald CJ, Schertzer WM, Spence C (2007b) The influence of lakes on the regional energy and water balance of the central Mackenzie River Basin. (Vol. I, this book)

Rouse WR, Blanken PD, Duguay CR, Oswald CJ, Schertzer WM (2007a) Climate-lake interactions. (Vol. II, this book)

Schertzer WM, Rouse WR, Blanken P, Walker AE, Lam DCL, León L (2007) Interannual variability of the thermal components and bulk heat exchange of Great Slave Lake. (Vol. II, this book)

Schuster M (2007) Characteristics of the moisture flux convergence over the Mackenzie River Basin for the 1990-2000 water-years. (Vol. I, this book)

Serreze MC, Walsh JE, Chapin III FS, Osterkamp T, Dyurgerov M, Romanovsky V, Oechel WC, Morison J, Zhang T, Barry RG (2000) Observational evidence of recent change in the northern high-latitude environment. Climatic Change 46:159–207

Smith CD (2007) The relationship between monthly precipitation and elevation in the Alberta Foothills during the Foothill Orographic Precipitation Experiment. (Vol. I, this book)

Smith LC, Sheng Y, Macdonald GM, Hinzman LD (2005) Disappearing arctic lakes. Science 308:1429

Smith SB, Yau MK (1993) The causes of severe convective outbreaks in Alberta. Part I: a comparison of a severe outbreak with two nonsevere events. Mon Weather Rev 121:1099–1125

Soulis ED, Seglenieks FR (2007) The MAGS integrated modeling system. (Vol. II, this book)

Spence C, Woo MK (2002) Hydrology of subarctic Canadian Shield: bedrock upland. J Hydrol 262:111–127

Spence C, Woo MK (2003) Hydrology of subarctic Canadian Shield: soil-filled valleys. J Hydrol 279:151–166

Spence C, Woo MK (2007) Hydrology of the northwestern subarctic Canadian Shield. (Vol. II, this book)

Szeto KK (2007) Cold-season temperature variability in the Mackenzie Basin. (Vol. I, this book)

Szeto KK, Liu J, Wong A (2007b) Precipitation recycling in the Mackenzie and three other major river basins. (Vol. I, this book)

Szeto KK, Stewart RE, Yau MK, Gyakum J (2007a) The Mackenzie climate system: a synthesis of MAGS atmospheric research. (Vol. I, this book)

Szeto KK, Tran H, MacKay MD, Crawford R, Stewart RE (2007c) Assessing water and energy budgets for the Mackenzie River Basin. (Vol. I, this book)

Thorne R, Armstrong RN, Woo MK, Martz LW (2007) Lessons from macroscale hydrologic modeling: experience with the hydrologic model SLURP in the Mackenzie Basin. (Vol. II, this book)

Trischenko AP, Khlopenkov KV, Ungureanu C, Latifovic R, Luo Y, Park WB (2007) Mapping of surface albedo over Mackenzie River Basin from satellite observations. (Vol. I, this book)

Walker A, Silis A, Metcalf JR, Davey MR, Brown RD, Goodison BE (2000) Snow cover and lake ice determination in the MAGS region using passive microwave satellite and conventional data. Proc 5th Scientific Workshop, Mackenzie GEWEX Study, Edmonton, Alberta, Canada, pp 39–42

Wen L, Rodgers D, Lin CA, Roulet N, Tong L (2007) Validating surface heat fluxes and soil moisture simulated by the land surface scheme CLASS under subarctic tundra conditions. (Vol. II, this book)

Woo MK, Modest P, Martz L, Blondin J, Kochtubajda B, Tutcho D, Gyakum J, Takazo A, Spence C, Tutcho J, diCenzo P, Kenny G, Stone J, Neylle I, Baptiste G, Modeste M, Kenny B, Modest W (2007d) Science meets traditional knowledge: water and climate in the Sahtu (Great Bear Lake) region, Northwest Territories, Canada. Arctic 60:37–46

Woo MK, Mielko C (2007) Flow connectivity of a lake–stream system in a semi-arid Precambrian Shield environment. (Vol. II, this book)

Woo MK, Mollinga M, Smith SL (2007b) Modeling maximum active layer thaw in boreal and tundra environments using limited data. (Vol. II, this book)

Woo MK, Rouse WR (2007) MAGS contribution to hydrologic and surface process research. (Vol. II, this book)

Woo MK, Rouse WR, Stewart RE, Stone JMR (2007a) The Mackenzie GEWEX Study: a contribution to cold region atmospheric and hydrologic sciences. (Vol. I, this book)

Woo MK, Thorne R (2003) Streamflow in the Mackenzie Basin, Canada. Arctic 56:328–340

Woo MK, Thorne R (2006) Snowmelt contribution to discharge from a large mountainous catchment in subarctic Canada. Hydrol Process 20:2129–2139

Woo MK, Thorne R, Szeto KK (2006) Reinterpretation of streamflow trends based on shifts in large-scale atmospheric circulation. Hydrol Process 20:3995-4003

Woo MK, Thorne R, Szeto KK, Yang D (2007c) Streamflow hydrology in the boreal region under the influences of climate and human interference. Phil Trans Roy Soc B (in press)

Yang D, Ye B, Shiklomanov A (2004) Discharge characteristics and changes over the Ob River watershed in Siberia. J Hydrometeorol 5:596–610

Zhang X, Vincent L, Hogg W, Nitsoo A (2000) Temperature and precipitation trends in Canada during the 20th century. Atmos Ocean 38:395–429

Subject Index

A

adiabatic 2, 480
acrotelm 19, 21
active layer 18, 125, 126, 128, 132–136, 265, 266, 269, 277, 475, 487
advection, advective 2, 13, 15, 33, 81, 83, 87, 88, 97, 98, 105–108, 110, 111, 119, 169, 177, 202, 239, 481
albedo 4, 15, 41, 43, 101, 112, 142, 147, 439, 449, 455, 477
alpine 266, 374, 398, 459
AMSR (Advanced Microwave Scanning Radiometer) 68–71, 73, 74
anticyclone 2, 476, 478, 480
anthropogenic 1, 363
AO, Arctic Oscillation 2
application 2, 3, 29, 32, 39, 40, 55, 56, 63, 68, 72, 91, 133, 152, 198, 201, 216, 281, 286, 296, 322, 333, 357, 377, 382, 385, 388, 393, 400, 411, 414, 446, 465, 478, 489
Arctic 2, 10, 15, 23, 26, 28, 81, 82, 84–86, 88, 89, 97, 98, 117, 126, 141, 222, 227, 235, 252, 266, 286, 436, 463, 464, 476, 480, 483
ASA (Aggregated Simulation Area) 375–378, 398, 403, 407, 408
Athabasca 23, 26–30, 53, 54, 64, 72, 74, 140, 141, 143, 289, 290, 294, 298, 299, 307–313, 315, 318, 322, 346, 363–367, 368, 369, 380–382, 411–414, 425, 430–432, 463, 464, 477, 482, 486, 488
AVHRR (Advanced Very High Resolution Radiometer) 39–43, 45–53, 55, 56, 141, 414, 445, 457

B

bedrock 24, 25, 33, 162, 177, 221, 222, 224, 227, 235–244, 247, 248, 253, 482
bias 5, 50, 51, 56, 172, 408, 439, 440, 442, 491
blowing snow (see snow) 3, 4, 12, 13, 81–87, 97, 98, 478, 480, 481
bootstrap 363, 364
breakup 9, 10, 26, 28–30, 33, 53, 55, 144, 148, 201, 205, 223, 224, 281, 282, 285–287, 289, 291, 292, 294, 295, 297–301, 307–314, 316, 318–320, 322, 323, 337, 338, 345, 347, 348, 350, 351, 353–356, 405, 488, 489
– mechanical breakup 28, 287, 289, 295, 351, 352, 358, 482
– thermal breakup 28, 291, 295, 351, 358, 482
bog 22, 237, 257–266, 273, 274, 276, 277, 482
boreal 3, 5, 9, 10, 15, 16, 23, 43, 81, 82, 84, 89, 92, 98, 100, 125–127, 129–135, 151, 162, 192, 193, 221, 222, 252, 257, 258, 374, 459, 467, 480, 486
BOREAS (Boreal Ecosystem-Atmosphere Study) 43, 467, 468
boundary layer 10, 13, 86, 105, 163, 169, 175, 457
Bowen ratio, BREB 162, 166, 177
bulk density 115, 128, 132, 266, 267
BWI (Basin Wetness Index) 59, 61–64, 68–76

C

CAGES (Canadian GEWEX Enhanced Study) 39, 41, 101, 436, 479, 489
calibrate, calibration 41, 46, 60, 63, 64, 71, 72, 76, 118, 166, 296, 320, 327, 330, 332–336, 341, 352, 357, 401, 402, 406–408, 417, 419–422, 429, 431, 437, 448, 450, 452, 460–462, 469
Canadian Shield (*see* Shield) 24, 27, 31, 47, 140, 147, 161, 222, 231, 235, 236, 242, 243, 247, 250, 251, 253, 482, 488
canopy 10, 12, 14, 45, 69, 83, 84, 89, 144, 237, 265, 276, 403, 404, 437, 439, 449
catotelm 19, 21
CGCM2, GCMII 327, 337, 339, 355, 356, 358, 416, 423, 425, 426, 428, 430, 431
channel 19, 20, 22, 24, 41, 61, 62, 72, 97, 98, 100, 112, 114–117, 127, 201, 222, 224, 226, 227, 230, 231, 235, 249, 250, 257–266, 273, 274, 276, 277, 281, 283, 286, 287, 289, 291–294, 298, 308, 310, 328, 329, 341, 346, 348, 350, 373–376, 378–381, 384, 387–392, 398, 399, 405, 413, 449, 457, 459, 482
circulation 153–155, 298, 371, 447, 478, 485
- mountain-plains circulation 483
- thermohaline circulation 476
circumpolar 33, 98, 141, 475, 476, 490
CLASS (Canadian Land Surface Scheme) 5, 45–50, 56, 83, 84, 435–442, 445–453, 455, 459, 467, 469, 470, 478, 490
climate change 1, 33, 121, 136, 153, 156, 181, 194, 281, 282, 293, 300, 302, 327, 328, 336–341, 345, 347, 355, 356, 359, 363, 369, 411, 412, 426, 478, 486, 487, 490
climate variability 1, 139, 144, 151, 156, 157, 161, 211, 253, 327, 475, 478, 485, 488
climate warming 113, 136, 156, 158, 282, 327, 336, 338, 475, 487, 488
CLIMo 144, 145
cloud 3, 5, 31, 32, 39, 41, 42, 44, 45, 46, 51, 53, 56, 71, 107, 141, 192, 286, 477, 483, 484

CMC (Canadian Meteorological Centre) 5, 63, 117, 447, 478
Community Land Model 16
conduction 16, 18, 91, 132, 484
connectivity 24, 25, 33, 221, 222, 229–231, 235, 252, 274, 455, 484
convect, convection 18, 132, 432
Cordillera 2, 26, 27, 81, 82, 480
CRCM (Canadian Regional Climate Model) 3–5, 40, 45–52, 56, 435, 441, 447, 462, 464, 477, 490
CRHM (Cold Regions Hydrological Model) 272, 278
cyclogenesis 2, 192, 480, 483
cyclone 2, 237, 476, 478, 480, 484

D

Darcy, Darcian 450, 452
degree-day, degree day 126, 132, 286, 349, 356, 357, 402, 404
DEM (digital elevation model) 72, 76, 98, 296, 371–374, 377, 380, 382–387, 390–392, 394, 395, 398, 399, 414, 445
dimictic 143, 150, 182, 197, 206, 216, 485
discharge 4, 19, 26, 30, 60, 71–73, 99, 100, 114, 116, 120, 156, 213, 221, 224, 226, 229–231, 245, 250, 258, 274, 275, 287, 291–293, 296–298, 310, 332, 333, 338, 341, 354, 355, 358, 379, 402, 403, 405,–408, 421, 476, 482, 484, 485
downscale, downwscaling 59, 74, 75, 422, 430, 432
drainage 18–20, 22, 24–26, 33, 98, 99, 221, 222, 226, 229, 230, 235, 243, 244, 248–250, 252, 260, 263–265, 267, 271–274, 277, 312, 364–366, 371, 373–376, 378–380, 382–389, 391–394, 399, 400, 414, 416, 417, 440, 445, 449–452
DYRESM (Dynamic Reservoir Model) 152, 154

E

ECMWF (European Centre for Medium Range Weather Forecasts) 414, 462, 468

eddy covariance 162, 166, 178, 181, 183, 184
ELCOM 153, 155, 206, 207
element 45, 178, 247–250, 252, 283, 296, 327, 328, 333, 415, 447, 449, 450, 469
El Niño 182, 202
emissivity 61–63
energy and water cycle (*see* water and energy cycle) 1, 40, 56, 217, 372, 479, 488
energy balance, energy budget 1, 4, 31, 39, 40, 53, 83, 84, 94, 97, 103, 108, 110, 120, 129, 147, 157, 161, 162, 182, 282, 330, 404, 436, 437, 445–448, 475, 478, 484, 489
ERA-40 411, 414, 415, 416, 420–424, 430, 431, 432, 462, 464, 466, 478
error 4, 60, 75, 99, 100, 106, 129, 153, 161, 168, 170, 171, 174–178, 214, 224, 314, 380, 381, 383, 388–390, 394, 399, 407, 408, 420–423, 430, 437, 439, 440, 442, 458, 459, 466
Eulerian 327, 328
evaporation 2–5, 9, 22, 24, 25, 30–32, 52, 55, 56, 59, 67, 73, 75, 100, 116, 139, 142, 143, 147–149, 151, 152, 156–158, 181, 182, 184, 187, 188, 191, 193, 221, 223–226, 229–231, 235, 239–241, 243, 244, 250, 274, 403, 415, 431, 442, 446, 465–467, 481–485, 488
evapotranspiration 23, 39, 43, 52, 56, 151, 193, 244, 249, 274, 275, 377, 379, 402–404, 446–469
extreme 2, 33, 47, 60, 135, 142, 281, 285–287, 300, 307, 309, 313, 314, 320, 321, 336, 338, 341, 345, 423, 426, 450, 452, 475, 479, 480, 483

F

feedback 43, 52, 181, 194, 413, 435, 445–478, 481, 485, 487, 490
fen 22, 257–266, 273, 274, 276, 277, 449, 482
fetch 163, 208
fire 5, 237, 475, 477, 478, 483, 487, 490

fill-and-spill 25, 33, 227, 230–232, 244, 245, 248, 252, 484
finite element 327, 328, 333, 447
flood 9, 10, 26, 28, 29, 33, 59, 60,62, 70, 71, 75, 231, 274, 281, 282, 287, 289, 294,–301, 307–311, 314, 316, 318, 320, 322, 323, 345, 347, 348, 351–359, 432, 478, 482, 485, 488–490
flow 16, 18, 19, 21–26, 33, 71, 72, 85, 97, 100, 113, 115–117, 120, 154–156, 202, 221, 222, 224, 226–232, 235, 239, 242, 244–250, 252, 253, 262–264, 247–250, 252, 253, 262–264, 272, 274, 277, 283–287, 289–298, 307, 310, 314, 327–330, 338, 345, 346, 350–359, 363, 368, 369, 371–374, 377, 381–383, 386,–395, 398, 399, 403–406, 408, 412, 420–432, 435, 440–442, 445, 449–453, 455, 460, 461, 463, 467–469, 478, 480, 482, 484–486, 488
– baseflow 432
– high flow 21, 26, 250, 297, 351, 353, 355, 431, 440, 482
– interflow 413, 429, 450–452
– lateral flow 19, 24, 263, 435, 440, 441, 445, 449, 450, 467–469
– low flow 26, 224, 368, 369, 405, 406, 420, 440, 484
– matrix flow 19, 453
– peak flow 100, 405, 430, 459, 482, 484, 488
– pipeflow 19
– quick flow 21, 22, 23
– slow flow 21
– stream flow (*see* streamflow) 115, 228, 232, 264, 274, 292, 423, 426
– subsurface flow 19, 21, 25, 115, 116, 224, 244, 246, 272, 274, 282, 484
– surface flow (*see* overland flow) 24, 25, 116, 117, 221, 222, 244, 246, 248, 263, 277, 450, 482
– two layer flow system 21–23, 33
flow connection 22, 24, 221, 274, 484
flux
– energy flux 5, 16, 32, 92, 97, 98, 110, 119, 147, 161, 166, 177, 328, 477, 478, 481

- heat flux (*see* heat) 16, 17, 31, 82, 86, 91, 101, 102, 104–107, 129, 139, 142, 143, 147, 148, 150–153, 156–158, 161, 162, 166, 171, 173, 174, 178, 181, 183, 185, 186, 188, 197, 202, 203, 212, 213, 215, 216, 435–442, 469, 481, 482
- turbulent flux 16, 98, 100, 101, 107, 108, 110, 151, 161, 162, 174, 177, 178, 183, 184, 187, 188, 192, 440, 481
- vapor flux 2, 148, 166

forecast 5, 29, 33, 59, 60, 63, 72, 281, 282, 294, 298, 299, 302, 307, 308, 310–314, 318, 320, 321–323, 413–415, 422, 432, 446, 447, 460, 478, 490

forest 3, 5, 9, 10, 12, 14–16, 23, 33, 40, 81–84, 89, 90, 92, 97, 98, 100, 113, 125, 126, 130, 133, 134, 144, 162, 237, 374, 401, 413, 458, 459, 467, 468, 477, 478, 480, 481

frazil 281, 283, 284, 293, 328–331, 335, 485

freeze-thaw 9, 16, 132, 436, 478

freeze-back, freeze-up 26, 28, 30, 129, 130, 132, 142–144, 148, 151, 201, 205, 212, 269, 270, 281,–283, 285, 291, 292, 299, 300, 308, 312, 336–338, 345, 347–349, 351, 353–355, 358, 484, 485

freshet 222

frost
- frost table 20, 21, 23, 24, 266, 269,271, 273, 277
- seasonal frost 16, 24, 450

frozen ground 16, 91, 405, 479

frozen soil 9, 18, 21, 33, 81–83, 91, 92, 115, 135, 242, 243, 266, 402, 405, 455, 475, 478, 481

Fuzzy Expert System, fuzzy logic 29, 299, 307, 308, 314–318, 320–323

FWS, fractional water surface 59, 61–64, 66–68, 70, 71

G

GCM, General Circulation Model, Global Climate Model 5, 300, 355, 356, 371, 395, 399, 412–416, 423, 424, 426, 431, 432, 447, 486, 487

GEM (Global Environmental Multiscale Model) 63, 117, 411, 414, 415, 419, 420–424, 430–432, 447, 448, 460–464, 466–468, 478

Great Bear Lake 27, 30, 53, 64, 140, 141, 143, 149, 177, 477, 482, 485

Great Slave Lake 27, 30–32, 43, 47, 53, 55, 64, 139, 140, 141, 143, 149–157, 162, 163, 170, 178, 181–183, 185, 186, 188, 191, 193, 194, 197–199, 201–203, 206–210, 213, 214, 217, 346, 482, 485

greenhouse 356, 359, 486, 487

ground ice 16, 18, 21, 127, 450, 481, 484

groundwater 26, 60, 224, 239, 250, 253, 274, 413

GRACE (Gravity Recovery and Climate Experiment) 464, 466

GRU (Group Response Unit) 117, 390, 449

GT (soil and vegetation skin temperature) 45–52, 56

GTOPO30, GTOPO-30 387, 445, 457

H

Hadley Centre 84

headwater 23, 25, 221, 222, 230, 235, 237, 242, 246, 247, 249, 250, 252, 290, 338, 459, 484

heat
- heat budget 2, 202
- heat conduction 16, 91, 484
- heat exchange 9, 31, 151, 197, 198, 201, 202, 209, 212–216, 277, 283, 329, 330, 333, 334, 336
- latent heat, Q_E 153, 157, 161, 162, 167, 168, 172–174, 176–178, 183, 186, 188, 202, 212, 330, 435, 437–442, 467, 469, 482, 485
- sensible heat, Q_H 16, 17, 31, 97, 98, 101–103, 105, 106, 129, 139, 143, 147, 148, 150–153, 156–158, 161, 162, 166–168, 171–178, 181, 191, 202, 435, 437,–440, 442, 481–483, 485

hummock 19, 20, 115, 116, 120, 285, 435, 436, 440, 442

HYDAT 477

hydraulic conductivity 16, 19, 68, 266–269, 274, 412, 418, 455
hydrograph 21, 118, 230, 231, 250, 264, 273, 289, 291, 298, 300, 353, 357, 373, 379, 384, 386, 389, 392, 405, 411, 420, 421, 423, 426, 429, 431, 446, 449, 462, 469, 482
hydrologic cycle 60, 76

I

ice
- ice cover 26, 28, 30, 31, 139, 144–146, 157, 201, 205, 211, 213, 216, 223, 226, 229, 282–287, 289–295, 297, 299, 300, 308, 327–332, 336–340, 345, 349–351, 353, 355, 356, 358, 359, 479, 481–483, 485, 488
- ice floe 283–285, 288, 341, 349
- ice free, ice-free 30, 44, 54, 147, 149, 150, 153, 156, 163, 178, 182, 187, 188, 191–193, 197, 198, 201–203, 205, 206, 212, 216, 223, 229, 231, 283, 454, 485, 487
- ice front 284, 285, 327, 328, 332, 333, 335–341
- ice jam, ice-jam 26, 28, 29, 33, 281, 282, 285, 287–292, 294–302, 307–310
- lake ice 16, 30, 31, 64, 132, 139, 142, 144, 147, 156, 158, 182, 201, 205, 223, 224, 226, 229, 482, 485
- river ice 9, 10, 26, 28, 33, 281, 282, 284, 286, 290–294, 298, 300–302, 307, 308, 310, 311, 322, 327, 328, 330, 333, 341, 402, 405, 475, 477, 478, 482, 485, 488, 489
- sea ice 13, 293, 476
ice content 24, 243, 450, 453, 454
icing 286, 484
IKONOS 260–262
impact 3, 5, 33, 92, 120, 125, 135, 139, 147, 156, 192, 197, 209, 216, 253, 281, 282, 284, 300, 327, 328, 337–341, 345, 348, 356, 358, 363, 364, 371, 379, 385, 399, 400, 402, 467, 469, 478, 487, 488, 490, 491
impermeable 21, 24, 239, 266
impervious 18, 24, 229, 405, 481

infiltrate, infiltration 9, 18, 19, 21, 23–25, 33, 59, 81–83, 85, 87, 89, 91–93, 115, 229, 239, 241, 243, 244, 247, 249, 277, 375, 402, 404, 405, 418, 475, 478, 481, 482
- frozen soil infiltration 18, 33, 81, 82, 475, 478
- meltwater infiltration 18, 243
inflow 22, 24, 25, 155, 206, 221, 223, 225, 226, 230, 231, 243, 244, 250, 333, 335, 336, 341, 364, 442
initialization 5, 397, 406
Interior Plains 22, 26, 27, 30, 140, 483, 484, 487, 488
interception 81–84, 243, 480, 481
inversion 2, 31, 148, 480
ISBA (Interactions Sol-Biosphère-Atmosphère) 411, 414–417, 419, 420, 427, 431, 432

L

lag time 129
lake
- lake ice 30, 31, 64, 132, 139, 142, 144, 147, 156, 158, 182, 205, 223, 224, 226, 229, 482, 485
- lake level 25, 221, 223, 224, 226, 228, 229, 249, 484
- lake size 53, 139, 140, 142, 144, 147, 158, 162, 177, 250
Lake Athabasca 27, 30, 53, 72, 140, 141, 143, 309, 346, 482
Landsat 85, 260, 264, 458
land surface scheme, land surface model, LSM, LSS 5, 16, 40, 45, 46, 83, 84, 87, 98, 117, 250, 411, 412, 414, 431, 435, 445–447, 469, 490
lapse rate 403, 404
latent heat 9, 16, 17, 18, 82, 86, 91, 92, 101–104, 107, 119, 132, 134, 147, 148, 157, 167, 168, 183, 186, 188, 212, 330, 343, 435, 437–442, 467, 469, 482, 485
LAI (leaf-area index) 69, 70, 72, 84
Liard 23, 26, 27, 28, 258, 259, 260, 363, 364, 366, 367, 368, 369, 380, 382, 401, 405, 406, 407, 408, 461, 463, 464, 486, 488

L

LiDAR 98
lightning 5, 483
linear 3, 41, 43, 61, 64, 72, 129, 153, 161, 172, 174, 181, 186, 188–190, 193, 194, 299, 312, 313, 319, 322, 330, 356, 411, 417, 419

M

macropore 25, 244
Mackenzie River 1, 4, 9, 26, 27, 28, 39, 59–61, 66, 81, 98, 125, 126, 139, 140–142, 145, 155, 161, 181, 198, 206, 235, 257–259, 276, 282, 286, 290, 346, 363, 369, 372, 374, 381, 382, 386, 388, 393, 436, 445, 458, 459, 463–467, 475, 476
management 33, 59, 125, 126, 232, 397, 491
Manning 274, 292, 336, 375, 404, 407, 450
Mann-Kendall 363, 364
mass transfer 81, 91, 139, 152, 153, 158, 161, 162, 168, 170, 172, 174, 178, 182, 193, 194

MAUP (Modifiable Area Unit Problem) 372, 373
melt 3, 9, 10, 14, 15, 16, 21, 26, 28, 30, 31, 33, 48, 50, 73, 81, 85, 88–90, 97, 98, 100–103, 105, 107–120, 129, 132, 134, 142, 147, 150, 156, 203, 212, 224, 226, 235, 249, 250, 269, 276, 277, 283, 287, 289, 291, 299, 329, 330–332, 345, 350, 355, 402, 404, 407, 478, 480–482, 485
meltwater 18, 21, 24, 82, 91, 92, 97, 98, 113–115, 117, 120, 156, 229, 243, 249, 269, 459, 482
modeling 3–6, 9, 13, 24, 46, 82, 98, 105, 117, 125, 128, 152, 161, 177, 198, 216, 221, 230–232, 257, 271, 277, 296, 298, 299, 302, 307, 315, 328–330, 333, 334, 341, 358, 359, 363, 369, 371, 377, 391, 394, 397, 399, 400, 402, 408, 411, 439, 442, 445–451, 456, 457, 467, 469, 475–478, 490
MODIS (Moderate-Resolution Imaging Spectroradiometer) 39, 40, 46, 49, 51, 52, 56, 68, 70–73, 407

moisture 2, 4, 22, 23, 31, 47, 59,–61, 67–72, 74, 76, 77, 92, 93, 100, 125, 132, 142, 143, 148, 212, 239, 241, 243, 244, 246, 248–272, 299, 313–315, 318–321, 323, 405, 412, 416, 417, 419, 431, 435–437, 439–442, 448, 450, 452, 464, 466, 478–481, 483–486, 488
moisture content 16, 18, 132, 267, 269–272, 435–437, 439–442, 483, 484
Muskingum 407
MWT, mid-winter thaw 30, 281, 289, 300, 345, 356, 357, 359, 488

N

NASA (National Aeronautics and Space Administration) 46, 56, 68, 195, 477
Nash, Nash-Sutcliffe 401, 402, 459–464, 469
NCAR (National Center for Atmospheric Research) 144, 430, 462
NCEP (National Center for Environmental Prediction) 14, 462, 464, 477
Nival 26, 235, 2590, 252, 482
NOAA (National Oceanographic and Atmospheric Administration) 39, 41, 48, 50
NSIDC (National Snow and Ice Data Center) 56, 63, 477
Network 20, 22, 33, 99, 128, 133, 136, 195, 216, 222, 229, 230, 235, 248, 250, 252, 260, 271, 273, 277, 291, 299, 324, 327, 337, 350, 371, 373–388, 390, 392, 394, 398–400, 414, 449, 456, 458, 488–490

O

open water, open-water 29, 43, 53, 59, 64, 70–72, 139, 142, 143, 147, 150, 157, 188, 191, 201, 205, 235, 257, 266, 281, 287, 291–294, 297, 307, 330, 347, 487
optimization 393, 397, 404–408, 440
organic 268, 277, 435–439, 442, 481, 482, 487
overland flow 25, 244–247, 482

outflow 25, 26. 155, 158, 206, 221, 223–231, 235, 244, 248–250, 313, 339, 408, 450, 484

P

Peace–Athabasca Delta, PAD 29, 59, 72–76, 282, 295, 296, 300, 345–347, 353, 355, 356, 359
parameterization 3, 15, 121, 371, 373, 379, 395, 397, 408, 435, 436, 442, 447, 450
passive microwave 53, 55, 59, 68–71, 76
PBSM (Prairie Blowing Snow Model) 12, 13, 86, 87
PDO, Pacific Decadal Oscillation 3, 485
Peace 26–30, 59, 72, 73, 75, 258, 288–290, 293, 295, 296, 300, 307, 327,–330, 333–341, 345–353, 355, 357–359, 363–365, 367–369, 380–382, 414, 432, 459, 463, 464, 486, 488
peat 18–23, 115–117, 120, 125, 134, 151, 188, 236, 237, 257–260, 262, 266–268, 413, 436
peatland 257, 258, 260, 263, 277, 435,–442
peat plateau 22, 236, 257–274, 276, 449
perched 21, 247, 295, 347
percoate, percolation 14, 21, 24, 97, 98, 113, 115
permafrost 16, 18, 19, 21–23, 33, 115, 125, 126, 132, 135, 136, 156, 236, 258, 261, 266, 272, 276, 450, 487
– continuous permafrost 98, 435, 436, 442, 459, 487
– discontinuous permafrost 22, 125, 162, 236, 257, 258, 274, 487
pervious, perviousness 3
PIEKTUK 3, 12, 13, 480, 481
pixel 40–44, 46, 51, –64, 70, 71, 74, 75, 402, 458
PNA, Pacific-North American 2
pond 10, 30, 114, 115, 140, 142, 239, 295, 300, 345–347, 436, 449, 485
porosity 19, 115, 128, 132, 266–268, 271, 272, 310, 331, 335, 405, 439, 453
Prairies 2, 88, 289, 300, 377, 486

precipitation 2–5, 10–12, 31, 44, 59, 64, 67, 73–75, 100, 103, 117, 139, 141, 156, 157, 221, 222, 235, 237, 245, 258, 274–276, 300, 311, 313, 335, 356, 375, 377, 379, 400, 403–405, 410–413, 415, 416, 423, 425, 427, 429, 431, 432, 437, 439–441, 461, 462, 466, 467, 469, 479, 480, 483–486, 488
predict, prediction 4–6, 29, 33, 36, 68, 71, 77, 78, 84, 111, 114, 117, 136, 153, 181, 190, 232, 286, 289, 292, 296, 298–300, 302, 307–309, 313, 323, 327, 336–340, 347, 350, 352, 353, 355, 357, 380, 411, 415–417, 419, 423–426, 429, 431, 432, 439, 443, 447, 448, 451, 459, 475, 477, 478, 482, 487, 490
pressure 2, 43, 117, 147, 148, 152, 166–168, 181, 184, 188, 189, 191, 192, 195, 209, 223, 311, 415, 437, 480, 483, 484, 486
Priestley 224, 239, 403

Q

Q_b^* (net flux of water) 224, 225

R

radar 3, 28, 293, 477, 484, 489
RADARSAT 75, 293, 294
radiation
– longwave, long-wave 5, 14, 43, 100, 200, 202, 283, 437, 439, 483
– net radiation, Q^* 147, 149, 156–158, 404, 437, 439, 480, 481
– shortwave, short-wave 4, 5, 14, 89, 107, 192, 349, 437, 439, 481, 483
– solar radiation 202, 211, 212, 216, 282, 286, 311–313, 333, 349, 481, 483
rain shadow 480
rainfall 417, 418, 424, 427, 429, 431, 478, 483, 484, 486
rainwater 24, 243, 267
reanalysis, re-analysis 12, 144, 403, 407, 413, 414, 431, 432, 462, 469, 477, 478, 489

recycling 2, 479, 483, 488
regime
– hydrologic regime 16, 75, 139, 143, 363, 369
ice regime 29, 144, 291, 327, 329, 331, 333, 335–337, 339–341, 345, 355, 357, 476, 488
thermal regime 18, 33, 125, 150, 300, 327
relative humidity 12, 13, 162, 166, 184, 200, 217, 286
remote sensing 29, 40, 56, 60, 75, 139, 281, 282, 293, 301, 402, 464, 475–477
reservoir 31, 59, 60, 152, 154, 230, 244, 290, 291, 338, 339, 341, 355, 358, 363, 398, 403, 417, 429, 450, 488
RFE (Regional Finite Element) 447, 460–463, 466, 467
Richard's equation 450
River1D 328, 333, 336, 339, 340
RIVJAM 296, 348, 352
roughness 101, 112, 162, 168, 274, 277, 285, 287, 297, 375, 404, 407, 437
routing 23, 230, 250, 257, 274, 277, 281, 286, 328, 371, 374, 386–392, 395, 398, 399, 407, 408, 412, 415, 447, 449, 457, 458, 469
runoff 4, 9, 19, 14, 15, 19, 21–26, 28, 33, 59, 60, 81, 97–101, 103, 105, 107, 109, 113–115, 117–222, 226, 229, 230, 235, 241,–252, 257, 258, 262–265, 267, 271, 274–276, 281, 286, 287, 289, 300, 350, 355, 357, 359, 364, 397–400, 403–405, 408, 411–419, 423–427, 429, 431, 432, 446, 449, 450, 462, 466, 467, 469, 482, 484

S

SAR (Synthetic Aperture Radar) 28, 293
atellite 3–5, 28, 39–41, 43, 45–48, 50, 51, 53, 55, 56, 59, 60, 62, 63, 68, 70, 73, 75, 76, 103, 110–112, 141, 149, 152, 260, 274, 277, 281, 293, 294, 301, 401, 407, 414, 445, 464, 477, 489, 491
scale
downscale 59, 74, 75, 422, 430, 432

large-scale 1, 2, 31, 94, 371–373, 399, 415, 485, 491
local scale 9, 68, 97, 105, 107, 111, 119, 253, 371, 415, 481
macroscale, macro-scale 33, 397, 408, 478
– synoptic scale 32, 151, 193
scaling 15, 48, 76, 83, 86, 371, 374, 399, 491
scenario 30, 113, 136, 144–146, 285, 300, 327, 337–339, 341, 347, 355–359, 411, 412, 416, 423–426, 428–432, 478, 487, 488
seepage 25, 117, 224, 243, 244, 286, 451, 452, 484
semi-arid 25, 221, 222, 226, 229–231, 237, 244, 250, 252, 479, 484
sensitivity 4, 18, 60, 62, 63, 68, 83, 136, 181, 182, 193, 194, 213, 216, 317, 321, 355, 404, 452, 453, 567
Shield 9, 24, 25–27, 30, 31, 47, 140, 147, 161, 162, 178, 221, 222, 227, 229–231, 235–239, 242–244, 246, 247, 250–253, 449, 482, 488
shift 76, 151, 187, 192, 194, 250, 301, 415, 475, 485, 486
shrub 13–15, 24, 86, 89, 101, 102, 111, 113, 114, 117, 126, 258, 259, 413, 437, 459
simulate, simulation 3, 5, 15, 16, 18, 33, 47, 56, 72, 86, 88, 90, 91, 93, 107, 110, 112, 113, 117–121, 125, 126, 128, 133, 135, 152–155, 197, 206, 207, 216, 253, 298, 327, 330, 332–339, 341, 357, 375, 393, 397, 398, 400, 402–408, 411, 412, 415, 416, 419–424, 426, 427, 430–432, 435–442, 445, 446, 450, 453, 458–467, 469, 477, 487–489
SLURP (Semi-distributed Land Use-based Runoff Processes) 374, 375, 393, 397, 400–408, 409, 447
slush 283–285, 329, 331, 349
SMMR (Scanning Multichannel Microwave Radiometer) 76
snow
– blowing snow 3, 4, 12, 13, 81–87, 97, 98, 478, 480, 481

- distribution 10, 15, 40, 60, 68–70, 72, 85, 86, 88–90, 97, 101, 105, 108, 114, 119, 125, 126, 135, 153–155, 168, 206, 214, 216, 235, 241, 268, 272, 273, 297, 319, 380, 398, 411, 415, 416–419, 431, 449, 450, 457, 478
- drifting 10, 126, 237, 329
- intercepted snow 10, 82, 83, 85, 480
- redistribution 10, 12, 15, 33, 81, 86, 127, 277, 480
- undercatch 10

snow ablation 10, 83, 92
- snow accumulation 10, 15, 81–86, 97, 98, 100, 112, 113, 129, 236, 237, 290, 346, 355, 424, 431, 488
- snow cover, snowcover 3, 10, 11, 15–17, 51, 81, 82, 87–90, 97, 98, 101–105, 107, 110, 111, 113, 114, 117, 119, 120, 126, 128, 144, 147, 237, 272, 285, 341, 407, 412, 418, 455, 456, 464, 477, 481, 486, 487, 489
- snow covered area, SCA 81, 82, 88–90, 97, 102, 104, 110, 111, 113, 114, 117, 120, 418, 455, 456, 464
- snowfall 3, 10–12, 83, 84, 100, 125, 158, 222, 355, 424, 427, 480, 484, 487, 490
- snow free, snow-free 15, 51, 87, 97, 98, 101, 103, 105, 106, 110, 111, 129, 144, 274, 407
- snow interception (see interception) 81, 83, 84, 480, 481
- snow melt, snowmelt 10, 12, 14–17, 21,–26, 33, 48, 50, 51, 56, 67, 73, 81–83, 85, 87–89, 91–93, 97–99, 101, 103, 105, 107–113, 115–117, 119, 120, 129, 221, 222, 224–227, 229, 231, 235, 237, 239, 241, 243, 244, 249, 250, 252, 258, 265, 270, 276, 286, 289, 299, 307, 355, 356, 359, 375, 402, 404, 405, 407, 408, 415, 418, 426, 429–432, 462, 475, 481, 482, 486, 488
- snowpack 97, 98, 106, 113, 115, 120, 299, 300, 318, 319, 321, 323, 345, 356, 357, 359, 405, 411

snow-free patch, snow patch 15, 97, 101, 105, 106, 112, 481
- snow sublimation 4, 9, 10, 13, 84, 86, 87, 480

soil moisture 16, 22, 23, 59–61, 68–72, 76, 77, 92, 93, 100, 132, 243, 244, 270–272, 313–315, 318–321, 323, 405, 412, 417, 419, 431, 436, 437, 439–442, 448, 450, 452, 464, 481, 484
soil temperature 91, 92, 100, 236, 270, 271, 437, 465
soil water 60, 68, 244, 267, 411, 414, 416,–418, 431, 445, 446, 448–450, 469
soil wetness 59–62, 69, 71, 76, 243, 477
source area 21, 117, 230, 246, 250, 374, 376, 484
specific yield 19
spin up, spin-up 406
spruce 12, 14, 23, 90, 127, 237, 413, 459
SRES (Special Report on Emissions Scenarios) 411, 416, 423, 425, 431, 432
SSM/I (Special Sensor Microwave/Imager) 53, 59, 61, 63, 64, 68, 70, 76, 77, 201, 205
stage 19, 32, 71, 100, 223, 282, 284, 292–294, 296–298, 345, 346, 348–350, 353–355, 358, 447
Stefan 16, 18, 97, 125, 128, 132, 135, 286, 359
storage
- antecedent storage 25, 221, 223, 229, 231, 249
- change in storage, ΔS 228, 229, 249, 445
- lake storage 25, 177, 221, 222, 224, 226–231, 250, 449, 484
storage capacity 21, 22, 139, 144, 231, 235, 238, 239, 241, 242, 244, 247, 248, 277, 418
storm 2, 5, 21, 117, 148, 150, 197, 203, 208, 209, 213, 226, 241, 245, 250, 273, 415, 423, 430, 432, 478, 484, 487

streamflow 21, 22, 26, 98, 114, 221, 222, 227, 229–231, 235, 246, 248, 25–253, 281, 282, 290, 292, 293, 301, 345, 363, 364, 369, 393, 400, 402, 403, 406–408, 411, 412, 415, 421–423, 426, 429, 431, 432, 445, 448, 449, 458–462, 464, 475, 476, 478, 482, 484, 486, 488, 489

subalpine 23, 24

subarctic 5, 14, 15, 21, 22, 30, 33, 82, 85, 162, 178, 221, 222, 227, 229, 235, 237, 239, 243, 250, 253, 277, 402, 403, 405, 407, 435, 436, 442, 478

sublimation

subsurface 19, 21, 25, 115, 116, 150, 209, 224, 226, 244–246, 265, 267, 271, 272, 274, 277, 416, 417, 429, 431, 482, 484

supercool 283, 284

SWE, snow water equivalent 3, 12–15, 21, 84–90, 100, 107, 108, 113, 114, 117, 275, 312–314, 318–323, 355, 426, 464, 465, 480

T

temperature
- air temperature 3, 12, 43, 63, 73, 83, 84, 100, 102, 103, 109, 111, 116, 117, 125, 127–131, 133–135, 139, 141, 144, 147, 148, 150, 162, 166, 167, 177, 182–184, 192, 193, 197, 200, 202–205, 215, 258, 286, 290, 299, 311, 327, 330, 333, 334, 338, 346, 349, 437, 481, 484, 486, 487
- brightness temperature 59, 60–63, 70, 74, 76
- cloud top temperature 39, 44, 45
- ground temperature 125–129, 131, 132, 135, 476, 477

land surface temperature, LST 39–52, 55, 56

temperature inversion 2, 31, 148, 480
- surface temperature 5, 15, 18, 39, 40–44, 47, 50, 51, 55, 62, 63, 101, 103, 129, 130, 133, 136, 148, 150, 152, 178, 184, 197, 200, 202–204, 206, 207, 209, 212, 216, 477

water temperature 39, 43, 44, 53–55, 153, 161, 162, 166, 182, 200–203, 207, 212, 215, 216, 224, 283, 284, 290, 327–331, 333–336, 338, 341, 487

thaw
- thaw depth 126, 128, 133–136
- thaw front 126, 132, 134

thermal

thermal conductivity 16, 18, 132, 134, 330, 481

thermal properties 18, 132, 481

thermal stratification 153, 182, 197, 206, 207, 209

thermocline 150, 182, 209

threshold 25, 28, 41, 83, 84, 87, 221, 227–229, 231, 235, 244, 247–250, 252, 295, 299, 312, 314, 318, 352, 374, 382, 484

tile 412, 414, 449

TOPAZ (Topographic Parameterization) 371, 373–377, 379, 383, 38–388, 390–393, 398, 399, 408, 458, 478

TVC, Trail Valley Creek 12, 13, 15, 85, 86, 98–104, 106–108, 110, 113, 114, 116–118, 120, 121, 436

tree 5, 10, 12, 14, 82, 83, 85, 126, 258, 259, 265, 276, 435, 436, 449, 459, 461, 481

treeline, tree-line 86, 97, 108, 119, 236

trend 2–4, 44, 51, 63, 64, 169, 216, 300, 345, 348, 355, 363–369, 477, 478, 485, 486, 489

tundra 3, 9, 12–17, 19, 20, 30, 33, 48, 49, 81, 85, 86, 89, 97, 98, 100–102, 107, 111–115, 117, 125–127, 129–135, 144, 252, 266, 401, 435, 436, 442, 458, 459, 480, 481, 487, 488

Twin Otter 48, 49, 101, 477

U

upland 13, 21, 23–25, 111, 113, 116, 117, 147, 156, 157, 177, 221, 222, 229, 235–237, 242–244, 248, 249, 264, 413, 436, 437–442, 482

upscale, upscaling 399

V

validation 50, 61, 133, 313, 314, 323, 327, 333–336, 339, 341, 397, 407, 446, 459, 477
  vapor 2, 21, 43, 76, 84, 147, 148, 152, 166–168, 181, 183, 184, 188, 189, 191, 192
variability
interannual variability 4, 150, 197, 198, 199, 206, 210, 213–216, 237, 486
– seasonal variability 31, 72, 73, 182
– spatial variability 16, 17, 39, 68, 75, 97, 98, 101, 104, 107, 108, 117, 120, 198, 206, 212, 215, 277, 373, 377, 400, 415, 432
– temperature variability 2
– thermal variability 33
variation 2, 4, 18, 40, 44, 47, 48, 55, 61, 63, 64, 66, 69, 72, 73, 88, 90, 91, 100, 101, 104, 105, 107, 113, 115, 126, 133, 144, 148, 150, 152, 153, 182, 197, 207, 209, 211, 213, 225, 235, 247, 252, 267, 268, 270, 271, 274, 277, 292, 320, 349, 350, 373, 377, 379, 380, 401, 402, 407, 411, 412, 414, 417, 421, 431, 432, 446, 450, 452, 466, 481
– inter-annual 18, 207
vegetation 10, 23, 33, 40, 45, 47, 52, 59, 60–64, 68–70, 72, 76, 85, 97, 98, 100, 101, 112, 117, 126, 136, 144, 151, 177, 193, 236, 237, 243, 258, 259, 277, 374, 401, 413, 417, 436, 437, 439, 450, 481, 483

W

water
– liquid water 62, 63, 73, 114, 115, 223, 329, 453
– meltwater 18, 21, 24, 82, 91, 92, 97, 98, 113–115, 117, 120, 156, 229, 243, 249, 269, 459, 482
– pore water 331, 332

water and energy balance, water and energy budget 1, 94, 445
water and energy cycle 82, 181, 194
water balance, water budget 4, 5, 9, 15, 23, 31, 33, 52, 100, 114, 116, 144, 221, 224–231, 235, 244, 247–249, 257, 274, 275, 377, 379, 393, 398, 403, 408, 445, 446, 448–450, 456, 465–467, 469
water table 21, 23–25, 115, 117, 241, 243, 244, 247, 266, 267, 269, 413, 469, 484
WATCLASS 77, 117, 118, 120, 445, 447–450, 452, 453, 455, 456, 458–469
  WATFLOOD 348, 357, 358, 371, 390, 395, 399, 445–449, 456, 458–470, 478, 490
WATPAZ 390–395, 399
wave 4, 5, 14, 43, 100, 107, 192, 200, 202, 274, 283, 286, 287, 289, 295, 298, 349, 351, 353, 358, 417, 437, 439, 450, 481
wetland 9, 10, 19, 22, 24, 26, 27, 29, 33, 52, 72, 127, 147, 156, 157, 162, 221, 222, 235–237, 243, 250, 257, 258, 263, 274, 276, 277, 282, 413, 449, 458, 468, 482–484, 487
Whaleback Is 163, 164, 197, 202, 212, 213, 216, 217
wind speed 13, 84, 87, 102, 105, 107, 109, 152, 161, 162, 166, 167, 181, 183, 184, 188, 189, 192, 197, 200, 203, 204, 286, 415, 437
Wolf Creek 14, 18, 19, 21–24, 26, 43, 82, 372, 374–377, 379, 400–402, 457, 459, 461, 464, 465
woodland 14, 23, 24, 436, 481

X

Xinanjiang 411, 412, 416, 417, 419, 431